Resilient and Adaptive Tokyo

Wanglin Yan • William Galloway • Rajib Shaw
Editors

Resilient and Adaptive Tokyo

Towards Sustainable Urbanization in Perspective of Food-energy-water Nexus

Springer

Editors
Wanglin Yan
Facultiy of Environmental Information Studies
Keio University
Fujisawa, Kanagawa, Japan

William Galloway
Department of Architectural Science
Toronto Metropolitan University
Toronto, ON, Canada

Rajib Shaw
Graduate School of Media and Governance
Keio University
Fujisawa, Kanagawa, Japan

ISBN 978-981-99-3833-9 ISBN 978-981-99-3834-6 (eBook)
https://doi.org/10.1007/978-981-99-3834-6

This Springer imprint is published by the registered company Springer Nature Singapore Pte Ltd.
The registered company address is: 152 Beach Road, #21-01/04 Gateway East, Singapore 189721, Singapore

Paper in this product is recyclable.

Preface

The Tokyo-Yokohama region in the Kanto plain in Japan is the largest urban conglomeration in the world. Urban land use in this region expanded rapidly in the 1960s and sprawled widely to create suburbia along with the development of transport infrastructure. Many urbanized areas were built during the postwar period of high economic growth and spread into suburbia and extra-urban areas along a radial system of railways. Workers living in outer communities endured long commutes to city centers that had limited access to local food, energy, and water (FEW) systems. Many suburban communities are now facing declining birth rates and an aging population.

Governments are now trying to remake cities, designing them to be more compact by attracting services and residents to the walkable areas near railway stations, amid nagging fears of an aging society. Meanwhile, built-up areas are approaching a time when infrastructure and other upgrades will be needed, especially as vulnerabilities have been revealed by recent disasters, such as the Great East Japan Earthquake in 2011, the heat wave of 2018, and super typhoons like Hagibis in 2019. Also, the Japanese government initiated an action plan on climate change adaptation in 2015. It has also launched an SDGs (UN Sustainable Development Goals) future city program in 2018 to accelerate the transition and transformation to a carbon neutral and sustainable society. In this context, the food-energy-water nexus plays a key role in achieving the aims of all these initiatives. The production, transportation, and consumption of FEW resources and services typically account for more than 70% of CO_2 emissions in cities.

Although the FEW nexus presents major physical challenges, it is usually addressed with unilateral policies and individual or isolated management tools. The potential synergies between food, water, and energy are not fully realized in many cases. Instead, it is not uncommon to see one depleting the other. In many cases, our cities and the systems that support them were not designed to address the FEW nexus. Consumers in cities are generally unaware of the interrelations of food, water, and energy, and therefore not motivated to change their behavior. Farmers in rural areas produce food under increasing stresses of water and energy. There are also gaps in awareness of the roles and impacts of climate change. Improving communication among stakeholders, with the support of scientific evidence, is a keyway to narrow these gaps.

To date, some FEW research has resulted in assessment tools and policy proposals, but it is still rare to see current scholarly literature addressing the relationship between design and physical environment and FEW nexus problems.

The Sustainable Urbanisation Global Initiative (SUGI) project of the Belmont Forum approached this issue with a multidisciplinary and transdisciplinary moveable nexus approach examining six cities around the world, through the lens of the food-energy-water nexus. The methodology developed and tested three modules for FEW management in cities. These consist of a design method, evaluation tools, and a mechanism for participation.

This book is a compilation of the latest research on the food-energy-water nexus, mainly using Tokyo as the focal area. It examines the FEW stocks and flows that support the world's largest metropolitan area, explores how actors have worked together to secure the resilience and sustainability of resources, and demonstrates the potential to use the resources to make the city more adaptive to climatic and social change.

The volume consists of 16 chapters. Chapter "Understanding Change in Tokyo Through Food, Energy, and Water Security" introduces the landscape and history of urban development in Tokyo, and policies relevant to securing the supply of food, energy, and water to create resilient and adaptive cities. Chapter "Design-Led Nexus Approach for Sustainable Urbanization" reviews principles relating to sustainability and provides the methodology of a design-led nexus approach followed by the M-NEX project funded by SUGI to examine the food-water-energy nexus. Chapter "Climate Change in Global Cities" introduces a quantitative method to predict global warming impacts in six target cities (Amsterdam, Belfast, Detroit, Doha, Sydney, and Tokyo).

One of the chief challenges of FEW is to scale the myriad solutions to have significant beneficial impacts on communities, cities, and even globally. Tokyo has witnessed rapid growth and dramatic socio-economic transformation in the last half century. Remarkable land use change is evident, from the city center to suburban areas, from the decrease of greenery and agricultural land uses in the transformation to residential and industrial urban land uses. In many areas the change was totally unplanned. In the case of Tokyo, the natural pace of change is rapid enough that we might imagine the intentional transformation of the entire metropolitan area with the use of policy as well as a group of technical solutions and building typologies.

In that context, Chapters "Scaling the Food–Energy–Water Concept in Tokyo" to "Green Infrastructure in Tokyo" discuss the patterns of those changes. Chapter "Scaling the Food–Energy–Water Concept in Tokyo" presents a detailed introduction to a conceptual methodology and possible outcomes in terms of design that builds directly on this observation. Chapter "Land Use Planning and Conservation Policy in the Tokyo Metropolitan Area" explores the volumes and patterns of land use in the fringe areas of the metropolis and discusses the driving forces and impacts of change on sustainable development. Chapter "Green Infrastructure in

Tokyo" visualizes the green space lost and found in Tokyo through the process of urbanization and identifies lessons for how metropolitan areas can accelerate the creation of green and blue infrastructure.

Urbanization has improved the quality of life for most citizens, while the environmental load, as evidenced by CO_2 emissions, has also increased correspondingly. Previous research has revealed that more than 80% of CO_2 emissions come from FEW consumption in daily life, although geospatial differences at the neighborhood level could not be identified because of the limited availability of data and methods.

In that context, Chapter "Calculating the Demand for Food, Energy, and Water in the Spatial Perspective" develops a simplified tool to measure the ecological footprint in order to assess the environmental load caused by demand for and supply of FEW in cities. Tokyo is promoting compact city policies to smooth the adaptation to future depopulation. While the policies may be favorable for the management of urban infrastructure and private services, they could also create "food deserts" that are remote from retail centers. Chapter "Identifying Gaps Between Food Supply and Demand Under Compact City Policies" quantitatively identifies the gaps between food supply and delivery while considering neighborhood walkability for an aging community.

Tokyo urbanized during the rapid economic growth of the twentieth century, growing along railways radiating from the city center to the suburbs. However, some farmland remains on the fringes of the metropolitan area and its value has been more recognized lately, so some has been conserved for allotment gardens. Chapter "Urban Agriculture as a Tool for Adapting Cities for the Future" investigates the geographic distribution patterns and products and examines the nexus effects of agricultural activities and the performance of ecosystem services by agricultural land.

Chapter "Assessing Urban Resource Consumption and Carbon Emissions from a Food–Energy–Water Nexus Perspective" qualifies the intensity of resource use and CO_2 emissions in food, energy, and water sectors in Tokyo based on input-output analysis. Using the monetary input-output table from the Statistic Division, Tokyo Metropolitan Government Bureau of General Affairs, this chapter quantifies FEW demand and supply and the flow of water and energy in FEW sectors, assesses resource-use efficiency in different sectors, and helps to visualize the energy and water consumption intensity in each city.

Chapter "Impacts of the 2011 Disaster on Food–Energy–Water Material Flows and Resource Use Efficiency in Yokohama City" takes Yokohama City as the research area to investigate the impacts of the 2011 Tohoku earthquake and tsunami disaster on material flows and resource-use efficiency of urban water, energy, and food systems, by using a series of input-output tables. The chapter examines the increase of coal consumption and consequently CO_2 emissions, as well as improvements in resource-use efficiency in water- and food-related industries. Japan has created a national hydrogen development strategy and road map, with designated targets and schedules up to 2050. However, a lack of proper business models hinders the acceptance of new technology in public and private sectors.

Taking hydrogen energy as an ultimate form of energy for the food-water-energy nexus, Chapter "The Potential of Hydrogen Energy and Innovative Diffusion Models in Japan" envisages lifestyles supported by hydrogen-based household FEW units in a future city, and proposes possible business models toward local production and local consumption. Along these lines, Chapter "Hydrogen Refueling Station Siting and Development Planning in the Delivery Industry" looks at the viability of installing refueling stations at logistics service centers and converting delivery fleets to hydrogen-powered fuel cell trucks, using one company in Kanagawa Prefecture as a case study. Each chapter demonstrates the unique potential of transportation and delivery industries in promoting hydrogen energy.

A large number of suburban cities developed around Tokyo during the period of rapid economic growth in the last century. Many of those areas are now facing an aging population and deterioration of urban infrastructure. Meeting these challenges requires cooperation among a broad range of stakeholders. Chapter "Visualizing Social Capital and Actor Networks for Sustainable Suburban Areas" introduces a success story of the "Next Generation of Suburban Development" led by Tokyu Corporation and Yokohama City. The chapter first reviews the activities that unfolded at the WISE Living Lab since 2012. Then it visualizes the actor network using the KUMU tool and finds that various actors were connected with each other through the living lab, and that the activities contributed significantly to nine of the UN Sustainable Development Goals.

Policy interventions for resilience and adaptation at the local government level are challenging for many reasons, such as national policy making schemes, lack of resources, and changes in the risk landscape. Chapter "Policy Interventions for Resilience and Adaptive Cities" provides ten specific policy measures for adapting to increasing systemic risks in the complex urban context.

Chapter "Toward a New Resilience" concludes that the Tokyo experience shows that resilient and adaptive cities rely on interactions involving human stakeholders and non-human factors at many levels.

We hope that the knowledge compiled in this volume will be valuable for growing cities in emerging countries, and that it will provide a basis for further discussions to advance resilience and adaptivity.

Fujisawa, Kanagawa, Japan — Wanglin Yan
Toronto, ON, Canada — William Galloway
Fujisawa, Kanagawa, Japan — Rajib Shaw

Acknowledgments

This publication was first proposed in 2021, in the midst of the COVID-19 pandemic. It was intended to provide scientific support for research and practice relating to the M-NEX project (the Moveable Nexus: Urban Food-Energy-Water Management Innovation in the New Boundaries of Change), which had received grant support from the Belmont Forum/JPI Europe under the "Sustainable Urbanisation Global Initiative (SUGI) Food-Water-Energy Nexus" (Project No. 11314551).

In the course of this work, we had to deal with unexpected inconveniences in field investigations and constraints on face-to-face communications, so the writing and editing ended up taking more time than we originally planned. On the other hand, the unprecedented outbreak of the coronavirus catalyzed an opportunity for us to experience the resilient and yet vulnerable aspects of the megalopolis of Tokyo, especially in the dimensions of essential demand and supply of food, water, and energy. The large population was able to survive the pandemic thanks to the cumulative work of generations in the construction of urban infrastructure and essential services relating to food-water-energy, which are the focus of this volume. I hope that the chapters of this book will provide some historical clues for other global cities to build resilience and adaptivity.

I am grateful that the SUGI project provided us a precious opportunity to examine the urban nexus through an international research project. The M-NEX consortium partners have inspired may new observations and research themes. The original plan was to deliver the book at the closing symposium of the M-NEX project in Tokyo in March 2022, and it was unfortunate that we were unable to hold the event due to COVID-19 restrictions.

I must thank the chapter contributors for their dedicated efforts to implement the proposal, reorganize their material, and remake their narratives to consider the nexus perspective. Most of the contributions were the latest research results from master's and doctoral students. The results and conclusions are contextualized to Tokyo only because of limits on time and human resources.

I am grateful to the Japan Science and Technology Agency (JST), partner of the SUGI project, for financial and project management support for M-NEX (Project

No. 1298) over the past 4 years. With JST's systematic management we were able to complete the M-NEX project despite the difficult conditions and to achieve more than we had expected.

I would also like to express my sincere thanks to Randal Helten for his professional and devoted editing efforts to make the chapters deliverable.

Finally, I must thank Springer for the kindness and patience shown while waiting for the late arrival of manuscripts.

Because of constraints on time and capacity, the chapters and volume may contain many flaws. Comments, suggestions, and discussion are welcome at any time.

Wanglin Yan
April 2023, Tokyo

Contents

Understanding Change in Tokyo Through Food, Energy, and Water Security 1
Wanglin Yan and Keidai Kishimoto

Design-Led Nexus Approach for Sustainable Urbanization 17
Wanglin Yan and Shun Nakayama

Climate Change in Global Cities 47
Yoshiaki Miyamoto

Scaling the Food–Energy–Water Concept in Tokyo 59
William Galloway

Land Use Planning and Conservation Policy in the Tokyo Metropolitan Area 79
Ruiyi Zhang, Wanglin Yan, and Rajib Shaw

Green Infrastructure in Tokyo 105
Keidai Kishimoto and Wanglin Yan

Calculating the Demand for Food, Energy, and Water in the Spatial Perspective 123
Shun Nakayama, Wanglin Yan, and William Galloway

Identifying Gaps Between Food Supply and Demand Under Compact City Policies 135
Shun Nakayama, Wanglin Yan, and Rajib Shaw

Urban Agriculture as a Tool for Adapting Future Cities 149
Keidai Kishimoto and Wanglin Yan

Assessing Urban Resource Consumption and Carbon Emissions from a Food–Energy–Water Nexus Perspective 171
Xujie Hu and Wanglin Yan

Impacts of the 2011 Disaster on Food–Energy–Water Material Flows and Resource Use Efficiency in Yokohama City 189
Liu Hanyu and Wanglin Yan

The Potential of Hydrogen Energy and Innovative Diffusion Models in Japan 211
Baizi Zeng and Wanglin Yan

Hydrogen Refueling Station Siting and Development Planning in the Delivery Industry 231
Lu Ziyu and Wanglin Yan

Visualizing Social Capital and Actor Networks for Sustainable Suburban Areas 253
Sumiko Asanoumi, Ayase Yonomine, Shun Nakayama, and Wanglin Yan

Policy Interventions for Resilience and Adaptive Cities 273
Rajib Shaw

Toward A New Resilience 285
Wanglin Yan and William Galloway

Correction to: Design-Led Nexus Approach for Sustainable Urbanization C1
Wanglin Yan and Shun Nakayama

Editors and Contributors

About the Editors

Wanglin Yan Prof. Yan is a distinguished scholar in Environmental and Geo-informational Science for urban and regional planning by multi-disciplinary and transdisciplinary approaches. He is currently the president of Geographic Information System Association, Japan. He is also serving the sub-regional node of Asia-Pacific Adaptation Network of UNEP and devoted to climate change adaptation and disaster risk reduction in Japan and northeast Asian countries. He has recently led the international consortium for the SUGI Nexus Program/M-NEX project and contributed the co-creative design-led approach for urban food-energy-water management.

William Galloway William is an architect and urban planner, trained in Canada, the UK, and in Japan. He taught at several universities in Japan, including Keio University and Waseda University, and currently teaches at Toronto Metropolitan University, Department of Architectural Science. In addition to research and teaching he is the co-founder and director of the architecture office, frontoffice tokyo. His research focuses on resilience and adaptation, often making use of lessons learned from the examination of growth and change in cities.

Rajib Shaw Shaw is the Co-chair of the United Nations Asia-Pacific Science Technology Advisory Group (AP-STAG) and Coordinating Lead Author (CLA) IPCC's sixth Assessment report. He is the editor of a book series on disaster risk reduction, published by Springer, and the Chief Editor of an academic journal *Progress in Disaster Science* by Elsevier. He is the recipient of the Pravasi Bhartiya Samman Award (PBSA) of 2021 in the Education Sector from the President of India. He also received the United Nations Sasakawa Award for disaster risk reduction as a lifetime achievement and for his contribution to global disaster resilience initiatives. More of his work can be seen at www.rajibshaw.org.

Contributors

Sumiko Asanoumi SFC Institute, Keio University, Fujisawa, Kanagawa, Japan

Ishitsuka Studio for Planning and Design, Shibuya, Tokyo, Japan

William Galloway Department of Architectural Science, Toronto Metropilitan University, Toronto, ON, Canada

Liu Hanyu Graduate School of Media and Governance, Keio University, Fujisawa, Kanagawa, Japan

Xujie Hu Graduate School of Media and Governance, Keio University, Fujisawa, Kanagawa, Japan

Keidai Kishimoto Graduate School of Media and Governance, Keio University, Fujisawa, Kanagawa, Japan

Yoshiaki Miyamoto Faculty of Environment and Information Studies, Keio University, Fujisawa, Kanagawa, Japan

Shun Nakayama Graduate School of Media and Governance, Keio University, Fujisawa, Kanagawa, Japan

Rajib Shaw Graduate School of Media and Governance, Keio University, Fujisawa, Kanagawa, Japan

Wanglin Yan Faculty of Environmental Information Studies, Keio University, Fujisawa, Kanagawa, Japan

Ayase Yonomine Graduate School of Media and Governance, Keio University, Fujisawa, Kanagawa, Japan

Baizi Zeng Graduate School of Media and Governance, Keio University, Fujisawa, Kanagawa, Japan

Ruiyi Zhang Graduate School of Media and Governance, Keio University, Fujisawa, Kanagawa, Japan

Lu Ziyu Graduate School of Media and Governance, Keio University, Fujisawa, Kanagawa, Japan

Understanding Change in Tokyo Through Food, Energy, and Water Security

Wanglin Yan and Keidai Kishimoto

Abstract Japan, one of the most developed countries in Asia, is recognized for having well-managed infrastructure and for its living environment. This apparent utopia is collapsing amid the ever-growing risks posed by climatic disasters and the unsustainability of natural resources. This chapter looks back at the development path of urbanization in Tokyo, reveals the vulnerabilities of massive urban development, and raises concepts that could help remake the city to be more resilient and adaptive to climatic and social changes in the twenty-first century.

Keywords Urbanization · Food–energy–water security · Resilience · Natural resources

1 The Urban Systems of Tokyo

Depending on the definition, Tokyo has different metrics for geographical coverage and population. Administratively, Tokyo-to (Tokyo Metropolis) is one of the 47 first-class administrative prefectures in Japan, with a population of approximately 13.7 million and a land area of over 2200sq. km. It includes 23 special districts (in terms of jurisdiction, equivalent to cities) in its eastern part, the Tama region (26 cities, 2 towns, and 1 village) in the western part, and a remote island region in the southeast. The 23 special districts are undisputedly the central core of the metropolitan area, with a population of nearly ten million. The remote island region is unlike the rest of Tokyo in terms of urbanization, so it will not be covered further in this book.

W. Yan (✉)
Faculty of Environmental Information Studies, Keio University, Fujisawa, Kanagawa, Japan
e-mail: yan@sfc.keio.ac.jp

K. Kishimoto
Graduate School of Media and Governance, Keio University, Fujisawa, Kanagawa, Japan
e-mail: keidai@keio.jp

W. Yan et al. (eds.), *Resilient and Adaptive Tokyo*,
https://doi.org/10.1007/978-981-99-3834-6_1

The Tokyo Metropolitan Area (TMA) refers to Tokyo-to and three neighboring prefectures (Chiba, Kanagawa, Saitama) and has a population of about 35,154 million and 12,000 sq. km in total land area.

The Greater Tokyo Area (GTA) contains Tokyo-to and the prefectures of Chiba, Kanagawa, and Saitama as well as parts of Gunma, Ibaraki, and Tochigi prefectures. It is statistically defined by the administrative area in which more than 1.5% of the working-age population is commuting to the TMA for work and school. The population of this region is about 36.44 million and the land area is about 13,565 sq. km.

The Greater Capital Area (GCA) covers Tokyo and seven prefectures (Chiba, Gunma, Ibaraki, Kanagawa, Saitama, Tochigi, Yamanashi). It was politically designated by the National Capital Region Planning Act in 1956 (https://en.wikipedia.org/wiki/Greater_Tokyo_Area#National_Capital_Region). It has a population of approximately 44 million and a land area of 36,898 sq. km.

The Greater Kanto Region (GKR) is the concept in the National Territorial Planning. It includes Tokyo and six prefectures in the Kanto plain (Chiba, Ibaraki, Gunma, Kanagawa, Saitama, Tochigi), three prefectures in the Koshin-etsu region (Nagano, Niigata, Yamanashi) and Shizuoka Prefecture, for a total of one capital and ten prefectures, with a total population of 52 million and a land area of over 70,000 sq. km.

According to the national census in 2015, TMA consists of 105 cities, 35 towns, and 3 villages (townships). These cities are systematically divided into "city center, sub-center, new urban center, central cities, stronghold cities, and general towns" by the National Capital Region Planning Act based on population and functions. Except for the Tokyo 23 special cities, all the cities regardless of size are equivalent to local governments in the political system.

In addition to the 23 special districts of Tokyo, a municipality with a population of over 500,000 can apply to be a designated city by decree. These municipalities within TMA include, in descending order of population: Yokohama City, Kawasaki City, Saitama City, Chiba City, and Sagamihara City, of which the first three have populations over one million. Chiba city is close to one million. The rest are medium sized, with populations of around 500,000 or less. The urban composition of the TMA is illustrated in Fig. 1.

2 Development Path of the Tokyo Metropolitan Area

Tokyo emerged as a city during the Edo era (1650–1868), took off during Meiji (1868–1912), and Taisho (1912–1925), developed further during Showa (1925–1989), and matured in the Heisei era (1989–2018). The Reiwa era began in 2019. The progress of urbanization in Tokyo is shown in Fig. 2, which illustrates the sprawl of Tokyo since the seventeenth century. Edo was a small castle at the beginning. It started to grow in Meiji, the later of the nineteenth century. The remarkable sprawl was observed in Showa before and after WWII.

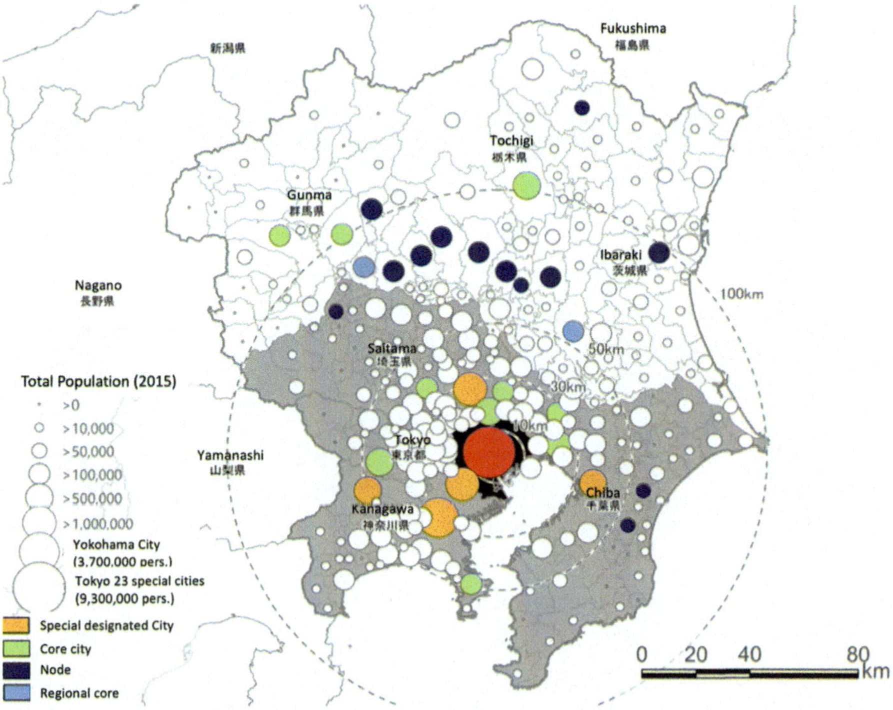

Fig. 1 The urban composition of the Tokyo Metropolitan Area. (Data source: e-Stat)

At the end of the Edo era, Western ideas and technologies began to enter Japan. The Edo military government started to construct roads, railways, and other urban infrastructure under modern concepts of urban planning. The Meiji government expropriated land from the samurai class for universities, government offices, and public parks. Since this conversion happened on a parcel-by-parcel basis, the land shapes and forms of use were mixed everywhere (resulting in the complex and diverse urban landscapes we see in Tokyo today.

The Great Kanto Earthquake of 1923 caused severe damage to the city center, with 40% of houses burned down and the streets demolished. The post-earthquake reconstruction triggered the first modern wave of urbanization in Tokyo and had significant impacts on later generations. The state and influential industry leaders strongly advocated for the construction of railways. By about 1930, a major railway network was almost completed, stretching from the terminal stations of the Yamanote ring railway to the hinterlands of the suburbs. Countless railway stations dot the metropolitan area today, serving as the most important transportation infrastructure system of the metropolis. Most of the development first happened around the railway stations, then expanded the outward as the stations became the centers of communities. Some of those residential communities were developed in a planned manner as the so-called garden cities, such as Tamagawadai and Denenchofu, in the

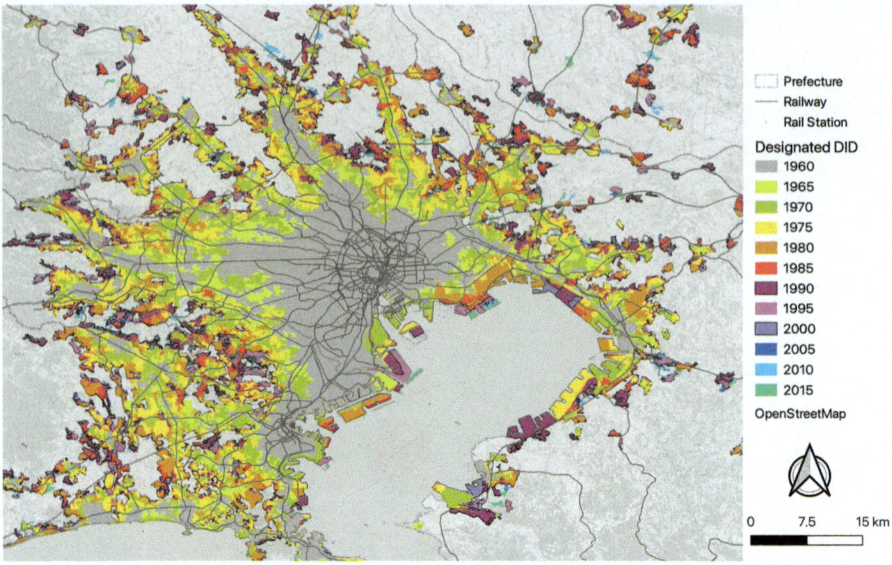

Fig. 2 The progress of urban development in Tokyo. (Data source: Digital land numeric information, MLIT)

1930s. However, those planned developments were very limited. Many neighborhoods were constructed on the agricultural infrastructure without any city planning, resulting in small properties, narrow streets, and inadequate public spaces.

The air strikes of the Second World War damaged a wide area of the city. The post-war reconstruction was designed under the General Headquarters Supreme Commander for the Allied Powers (GHQ/SCAP). Ambitious reconstruction plans were made, city roads were to be widened in order to meet the expected demand for automobiles. However, some of the plans were never completed due to financial constraints and difficulties in land acquisition. In 1958, the First Metropolitan Area Basic Plan was formulated with reference to the Greater London Plan, intended to control population growth in the city center, reduce congestion, and conserve a green belt to prevent urban sprawl. However, this vision had to be abandoned in the Second Metropolitan Area Basic Plan due to pressure from the urban development industry and a metropolitan population that was burgeoning beyond expectations (Takeuchi and Ishikawa 2008).

Changes in land use and built-up areas in suburban areas of the TMA mostly happened after 1960, along with the developed private railways. In response to urban problems such as overpopulation, environmental pollution and rising land prices, growth in the suburbs became the main coping strategy during the period of rapid economic development. People were eager to have bigger homes and better living environments, which resulted in a large increase in demand for housing. This brought a second wave of urbanization to Tokyo, eventually feeding an asset price bubble that became known as the "bubble economy." During this period, developments of various scales were carried out, greatly expanding the urban area. By the end of the 1980s, the current spatial structure of Tokyo had basically taken shape. Subsequent large-scale developments became less common except for several government-led projects such as Tama New Town, Kohoku New Town, Chiba New Town, and Tsukuba Science City. The only private company-led large-scale development was the Tama Garden City project, by Tokyu Company along today's Denentoshi Line, which occupies 3000 ha of land and serves a population of 890,000 (Tokyu Corporation, https://www.tokyu.co.jp/global/).

In recent years, the maintenance and renewal of infrastructure has become a major issue (Tsuzuki et al. 2007). The renovation of many densely populated areas with wooden houses in central areas was never improved due to the shrunken post-war reconstruction. Factories have been relocated to remote regions even overseas due to soaring land prices and high labor costs leading to the hollowing out of urban centers.

Conversely, the collapse of the bubble economy in the early 1990s was the opportunity to launch renewal programs to promote the return of population to the urban center. The Act on Special Measures concerning Urban Reconstruction was enacted in 2002, and with it the government initiated special urban regeneration zones in Tokyo to vigorously promote the development of the urban center. The redevelopment has included projects based on urban planning as well as projects involving building reconstruction. These redevelopment projects can greatly improve the efficiency of land use and have a spillover effect on the transformation of the surrounding densely populated areas.

3 Securing Food–Energy–Water Resources in the Metropolis

3.1 Securing the City from Disasters

Tokyo is geographically located at the inner reaches of Tokyo Bay and extends to the hinterlands of the Kanto Plain (Fig. 3). This geographic location has brought great benefits to the city in transportation and land development but vulnerabilities to natural disasters. From the early days of urban development, the builders of Tokyo have paid much attention to protect themselves from natural disasters. The ancient Edo Castle was built on the edge of the Musashino Terrace at the mouth of the Ara and Tone Rivers, right on the edge of the terrace and wetlands (near the present Imperial Palace and Tokyo Station). This ensured that the castle was protected from flooding on the one hand and allowed the use of water transportation and land reclamation on the other. As Edo Castle was built, the Shogunate (Bakufu) initiated large-scale water conservation projects to ensure the safety of Edo Castle and to better develop land resources. The biggest project was to redirect the flow of the Tone River to the sea. From then on, the Kanto Plain was flood-and-drought-

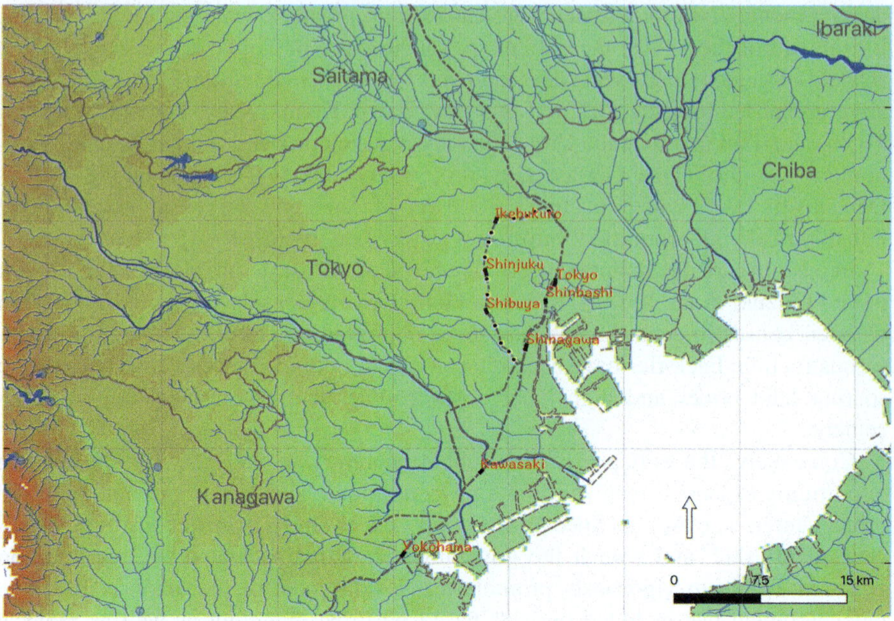

Fig. 3 The geographic layout and elevations in Tokyo. (Data source: Digital land numeric information, MLIT)

protected, and as commerce-flourished, it rapidly developed into the political and economic center of Japan.

In terms of topography, the eastern part is on an alluvial plain, while the western part of Tokyo is built on a hilly terrace. This provides easy transportation between the inland and the bay area. The terraces were favorable, first for flood and earthquake prevention, and second for superior landscape features. From Edo Castle, it was possible to reach a large area of the northern Kanto Plain inland via eight rivers, ensuring a constant flow of agricultural, forestry, and mining resources from the northern Kanto Plain. The spatial layout of TMA today is closely linked to these conditions.

3.2 Food Security

Food, energy, and water resources are indispensable for survival, and their conservation and efficient use are necessary for the sustainability of urban society. Food security as being when "all people, at all times, have physical and economic access to sufficient, safe, and nutritious food to meet their dietary needs and food preferences for an active and healthy life" (https://www.fao.org/3/w3613e/w3613e00.htm). The vast Kanto Plain ensured the production and provision of food along the

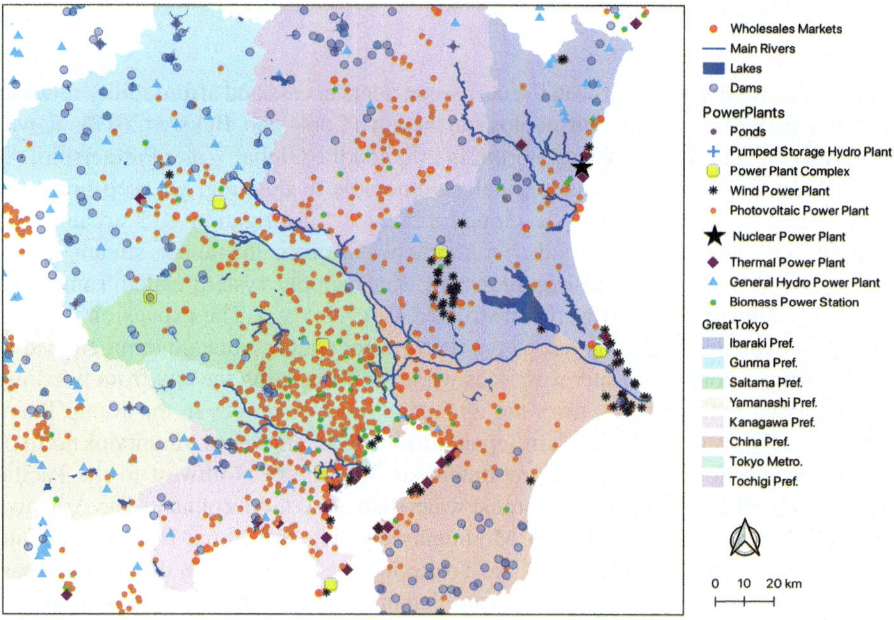

Fig. 4 Food–water–energy resources in Tokyo. (Data source: Digital land numeric information, MLIT)

well-established water transportation systems from the hinterland along rivers, helping the ancient Edo became a self-sufficient mega-city.

Modern cities are premised upon high population concentrations. Urban growth is accompanied by the conversion of land for food, energy, and water into industrial products and services. Tokyo has also converted massive agricultural and natural land to built-up areas, and food self-sufficiency is very low. The process of urbanization has resulted in a steady decrease in agricultural land and the agricultural labor force in Tokyo since 1955. Meanwhile, farmers are aging. Consequently, Tokyo is only 1% self-sufficient for food today. Kanagawa, neighboring prefecture to the West, with a population of eight million, is only 2% food self-sufficient. The Japanese food table depends mostly on food from overseas. The country currently provides only 37% of its own food calories (MAFF 2022).

To secure the food market, the Japanese government has designated 14 types of vegetables that account for 74% of the market including cabbage, spinach, Chinese cabbage, green onion, lettuce, potato, taro, Japanese radish, carrot, onion, cucumber, eggplant, tomato, green pepper. More than 70% of fresh meat, fish, and vegetables are distributed through wholesale markets to supermarkets and finally to consumers. E-commerce and direct trade between producers and consumers are growing in popularity, especially after the outbreak of the COVID-19. These factors make the regional food system very complicated. Figure 4 illustrates the distribution of land use and wholesale markets in the TMA. It clearly displays that food is produced and transported from the northern Kanto region.

3.3 Water Security

An integrative definition of water security considers access and affordability of water as well as human needs and ecological health (Cook and Bakker 2012). It was introduced at the Second World Forum in 2000 by the Global Water Partnership. As mentioned above, Tokyo was originally vulnerable to flooding, yet well-protected since the Edo era with well-developed channels for irrigation and urban water supply. The ancient city of Edo withdrew water from the upper stream of the Tama River. A complicated irrigation and water supply system called Tamagawa Josui and Shinagawa Josui were constructed 400 years ago. The expansion of built-up lands in modern times increased the demand for water, so the government had to look for more abundant water resources while still protecting the city from flooding. Fortunately, TMA benefits from the abundant water resources in the Kanto Plain. The Tone River, second longest in Japan, originates in the Kanto Mountains north of the plain and flows through the mountainous area in the northwest to the Pacific Ocean in northeast, forming the largest watershed area of the country. The Arakawa River originates in the Chichibu Mountains and like the Tama River flows into Tokyo Bay via the central cities of the Tokyo metropolitan area. In the southern part of the plain, the Sagami River flows into Sagami Bay from the Tanzawa Mountains through Kanagawa Prefecture. There are many flooded terraces in northern Chiba Prefecture, northern Kanagawa Prefecture, and southern Tokyo. This topography is the basis for a diverse urban landscape. Figure 4 shows that most of the water supply of the metropolis originates from outside the prefectural boundaries. The dams in the upper reaches of all these rivers are well managed and developed. Because of urban development in the western hilly terraces the water to serve that region must first be transported to higher land and then distributed to those consumers. Due to restrictions on the location of water intakes, about 80% of the total capacity of all water treatment plants is taken from points at an elevation of 5 m or less. However, about 70% of the water supply plants to which water is sent are located at elevations of 20 m or higher. Tokyo Waterworks is considering redesigning the water intake and distribution system by relocating water intakes from lowlands to highlands so that water can flow naturally with gravity while generating hydropower (https://www.waterworks.metro.tokyo.lg.jp/).

3.4 Energy Security

Energy security from the perspective of energy consumers is defined by the International Energy Agency as “uninterrupted physical availability at a price which is affordable while respecting environmental concerns” (IEA, http://www.iea.org/). This definition can be parsed into three indicators: availability, affordability, and acceptability (Hughes 2012). Despite consuming the most energy in Japan, the TMA itself does not produce any of its own primary energy. Oil, gas, and coal are imported from overseas, while electricity is sourced in neighboring regions. Figure 4 shows that the Fukushima Nuclear Power Plant is one of the providers of electricity to the region. The Tohoku earthquake, tsunami and nuclear accident of 2011 revealed the vulnerability of the TMA's energy systems (Vivoda 2012). With the temporary and/or permanent closure of many nuclear reactors, Japan has accelerated the installation of renewable energy in the process of reconstruction, while mainly coal-thermal power plants were also constructed in the coastal area of Tokyo Bay to meet urgent demand.

The 2011 earthquake gave more momentum to renewable energy, especially solar photovoltaic systems. It is interesting to note differences in the spatial distribution of power generation facilities. Hydropower plants are in the upper reaches of rivers, thermal power plants are along the coastline to access oil and gas imports from overseas, and nuclear power plants are in villages and towns far from the urban core. In contrast, many solar and biomass power plants were installed closer to the urban core. Although the capacity of these generators may currently be small, the patterns of decentralized distribution demonstrate the potential of new energy systems based on renewables. Progress with renewables in Japan has been driven by regulatory reforms of the electricity market. After the disasters of 2011, the monopoly previously held for decades by the Tokyo Electric Power Corp. (TEPCO) collapsed. Electricity retailers emerged in the market, propelled by subsidies under the feed-in tariff (FIT) system. The capacity of photovoltaic installations soared by a factor of 2.6 from 2011 to 2019 (http://www.jpea.gr.jp/). Photovoltaic power generation also presents opportunities everywhere in both urban and rural areas to utilize the ubiquitous solar resources. Nevertheless, the installation rate is still low, with only 3–5% of established houses equipped with solar panels in suburban Tokyo. The government is planning to accelerate solar power generators to 60% of the newly constructed homes by 2040.

4 The Legacy of Urbanization in Tokyo

Securing the supply of FEW is an arduous and long-term task for mega-cities. It requires not only quantity but also quality of resources.

As Tokyo's economy grew, the population shifted from rural areas to cities, and Tokyo became one of the most crowded cities in the world. This growth was accompanied by urban problems such as overpopulation, environmental pollution, and rising land prices. These pressures resulted in growth shifting from the city center to suburban areas to take advantage of lower housing costs but resulted in long daily commutes for many. In response, the government pushed hard to upgrade and expand railroads to increase train capacity, improve suburban-urban connections, and increase transportation efficiency. To some extent these efforts have led to improvements of commuter train congestion rates from more than 200% previously at the end of 1970s to 160% of load capacity just prior to the COVID-19 pandemic.

Along with population concentration and industrial development, water and air pollution from industrial and domestic sources became serious problems. The Tama River was considered a symbol of the problem of deteriorated water quality in those days. Urbanization triggered other environmental hazards from four major pollutants, including particulate matter (PM10 and PM2.5), ozone (O_3), nitrogen dioxide (NO_2), and sulfur dioxide (SO_2). Due to these pollution problems and the oil crises of 1973 and 1979, people began to pay attention to the ecological environment, urban infrastructure, and the concept of environment began to be introduced in development. In the 1970s, the governor of Tokyo had a mid-term and long-term comprehensive plan of "Public Square and Blue Sky Initiative" (Tokyo-to 1971), promising to improve the living environment in the city center. The central government subsequently initiated development plans for Tsukuba Science City (1961), Tama New Town (1963), Minato-Kita New Town (1965), and Chiba New Town (1966).

As a result, the urban problems in Tokyo were resolved gradually through the cooperation of government and industry and the involvement of citizens in the late twentieth century. Since the end of 1990s and after entering the new millennium, the sustainability of cities has risen to the top as an issue for urban management. Urban sustainability is typically expected to promise three things: stable provision of urban infrastructure including food, energy, and water (FEW) services, a constant improvement in quality of life, and the significant reduction of environmental impacts. For urban residents, FEW is the basis of life. Government and industry are responsible for securing the provision of resources and services.

Regarding sustainability, Japanese cities are facing three major crises today in the form of climate change, aging populations, and the deterioration of infrastructure. Japan has always been a country prone to various disasters, especially earthquakes and typhoons. Disaster management is always a key priority of the national and local governments. Severe damage from past disasters has left bitter memories. In recent memory, the Great East Japan Earthquake of 2011 resulted in power blackouts and severe shortage of food supplies. Typhoon Etau in 2015 flooded parts of northeastern suburbs of the metropolis and breached the levees of the Kinu River. Typhoon Hagibis in 2019 struck the central area of Tokyo and flooded several neighborhoods along the Tama River. More extreme weather due to climate change could make the situation worse in the Tokyo region by 2080 (Miyamoto, chapter "Climate Change in Global Cities", this volume).

The second crisis in Tokyo, and across Japan as well, is a shrinking population. Urbanization has pushed and pulled people to large cities. The more people moving to Tokyo, the fewer people will be producing food in rural areas. Meanwhile, the average age of agricultural laborers is ever increasing. These factors have enlarged the gap between domestic food production and consumption. One response might be to import more food from rural areas and even from overseas, but this is made more difficult with the gradual decline in the Japanese population, and declining birth rates particularly in rural areas. Tokyo itself is not yet at the stage of significant population decline, and many topics are still at the discussion stage, but population decline and land idleness are foreseeable in the future. The population has already peaked in some cities in the TMA in certain areas that are somehow inconvenient to live in. So one looming question is who will feed the urban population in future?

Third, much urban infrastructure was built during the period of high economic growth in the twentieth century, such as roads, bridges, waterworks, and tunnels. This infrastructure has aged and is approaching the time for renewal. The government is making efforts to improve efficiency and develop management methods to extend the lifespan of infrastructure, but this is a very costly challenge due to population decline, the aging of the labor force, and urban decay. This situation is particularly true in suburban areas (Ozeki et al. 2010) where the population is shrinking more quickly. The growth of suburban built-up areas will be sporadic, while the performance of maintenance gets worse. Who should pay for all of this and where can investment funds be secured?

One of the answers is the strategy of making cities more compact to optimize spatial configuration. The national government has initiated new urban zoning systems under the dual concepts of being compact and using network design. It aims to attract population to neighborhoods that are located near to railway stations while connecting those areas by public transportation. Areas with less population could be rezoned from urban promotion areas to urban control areas. More than two-thirds of municipalities have revised their city master plans and introduced this initiative since the enactment of the Act on Special Measures concerning Urban Reconstruction in 2002, though will still takes some time to see the effects.

Another response to this challenge is the concept of green infrastructure (GI), which has emerged in Europe and the USA. The basic concept is that the ecological functions of the natural environment can partially replace the functions of "gray" infrastructure. The definition of GI is still rather broad, but discussions are already active in Japan. Representative GI in Japan includes the Watarase Yusuichi (a natural river flood plain/wetland and water detention basin and levee system) created in the early twentieth century in the northern TMA, and kasumi levee systems built long ago in rural areas to retain overflow waters during flood seasons (Ichinose 2015). GI has attracted more attention in Japan since the Tohoku earthquake and tsunami disaster of 2011, when the nation was shocked by scenes of collapsed seawalls in the Sanriku coastal region. Each area has certain ecological resources that can be suitable for GI and serve various functions, such as disaster prevention, ecology, and soil and water conservation. Some local governments have moved quickly. For example, Setagaya City in Tokyo is planning to improve green cover from the current 25–33% in the future (https://www.city.setagaya.lg.jp/).

5 Perspectives About Resilient and Adaptive Cities in Terms of the FEW Nexus

Japan's highly urbanized and consumption-oriented society has seen big increases in food manufacturing and the demand for food and beverages. The convenience and security of life are based on the resources of food, energy, and water transported from distant places. Conversely, the impacts of human activities on the environment have increased dramatically, as measured through indicators such as food mileage (Specht et al. 2014; Murphy 2007), CO_2 emissions (Munksgaard et al. 2000), and virtual water (Allan 2003). This dilemma is illustrated in Fig. 5, which expresses environmental impacts conceptually with the FEW triangle (demand for food, energy, and water) and the quality of life is illustrated by the AHR triangle (accessibility, health, and resilience). In the ancient city of Edo, the intensity of resource use was limited, while the quality of life was low (in the sense of material consumption). The required provisions could be mostly met with local products. Modernization and urbanization have changed the landscape completely. A huge amount of farmland and forest has been converted to factories and residences, and the main industry has shifted from agriculture to manufacturing. Consequently, greenery disappeared, air was polluted, and water was contaminated, but through the determined efforts by the Japanese government and citizens, urban pollution problems had largely been mitigated by the end of the twentieth century, while the material satisfaction of life improved dramatically. Conversely, the convenience of life admittedly imposed costs on the natural environment. The comforts of urban life are built on the transboundary movement of FEW resources. Food, energy, and water are imported from remote regions. All of this has eventually been manifested

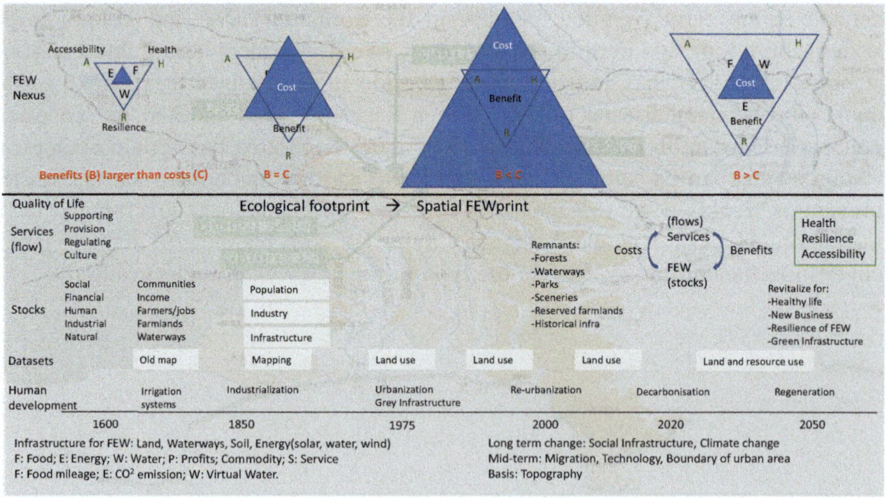

Fig. 5 Progress of urbanization and environmental costs through the lens of food–energy–water (FEW)

as pressure on the Earth system, resulting in ecological overshoot of planetary boundaries (Rockström et al. 2009).

To turn things around, significant improvements are needed in resource efficiency. The Paris Agreement was eventually adopted in 2015 after lengthy negotiations, with the international community agreeing on the goal to reduce GHG emissions by 50% by 2030 and by 80% by mid-century in order to keep the rise in mean global temperature to well below 2 °C from preindustrial levels and preferably limit the rise to 1.5 °C (Rogelj et al. 2016). Cities as major emitters have an obligation to be leaders in these efforts. The Japanese government has promised to reduce GHG emissions by 46% by 2030 relative to 2013 and achieve carbon neutrality by 2050. The Tokyo metropolitan government too has made a political decision to aim for net zero carbon emissions by 2050.

It is clear that we must think about how to reduce environmental impacts and make more room for the environment and nature and at the same time to rebuild our society to be smarter and more sustainable. This may be easy to say but difficult to do. Some leaders in Japanese industry believe that they have already done a lot and that there is not much more that needs to be done in cities. To make more progress we need to break through silo-based thinking. Nexus thinking is such an approach from the academic community. It involves connecting sectors and disciplines (Hoff 2011) in order to build a circular economy, realize the so-called donut economics (Raworth 2019), reduce our environmental load to within economic/ecological limits, and create a more just and health society.

6 Closing Remarks

The development story of Tokyo this chapter told us that the megapolis itself could be rebirthed through severe events. The first wave of urbanization occurred in the 1920s–1930s through the recovery from the Great Kanto Earthquake in 1923. Modern urban planning was introduced during reconstruction. The air strikes of the Second World War burned the city down again. The post-war recovery brought suburbanization along with a long period of economic growth. This recursive process is theorized by the notion of panarchy (Gunderson and Holling 2002), which posits that all societies repeat a circle of four phases: birth, growth and maturation, death, and renewal. *Change is normal through the process. The key for sustainability is to learn how to make our cities resilient and adaptive to change.* This was one of the lessons learned from the 2011 Tohoku earthquake and tsunami disaster (Yan and Galloway 2017).

The Japanese people have deeply reflected on the resulting nuclear accident and accelerated the energy transformation to renewables. TEPCO monopoly hold has been broken. Solar power generation as well as community-based renewable businesses have been growing in popularity. Ten years later Japan had largely recovered from the shock of 2011, but then the world became embroiled in the havoc of the COVID-19 pandemic. The unprecedented pandemic brought a big shock to the metropolis. The movement of migrants to the city center was suddenly redirected to the suburbs, for health and safety reasons from coronavirus. The volumes of commuters by train decreased by 20%, and the vacant rate of office buildings in the city center rises to historic high. People were requested to stay home and work remotely. Shops and restaurants were closed, or service hours were shortened, and alcohol sales were restricted. The pandemic provided time for us to rethink about the sustainability (Yan and Roggema 2017, 2019): the relationship between the places of production and places of consumption of FEW resources; the relationship between costs and benefits of investments; and the relationship between places of work and places of living. It is also an opportunity to translate many ideas into actions. During the pandemic, people discovered that they could work remotely while also being able to spend more time with family and community. This inspires us another vision of cities that can adapt to a new normal for managing natural and social resources of cities more effectively. Nexus thinking could be a promised approach to make the concept true (Hoff 2011). The design-led nexus concepts and practices in this volume are expected a reference for students and practitioners.

Acknowledgments This chapter was based on the work of the M-NEX project, under a grant from the JST/Belmont Forum Collaborative Research Area: Sustainable Urbanisation Global Initiative (No. 11314551). Local governments, companies, and communities were involved in the activities of the national teams. We are grateful to JPI Europe Urban for initiating the Sustainable Urbanization Global Initiative—Food-Water-Energy Nexus and making the M-NEX project possible.

References

Allan JAT (2003) Useful concept or misleading metaphor? Virtual water: a definition. Water Int 28(1):4–11

Cook C, Bakker K (2012) Water security: debating an emerging paradigm. Glob Environ Change 22(1):94–102

Gunderson LH, Holling CS (eds) (2002) Panarchy: understanding transformations in human and natural systems. Island Press, Washington, DC

Hoff H (2011) Understanding the Nexus. Background paper for the Bonn 2011 Nexus conference. Stockholm Environment Institute no. November, pp 1–52

Hughes L (2012) A generic framework for the description and analysis of energy security in an energy system. Energy Policy 42:221–231

Ichinose T (2015) Disaster prevention and reduction in time of population shrinking. J Rural Plann 34(3):353–356. https://doi.org/10.2750/arp.34.353

MAFF (2022). https://www.maff.go.jp/j/zyukyu/zikyu_ritu/011.html

Munksgaard J, Pedersen KA, Wien M (2000) Impact of household consumption on CO2 emissions. Energy Econ 22(4):423–440

Murphy AJ (2007) Grounding the virtual: the material effects of electronic grocery shopping. Geoforum 38(5):941–953

Ozeki Y, Togawa T, Suzuki Y, Kato H, Hayashi Y (2010) Evaluation method for maintenance and renovation costs of the future infrastructure in the metropolitan area. Infrastructure Planning Review 27:305–312. Released November 29, 2017, Online ISSN 1884-8303, Print ISSN 0913-4034. https://doi.org/10.2208/journalip.27.305

Raworth K (2019) Dollars to doughnuts: the shape of a new economy. GTI Interview, Great Transition Initiative. https://www.tellus.org/pub/Raworth-Dollars-Doughnuts.pdf

Rockström J, Will S, Noone K, Lambin E, Lenton TM, Scheffer M, Folke C et al (2009) Planetary boundaries : exploring the safe operating space for humanity. Ecol Soc 14:32

Rogelj J, den Elzen M, Höhne N, Fransen T, Fekete H, Winkler H, Roberto Schaeffer F, Sha KR, Meinshausen M (2016) Paris agreement climate proposals need a boost to keep warming well below 2 °C. Nature 534:631

Specht K, Siebert R, Hartmann I, Freisinger U, Sawicka M, Werner A, Thomaier S, Henckel D, Walk H, Dierich A (2014) Urban agriculture of the future: an overview of sustainability aspects of food production in and on buildings. Agric Hum Values 31(1):33–51

Takeuchi T, Ishikawa M (2008) Study on the greenery policy in the 1950s and 1960s of Tokyo special districts. J Urban Plann 43(3):199–204. https://doi.org/10.11361/journalcpij.43.3.199

Tokyo-to (1971) Public Square and Blue Sky Initiative of Tokyo, 251p. https://www.kosho.or.jp/products/detail.php?product_id=220534053 (in Japanese)

Tsuzuki M, Nakamura F, Okamura T (2007) Fundamental study on urban redevelopment and the impacts on neighbourhoods in Tokyo special districts with GIS. J Urban Plann 42(3):259–264. https://doi.org/10.11361/journalcpij.42.3.259

Vivoda V (2012) Japan's energy security predicament post-Fukushima. Energy Policy 46:135–143

Yan W, Galloway B (2017) Rethinking resilience: adaptation and transformation. Springer, 396p

Yan W, Roggema R (2017) Post-3.11 reconstruction, an uneasy mission. Tsunami and Fukushima disaster: design for reconstruction. https://doi.org/10.1007/978-3-319-56742-6_2

Yan W, Roggema R (2019) Developing a Design-Led Approach for the Food-Energy-Water Nexus in Cities Urban Planning 4(1):123–138. https://doi.org/10.17645/up.v4i1.1739

Design-Led Nexus Approach for Sustainable Urbanization

Wanglin Yan and Shun Nakayama

Abstract Modern cities are commonly seen as places where one can find a better life. At the same time, they face challenges in achieving sustainability, a goal crucial to the fundamental changes called for globally and locally. Food, energy, and water (FEW) resources are indispensable for human survival in general, while the importance of their conservation and efficient use is often underestimated. There exist big gaps in awareness of the existing threats and opportunities in cities in this regard. This chapter reviews the concept of nexus thinking, clarifies the features of the concept when applied in cities, and proposes design-led interdisciplinary and transdisciplinary methodology for innovative management of FEW resources in cities.

Keywords Sustainable urbanization · Urban nexus · Decarbonization · Design-led approach · FEWprint

1 Introduction

Modern society faces severe environmental problems as we approach or cross multiple thresholds of planetary boundaries (Rockström et al. 2015). The Paris Agreement on climate change, adopted in 2015, calls for a global temperature increase of no more than 2 degrees Celsius (°C) and preferably no more than 1.5 °C (Rogelj et al. 2016; Masson-Delmotte et al. 2021). In line with this, countries have declared

The original version of the chapter has been revised. A correction to this chapter can be found at https://doi.org/10.1007/978-981-99-3834-6_17

W. Yan (✉)
Faculty of Environmental Information Studies, Keio University, Fujisawa, Kanagawa, Japan
e-mail: yan@sfc.keio.ac.jp

S. Nakayama
Graduate School of Media and Governance, Keio University, Fujisawa, Kanagawa, Japan
e-mail: shun-nakayama@keio.jp

W. Yan et al. (eds.), *Resilient and Adaptive Tokyo*,
https://doi.org/10.1007/978-981-99-3834-6_2

Table 1 CO_2 emission reduction targets of some of the world's major economies

Nation/ region	GHG reduction targets		Share of global emissions (2018)
	Mid-term	Long-term	
USA	–26% to –28% (2005 base) by 2025	Carbon neutral by 2050	14.7%
	–≫–50% to –52% by 2030 (2005 base)		
Japan	–26% by 2030 (2013 base)	Carbon neutral by 2050	3.2%
	–≫–46% by 2030 (2013 base)		
EU	–55% by 2030 (1990 base)	Carbon neutral by 2050	9.4%
UK	–68% by 2030 (1990 base)	Carbon neutral by 2050	
	–≫–78% by 2035 (1990 base)		
Canada	–30% by 2030 (2005 base)	Carbon neutral by 2050	1.7%
	–≫–45 to –45% by 2030 (2005 base)		
China	–65% per GDP CO_2 emission by 2030 (2005 base)	Carbon neutral by 2060	28.4%
	–≫2026 to 2030 reduce coal consumption gradually		
India	–33 to –35% per GDP CO_2 emission by 2030 (2005 base)	No mention	6.9%
Russia	–30% by 2030 (1990 base)	No mention	4.7%
Korea	–17% by 2030 (2017 base)	Carbon neutral by 2050	1.8%
	–≫New target to be submitted		
Brazil	–43% by 2030 (2005 base)	Carbon neutral by 2050	1.2%
	–≫Prohibit illegal logging by 2030		

Source: Adapted from Ministry of the Environment, Japan, https://www.mofa.go.jp/mofaj/ic/ch/page1w_000121.html, 2022/1/11 present
Note: Bold font indicates new target announced at the UNFCCC climate summit in 2021

their greenhouse gas emission reduction targets for 2040 or 2050 (Table 1). Japan is accelerating its decarbonization efforts, aiming to reduce carbon dioxide (CO_2) emissions by 46% in 2030 from the 2013 level and to be carbon neutral by 2050.

The essence of decarbonization is about consuming fewer natural resources and reducing greenhouse gas emissions while creating more goods and services to put economic and social development on a sustainable track. Since the movement commenced in the early 1990s, Japanese industries and sectors have made great efforts and achieved remarkable success, despite the ups-and-downs of economic conditions in the last century: the rapid economic growth period in the 1960s, the oil crisis in the 1970s, the bubble economy in the 1980s, bursting of the bubble in the 1990s, economic stagnation in the 2000s, and the slow-paced recovery of the 2010s. Energy-use efficiency in gross domestic product (GDP) per kilogram of oil steadily improved in the 1960s and 1990s. Correspondingly, CO_2 emissions intensity also remained at a relatively high level. It seems we are hitting the ceiling of achieving further significant improvements (Fig. 1).

The rush to reach climate targets has often led to conflicting efforts between policies due to a lack of coordination between sectors. The conversion of cereals (e.g., maize) into biofuels has put pressure on food markets and raised food prices;

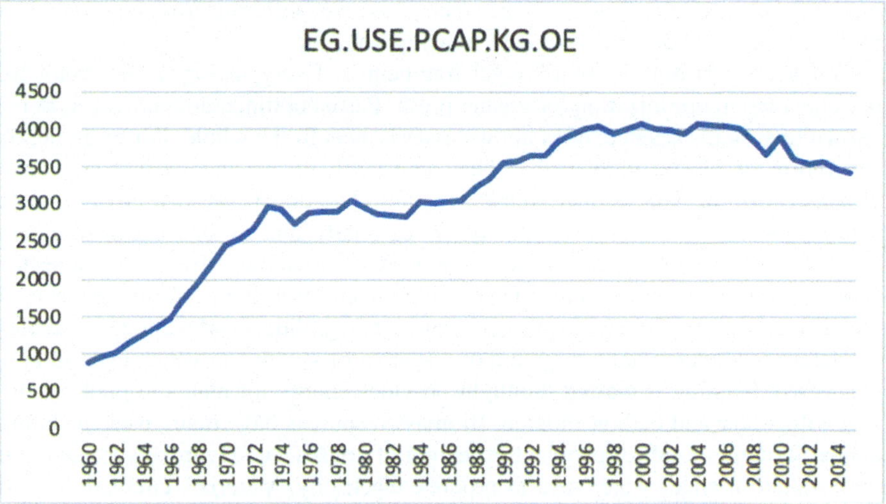

(a) Energy use (kg of oil equivalent per capita) 1960–2014

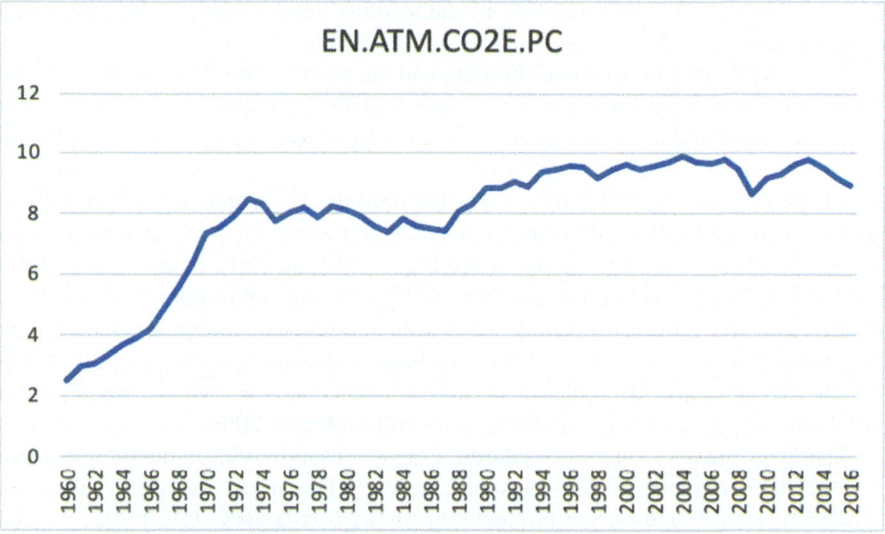

(b) CO_2 emissions (metric tons per capita) 1960–2016

Fig. 1 Economic development and CO_2 emissions, Japan. (Data source: World Bank, Development Indicators) (**a**) Energy use (kg of oil equivalent per capita) 1960–2014. (**b**) CO_2 emissions (metric tons per capita) 1960–2016

turning farmland into solar farms depresses food production and reduces food self-sufficiency; the need to increase food production has led to excessive consumption of water resources and degradation of the ecological environment in many countries; and acceleration of urbanization in developing countries focused on high efficiency of economic activities caused human suffering from environmental pollution, poor public safety, disease transmission, etc.

Our society is highly complex yet fragmented. Every sector or sub-sector has been devoted to optimization for greater profit. Partial optimization, however, as it is known in systems science, does not necessarily benefit the whole. For example, the generation of plastic garbage has been constantly increasing despite significant progress with the 3Rs (reduce, reuse, recycle). The effects of such environmental approaches are often confined locally or regionally due to data and institutional constraints. For a sustainable society, we need to move away from silo- or sector-based approaches toward an understanding of entire systems, from upstream to downstream, from home to town to city, and across political and natural boundaries.

Nature is fundamentally a system or a system of systems in which everything is connected. Nothing is wasted. Sunlight provides energy for plants to photosynthesize with water and carbon dioxide. Biomass resources have many uses, including animal foraging, food for people, the heating of houses, etc. Hydropower plants can be installed in small waterways to produce electricity. Materials circulate around, and energy cascades from higher to lower quality forms, and finally radiate out to the universe again. By connecting such systems, nature avoids trade-offs and creates synergies. Aware of the adverse impacts of human activities, scientists think about how to introduce the connectivity of nature into human systems for sustainable development.

This movement toward nexus thinking and the nexus approach was triggered by a Nexus Conference held in Bonn in 2011 (Hoff 2011; Bazilian et al. 2011). The term *nexus* first appeared in the scientific community in 1876 in the journal *Nature*, where it was used to describe “the organic connection between the nerves of the brain and the tactile centers” (Nature 2016). It was not until the 1970s that nexus thinking was further developed by the pioneering work of Ignacy Sachs on the food–energy nexus at the United Nations University in the late 1970s and early 1980s (Sachs 1980, 1988). The background at that time was that environmental problems around food, energy, and water were worsening worldwide and required integrated thinking and responses. The World Bank worked on the food–water–trade nexus (McCalla 1997), which was subsequently replaced by new concepts such as “virtual water” at the Kyoto World Water Forum in 2003 (Allan 2003; Merrett 2003).

The 2011 Nexus Conference caused a great sensation with its ambitious vision and acknowledgment of critical social needs. “The nexus focus is on system efficiency, rather than on the productivity of isolated sectors” (Hoff 2011), and it has become a popular research topic in the last decade. The number of nexus-related publications has increased exponentially, from several papers in 2009 to several hundred in 2018 (Newell et al. 2019). Many researchers focused on resource availability to secure supply and demand (Daher and Mohtar 2015). Others worked on urban metabolism in supply chains to improve resource-use efficiency (Bazilian et al. 2011).

The main goal of nexus research is to reduce inputs from outside a region by encouraging local production for local consumption (Siddiqi and Anandon 2011) and moving toward a circular economy (Bhaduri et al. 2015). As a critical global issue, research focused on the nexus of food, water, and energy, because the three are connected to each other in direct ways. In the age of global networks, food, water, and energy move across borders and regions, in theory and in practice, and the nexus is evident at various scales: global, continental, national, regional, community, and individual building.

The target of nexus research has expanded from nexus pairs of FE (food–energy), EW (energy–water) (Irena 2015; Varbanov 2014), and FW (food–water) (White et al. 2015), to include climatic impacts (Carpenter et al. 2015; Johansson et al. 2010), and now to the three pillars of FEW (Endo et al. 2014; Endo and Oh 2018). Inspired by the integrative concept (Cairns and Krzywoszynska 2016; Nature 2016), many researchers have started to rethink the physical and social dimensions through the nexus lens and apply the energy–water–food–X nexus approach (Sperling and Berke 2017), for example, FEW–health (Kumar et al. 2019; Ramaswami 2020), FEW–climate change, and FEW–SDGs (Yang and Goodrich 2014), etc.

Cities as centers of resource use have attracted particular attention for the complexity of urban issues. Urbanization displaces agricultural land, encourages people to leave farming, and reduces the local production of food. As a result, the gap between local food supply and demand widens, resulting in ever-increasing impacts on the local and global environment. Today, more than 56% of the global population lives in cities, which account for 80% of global energy consumption and 70% of global CO_2 emissions (Dhakal 2010). In the household sector, food, energy, water, and mobility are reported to account for 60% of the total emissions including shopping, eating, transport, heating and cooling, water, and sanitation and 90% in indirect home consumption (Ramaswami et al. 2021). The urbanization rate is predicted to be 60% by 2050, with drastic increases in consumption for food, energy, and water by 50%, 60%, and 55%, respectively.

Meanwhile, modern cities rely increasingly on advanced technologies, involving upstream resource extraction, intermediate transport by road, rail, port, and air, and end consumption by residences and businesses. Cities usually manage the demand and supply for water, electricity, gas, and food stuffs by sector. One product or service is often completed by many companies and sectors in a long process of supply chains, from production and transportation to consumption and disposal. Industrial sectors typically compete with each other. Nexus thinking is generally weak in urban planning and management as they are typically practiced, so there is room for urban nexus thinking to address the food–energy–water nexus in cities from a holistic perspective involving global and local human–environmental systems (Hoff 2011).

The urban nexus is emerging as a frontier of research on sustainable cities (Sperling and Berke 2017; Newell et al. 2019; Ramaswami 2020; Yuan et al. 2021). Soon after the 2011 conference, the EU launched the Urban Nexus Project, which published a series of reports (Urban Nexus 2013a, b, c). The ambition of the initiative was to break through silo-based technological and institutional ceilings. ICLEI (Local Governments for Sustainability) and GIZ (Deutsche Gesellschaft für Internationale Zusammenarbeit) took the baton on practical urban nexus projects in Southeast Asia (Lehmann 2018). These studies focused on the issues of FEW in emerging cities and investigated nexus effects through the shift to distributed urban infrastructure systems.

Nexus research so far has proposed a paradigm shift in the way services are provided and consumed for established cities, especially in developed countries, where people often take the established FEW services for granted. A common perception is that a city is already adequately resource-efficient and that there is little room for further improvement. Many consider it to be natural for developed cities to be on the receiving end of precious land resources from outside. Such thinking has limited our imagination of sustainable cities and biased urban nexus research to looking at how to secure resource supplies for cities from outside. This may have blinded researchers to the great potential of cities, not only as agglomerations of people but also productivity by the concentrations of buildings, floor space, and intelligent systems.

Cities are a complex three-dimensional space of natural and man-made systems with less land use and higher energy efficiency. The search for greater sustainability requires us not only to look outside of cities, but also to exploit the potential within cities. Technology is advancing quickly and making cities centers of innovation. Food, energy, and water can be produced in cities with high efficiency. Solar panels could be installed everywhere in cities. Urban agriculture is gaining attention as an emerging industry. The use of permeable pavement is on the rise and can transform infrastructure from gray to green. Such technologies are already available but not broadly deployed in cities. The implementation of such technologies could lead to competition with conventional uses of land and space. We do not yet know yet how to make them coexist harmoniously with each other. To this end, nexus thinking and nexus approaches can be powerful tools.

However, our understanding of the urban nexus is still at an early stage (Liu et al. 2017; Arthur et al. 2019). First, the nexus concept itself is esoteric and has no fixed definition; second, the subject of nexus thinking is complex and often unexplored in cities; third, the nexus approach is not well established and there is no commonly accepted nexus methodology; fourth, nexus effects are often context-dependent, and there is no single indicator to evaluate them. This immaturity of nexus thinking limits the feasibility of using the nexus approach in cities. This chapter attempts to clarify the complexity of urban systems generally and proposes design-led nexus approaches for urban sustainability based on the *moveable* nexus thinking for transferring knowledge and technology of food-water-energy nexus in research to practice in any cities through a series of design processes (Roggema, 2021).

2 Approaches Toward Urban Sustainability

2.1 Urban Systems and Research Approaches

Urban systems have traditionally been explored in different ways. Typically, an urban system can be classified as a geospatial system, a socio-ecological system, a metabolic system, and most recently, a cyber-physical system.

Classic approaches include geospatial systems in the agricultural land-use theory by Johan Heinrich von Thunen Tünen (Samuelson 1983) and central place theory by Walter Christaller (Agarwal 2007). These theories apply economic principles that govern human activity in cities to geographic space and structurally model the urban area in the geographic dimension. Urban sustainability built on these theories has been understood as a dynamic of growth and shrinkage driven by profit-seeking market powers. Limits of natural resources and biocapacity were generally left out of consideration or treated as givens. This oversight is said to be at the root of urban and global environmental problems.

To augment economic theories about cities, the concept of the socio-ecological system was introduced as applied theory of ecology and complexity (Frank et al. 2017) and presented. This theory recognizes that economic and social activities are built upon the natural environment, and that cities will not be sustainable if they cannot coexist with the ecological environment. It aims to model the dynamics between man and nature and seeks a balance between the demands of human society and the capacity of natural systems. "This theory posits a "theory-neutral framework" to analyze the linkages and relations of an ecological (non-human, physical, or material) system with one or more social systems while paying more attention to economic or quantifiable units (e.g., employment, tourist numbers, population, etc.) or to less quantifiable components such as social learning and land use" (Covarrubias 2019).

The metabolic system models the circulation of materials and the flow of energy, rationalizing the system in terms of material stocks and energy flows, as "natural resource systems supply urban areas in the form of flows" (Zhang 2013). It describes the city as an open socio-ecological system, the sustainability of which depends on a healthy metabolism. Flows are the continuous streams of objects, materials, resource units, ideas or information, or any other form that moves between at least two points. These can be either material flows (e.g., drinking water) or social flows (e.g., policies for drinking water provision). A material additions-and-withdrawals perspective is, therefore, insufficient. It needs to go further into a more sociology-based analysis of flows, which focuses on "the role of policies, institutional arrangements, networks, and social meanings shaping urban provisioning of resources" (Covarrubias 2019).

Finally, a recent trend is to consider the urban system as a fusion of cyber and physical systems, explicitly addressing the role of information, communication, and digital technologies (Yan and Sakairi 2019). "Cyber-physical systems (CPS) are complex systems with organic integration and in-depth collaboration of computation, communications, and control (3Cs) technologies". The proposition is that urban sustainability can be achieved using digital technologies to further improve resource efficiency. The concept has been incorporated into governmental policies like Industry 4.0 (Germany), China 2025 (China), Society 5.0 (Japan), etc. Under the support of those initiatives, "smart cities" projects have been implemented in local governments, industry, and communities (Cassandras 2016; Gassmann et al. 2019).

Depending on how each urban system is perceived, different methodologies and approaches have been applied in research and practice.

Perceiving cities as complex geospatial systems, planners have diligently sought for the best spatial form of cities to be buildable, manageable, livable, and sustainable. The utopian journey never ends, with aspirational concepts evolving from the garden city to compact city, then smart city to resilient city. Key issues within the discourses include, but are not limited to, the proper size of a city, built-up density, and allocation of urban functions. However, there are tensions between efficiency and effectiveness. Cities may be seen as economically efficient, but they may not be so effective in resource and environmental sustainability.

To address such tensions, various approaches have been developed under the umbrella of social-ecological systems, such as the ecosystem-based approach (Brink et al. 2016), river basin-based approach (Kaushal and Belt 2012), and nature-based approach (Scott et al. 2016; Roggema et al. 2021). Professional tools and models have been developed, including the social-ecological system framework (SESF), the vulnerability framework, and the driver-pressure-state-impact-response (DPSIR) framework. To foster a better understanding of the dynamics and complexity of social-ecological interactions, a variety of assessment methods including both quantitative and qualitative approaches (e.g., system dynamics modeling, network analysis, agent-based modeling, multi-criteria analysis/indicator-based aggregation, and integrated assessment/decision support systems/coupled model frameworks) are available. "Despite this progress, the operationalization of the conceptual frameworks through applying innovative methods and tools to allow for the sustainable development of SES is still an active field of investigation" (Gain et al. 2020).

The technical approach of integrated water resource management (IWRM) (Grigg 1999) is widely used. Discussions on IWRM have emphasized issues of integration through an analysis of (a) inter-sectoral competition for surface freshwater resources; (b) integration of water management at farm, system, and basin scales; (c) conjunctive use of surface and ground water resources; and (d) prioritizing water for human consumption and environmental protection (Khurian and Turral 2010). Some have pointed out that this approach neglects the political dimension through a focus on "natural boundaries" and "neutral planning and participation" (Wester and Warner 2002).

Urban metabolism analysis has become an important tool for the study of urban ecosystems, and diverse methods and tools have been developed for the analytic approach. It considers that "the problems of large metabolic throughput, low metabolic efficiency, and disordered metabolic processes are a major cause of unhealthy urban systems" (Zhang 2013). The concept of the urban metabolic system has evolved from a linear input–output system to a cyclical feedback system, and more recently, a complex network system. Typical methods often used include material flow analysis, energy flow analysis, ecological footprint analysis, life cycle assessment, and input–output analysis. With its quantitative perspective, this approach does not cope with social aspects directly. To improve the applicability of approach in policy making, political goals such as decarbonization "should be integrated explicitly with natural carbon metabolic processes to define carbon sources and sinks within an urban system, as well as the factors that influence the carbon-related processes at various spatial and temporal scales" (Zhang 2013).

Cyber-physical systems (CPS) aim to fully combine sensing, processing, and actuation of information with urban physical systems, through the smart use of sensors and data content. The prevalence of the Internet brings CPS from the laboratory out into society, as the name Geo CPS indicates (Yan and Sakairi 2019). This raises critical issues in the geographic space. For instance, smart electric power grids are transforming into open systems that can accept distributed renewable energy sources; water works may be integrated to include independent wells or rooftop tank systems; driverless vehicles must simultaneously respond to traffic conditions; healthcare robots must communicate with patients in a manner appropriate to their conditions. The synergy effects of position and attributes present both opportunities and challenges in Geo CPS.

Our collective understanding of urban systems has been evolving continuously, from human-centric geospatial systems to human-nature harmonized social-ecological systems, to functionally oriented material/flow metabolism, and finally to the fusion of cyber and physical spaces. Each of the theories tries to take the intrinsic attributes of cities from a specific perspective. Scientific and technological approaches were derived independently, so the connections and relationships between systems and approaches were not explored very well. A more prevalent view today is that cities are nodes where cross-sectoral actors, resources, infrastructures, policies, and utility services come together for the provisioning of water, energy, and food (Covarrubias 2019). Needless to say, understanding the characteristics of the nodes and managing their forms and functions are essential to achieving the sustainability of cities. The emergence of nexus thinking and approaches is a response to these social and urban calls for action.

2.2 *Nexus Thinking and Approaches*

Cities are driving forces of economic growth and cultural activities today, though they impose negative impacts on the natural environment. It is no exaggeration to say that the sustainability of our planet depends largely on how we build our cities. According to Sperling and Berke (2017), "a major challenge to achieving resilient and sustainable cities is increasing the areas that overlap each other and identifying most-critical uni-directional, bi-directional, and multi-directional dependencies, synergies, and feedback loops between systems."

While geospatial, socio-ecological, and metabolic systems mostly focus on physical dimensions of the urban system, nexus thinking deals with complex environmental and sustainable development issues. Trade-offs and synergies *between* elements and systems must be understood to make the whole system or society more efficient. According to Hoff (2011), "The nexus approach [is] an approach that integrates management and governance across sectors and scales." Another description is that "It is a form of systems thinking which focuses on the inter-linkages between natural resources and the ways in which the linkages are or could be managed and steered into more sustainable and integrated configurations" (Vogt et al. 2014). Sperling and Berke (2017) conceptualized the urban nexus science roadmap as a means of identifying and exploring synergies, trade-offs, and co-benefits across water, energy, and food (WEF) systems for building resilient and sustainable cities.

Research so far in the nexus arena has mostly focused on the physical aspects of the concept. Covarrubias thought the social dimension cannot be ignored, and defined nexus thinking as being "about the interaction between WEF provisioning in cities consisting of socio-material flows. . .Such a more socio-material informed flow analysis is, however, often ignored when analyzing interconnections between WEF in cities" (Covarrubias 2019). The material–social system is characterized by the fact that it also covers the norms and institutions of human society that manage the physical elements. He argues that "the analysis of the urban nexus should not just focus on creating cross-sectoral synergies or identifying cross-sectorial trade-offs, but also on bridging the material-versus-social divide that has long characterized systems of urban provisioning" (Covarrubias 2019). The material–social system covers the pitfalls of the urban metabolic system view by the introduction of the network governance approach. "Conducting a social flow analysis would go beyond the material aspects of flows and instead center on the social organization, actors, networks, policies, ideologies, discourses, and any kind of socio-cultural meaning that goes along with the material flows of water-food-energy" (Covarrubias 2019).

There are two ways in which material and social flows of WEF can come together and constitute a "nexus." The first is where the material flows are the main driver in creating a nexus between WEF provisioning in the city (Covarrubias 2019). The second way in which a nexus can come about is when cross-sectoral linkages are more socially driven; the nexus is then a result of social interventions, such as a new policy or strategy for the provisioning of energy to certain neighborhoods requiring changes in infrastructures and the circulation of material flows (Covarrubias 2019). "Such cross-sectorial modes of steering and planning of material flows in the city represent an approach to governance that was referred to as the 'urban nexus'" (Covarrubias et al. 2019). The idea was inspired by Castells' "network society" theory to describe the social dimensions of the urban nexus by scope, connectivity, density, function, and values. Covarrubias discussed the power of social actors in shaping social networks and creating additional values (Covarrubias et al. 2019). The series of papers by Covarrubias conceptualized the material–social system of the urban nexus in terms of three aspects (networks, functions and values, and power dynamism) and qualitatively verified the existence of the nexus in the Amsterdam context. However, how to quantify the effects and operate the connections and interactions quantitatively remains a topic to be studied further.

2.3 Recent Trends in Urban Nexus Practice

The essence of nexus thinking is to do more with less by improving the efficiency of investment in resources and land use (Martínez-Martínez and Calvo 2010; Kurian and Ardakanian 2015). "While nexus literature is long on determination and ambition, it is short on grounded evidence on the essential elements of FEW security, such as operational definitions that help to link research and practice, particularly within urban systems". This is because the concept of the nexus is difficult to understand, the findings of scientific research are not immediately available, there is no contact point in vertically divided administrative systems to receive cross-cutting results, and practical cooperation between departments is even more difficult than research. Scientific knowledge is not popularly applied in the design practices linked to the creation of the urban built environment. Similarly, with some notable exceptions such as the Living Building Challenge, architects design buildings to save and manage energy, or work on the redevelopment of urban neighborhoods to improve liveability. Landscape architects work on urban landscapes, urban agriculture, and green infrastructure to create the feeling of a greener urban life. City planners study land-use policy to improve efficiency of transportation and distribution. With regard to the FEW nexus, each profession has a kind of limitation of scope that needs to be bridged both in breadth and in scale.

To break this bottleneck, the National Science Foundation (NSF) in the USA began a focused research effort in 2017 with a large budget for the food–water–energy nexus. In response, the Belmont Forum launched the Sustainable Urban Global Initiative: Food–Water–Energy Nexus (SUGI-NEXUS) with three goals: (1) to integrate FEW nexus knowledge, (2) to promote technological innovation, and (3) to trigger stakeholder action. M-NEX is an initiative to integrate FEW knowledge and technologies into urban design. The M-NEX project was selected for its uniqueness of a design-led approach to urban design, introducing FEW knowledge and technologies and promoting co-creation through urban living labs (m-nex.net). The project has been conducted in six cities: Amsterdam, Belfast, Detroit, Doha, Sydney, and Tokyo. As a result of the project, the design support platform M-NEX was developed through a series of charrette design workshops. The following section of this chapter will introduce the essence of the design-led nexus approach and the case of practice for an adaptive Tokyo.

3 Design-led Nexus Approach

It is common understanding that sustainable cities must be constructed on robust land with resilient infrastructure and services. Meanwhile, every city is unique in terms of its land, people, and the systems for securing FEW in the bioregion. Therefore, solutions for sustainability will also be unique in different contexts, scales, and timing. To respond to these challenges, the moveable nexus and design-led approach was proposed by the M-NEX project of the SUGI-Nexus program. Instead of defining FEW as securities and risks, the moveable nexus considers that the FEW nexuses could be revitalized in an urban built-up context, and by doing so, turn problems into opportunities to create a supportive environment and sustainable services, even in built-up urban environments (Yan and Roggema 2019). The multiple facets of the moveable nexus have been highlighted by the project publication (Roggema 2021). The philosophy is implemented in the design-led nexus approach, which combines the physical world in situ and cybernetic knowledge in human brains or computers into design solutions that are adapted quickly to local contexts. The approach contains three operational techniques: the charrette design workshop, the iterative design process, and the evaluation indicator FEWprint (Yan et al. 2021), as described below.

3.1 Charrette Workshop

Design itself is a process to integrate knowledge and technology at the architectural, urban, and regional scale (Cairns and Krzywoszynska 2016). It takes design behavior as a kind of boundary object (Newell et al. 2019) through a sophisticated design process. The design-led approach fully embraces the essence of nexus thinking-integration and interaction. The M-NEX project uses charrette workshops to organize the relevant knowledge and technology through intensive communications (Yan and Roggema 2019). A charrette workshop is a participatory design activity that offers a demanding process with expectations about delivering innovative solutions within a tight time schedule (Roggema and Yan 2017). The step-by-step design process gives a clear guide for performance and turns inspiration and creativity into regional-to-local design propositions that belong together. Participation and communication provide opportunities for stakeholders to envision and create new ideas on the future of a geographical area from diverse perspectives. This is particularly true in participatory design, where design proposals are the result of professional design activities and communications with a variety of stakeholders. It is extremely useful, as it can create something out of nothing that existed before, presenting opportunities to be continuously and collaboratively adaptive as a city, as a landscape, and as a society (Roggema 2021).

3.2 Iterative Design Method

The M-NEX project developed an iterative design method built on successive charrette workshops. It consists of three distinct phases: (a) exploration, (b) iteration, and (c) communication. *Exploration* sets out the ambitions, visions, expectations, and formal program demands of the various stakeholders and partners—notably identified through the participation tools and techniques. The *iteration* phase consists of continual design iterations framed by radical hypothetical "what-if" questions. These intend to stretch propositional boundaries beyond traditional models to open up new, large-scale projects that stakeholders and partners might not have otherwise considered. Once the iterative process has concluded, the final stage is *communication* back to stakeholders and communities. This stage presents outputs of the design process in legible and understandable formats for all audiences. This method allows for transdisciplinary teams to engage with the contexts of each workshop and to understand the perspectives of stakeholders that directly feed into the participation platform. Similarly, the method is "moveable" (transferrable), for application in other cities around the world where debates on the future of urban regions are required to address climate change, for example. The process is flexible with respect to the composition of a given interdisciplinary team, allowing different groups of participants to engage with various solutions to the challenges of a particular place.

3.3 Nexus Tool: FEWprint

Making the workshops workable requires the support of data, information, and communication tools. The evaluation of design solutions is a tricky issue. There exists a long list of indicators to assess the impact of human activities on the environment, such as the most typical ones, which include food mileage, CO_2 emissions, virtual water use, and the ecological footprint (EF) (Wackernagel and Rees 1998). None of them can be used easily for the trial-and-error process of participatory design. Inspired by the EF, M-NEX developed an indicator called the FEWprint in terms of equivalent land area as environmental load by the sum of (1) the land area needed to meet the supply and demand for food, energy, and water and (2) the forest area to absorb the corresponding CO_2 emissions related to FEW resources and services. The FEWprint is applied iteratively through the design process. The output can express the existing environmental load of the FEW demands as a baseline, or the effects of FEW production and creative design for FEW supply at the local level within a household, a street block, a neighborhood, or a city. Such a simplified indicator is extremely useful to assess the performance of design work under different scenarios and alternatives. Stakeholders can understand the environmental costs, the trade-offs, and the synergies of different solutions and eventually rethink the inter-relationships in their behaviors. A simple example of the

FEWprint was provided by Yan and Nakayama (2021) for the environmental load of vegetables in suburban Tokyo.

4 Practice of the Design-led Nexus Approach

4.1 International Design Workshops

The M-NEX project organized international workshops bi-annually over the project period, from 2018 to 2021, at six living labs, including Amsterdam, Belfast, Doha, Detroit, Sydney, and Tokyo-Yokohama. Yan and Roggema (2019) reported on the geographical features, bioregional differences, and social themes of each study area. The key advantage of the design-led nexus approach was to harvest knowledge through a series of local and international activities and to implement the moveable nexus method incrementally. Each city team hosted at least one international workshop over the period. Consequently, the knowledge obtained at each workshop was integrated and provided as expertise and solutions from the M-NEX Project at multiple levels, from building to neighborhood, city, and region. The process and results of the workshops are described in which covers how the design-led approach works over time and the main products of the project. Here we highlight the Tokyo contributions and concept applications.

4.2 M-NEX Living Lab Tokyo

The aim of the M-NEX project in Tokyo was to redesign the FEW management systems for stable provision of FEW services and improvement of quality of life. Typical neighborhood areas were selected according to their location from the city center to the suburbs (Yan and Nakayama 2021). The transition of urban forms and land-use patterns in the transects studied presented a perspective to examine the progress and impacts of urbanization on local food systems. The project was promoted and integrated with the WISE Living Lab (WLL), one of the key stakeholders of this project. WLL was originally a joint facility established by Tokyu Corporation for urban revitalization in 2017 as a follow-up project of the 2012 initiative of the "Next-Generation Suburban Town Project (NST)" http://jisedaikogai.jp) between the company and Yokohama City. Over time, WLL has engaged in many activities, such as investigation of resources in the community, future vision design, experiments in mobility, and women's participation. It provides a solid platform for any stakeholders to cooperate with the government (Yokohama City), companies (Tokyu Corporation), the community (neighborhood associations), and residents (civil society groups). The M-NEX project joined WLL in 2018 and worked with the lab through the project period. This was the first attempt for the lab to think about sustainability through the food–energy–water nexus.

4.3 Implementing the M-NEX Nexus Approach in Tokyo

The M-NEX method has been developed to be applicable in any city, although solutions may differ depending on context. To apply the method it is crucial to understand the context and customize the methodology. It may be easier in the Netherlands to propose an all-vegan diet in conversation with stakeholders than in the USA, for example, or practices will differ at the individual and cultural level in Qatar versus Japan (Yan et al. 2021). Following the principles of the design-led approach (Yan and Roggema 2019), the M-NEX Tokyo project worked with WLL on the sustainability of suburban housing, in particular for Sustainable Development Goal (SDG) 11 (Yan and Nakayama 2021).

4.3.1 Reframing the M-NEX Design Method

Generally, the M-NEX method includes three phases: exploration, iteration, and communication. This is coupled with the process of application development by design, evaluation, and participation. The process is reframed in Fig. 2 for a specific target site. It includes the typological assessment of the spatial conditions of a given site or location. It also provides a baseline for assessing how the neighborhood is adapted physically, but also how activities could alter FEW resource use through FEWprint calculations. Similarly, it allows for a generalizable approach that can be employed across a larger area with similar spatial, social, and economic conditions. The process often examines how to integrate urban growing systems, techniques, and technologies into a given site while exploring the resultant programs, economies, and opportunities it creates for residents and/or communities. It prioritizes

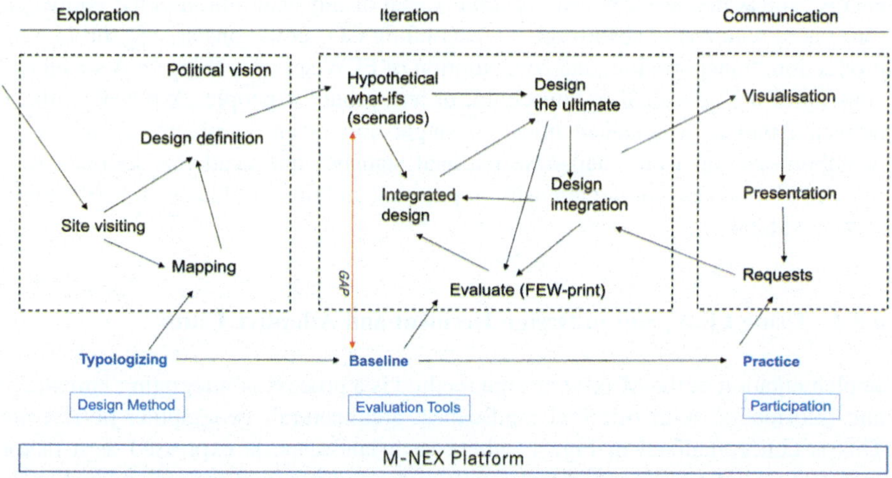

Fig. 2 Reframed M-NEX design method for activation

ideas of circular resource use and proximity of complementary activities or processes. The FEWprint indicator, developed through the evaluation platform, is integrated at this stage to assess the effects, and the design process continues until the parties involved are satisfied with the results. Iteration takes place qualitatively (descriptive of what-if possibilities, site conditions, stakeholder, and partner feedback), quantitatively (FEWprint assessment), and spatially (modeling, mapping).

4.3.2 Measuring Nexus Effects Using FEWprint

Urbanization broke up FEW nexuses by creating spatial distance between parts, and it also created sectoral segregation. The distance between demand and supply of goods and services at the urban scale depends on context, such as the size of a city, and the location of a site within the region. Generally, there is very limited space reserved for producing or managing FEW components locally within densely built-up environments like Tokyo, while more opportunities are available in suburban or extra-urban zones. However, the situation is changing because of a revolution in the production of renewable energy such as photovoltaic panels and vertical agriculture. Self-produced or locally produced and harvested food, energy, and water could shorten the distance between demand and supply, eventually reducing the FEWprint where that gap is shortened. Therefore, the FEWprint is useful to evaluate not only the consumption intensity of FEW but also the effects of efforts to reduce the imported resources.

By using FEWprint effectively, the M-NEX method can (1) redesign the relationship between demand and supply, (2) reassess the costs and benefits of FEW resources and services, and (3) rediscover opportunities for innovative FEW management in cities.

The FEWprint indicator is calculated as the land area needed to produce food, energy, and water services for a specific social or physical unit, and the equivalent forest area required to absorb the corresponding CO_2 emissions arising through the production, transportation, and consumption of FEW services (Fig. 3). A social unit consists of an individual, a household, or any group of people. A physical unit is defined either as a detached house, a single unit in an apartment, a building, a neighborhood, or a city. Japanese national statistics and academic journals were referred to as a basis for the intensity of personal and household use on FEW supply and consumption.

4.3.3 Using FEWprint to Design Resilient and Adaptive Cities

Implementation of the M-NEX design method is a process of integrating knowledge and information with different landscapes: architectural, geographic, policy, etc. This is conceptualized in Fig. 4, where each landscape is expressed as a factor, each one including various objects, with social and physical features related to the supply and demand of FEW components. The demand for FEW can be assessed at

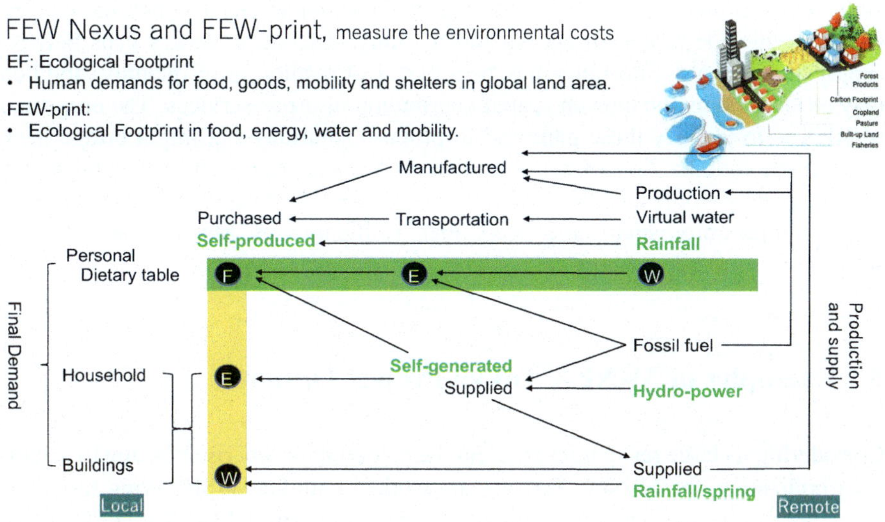

Fig. 3 FEW nexus and the FEWprint

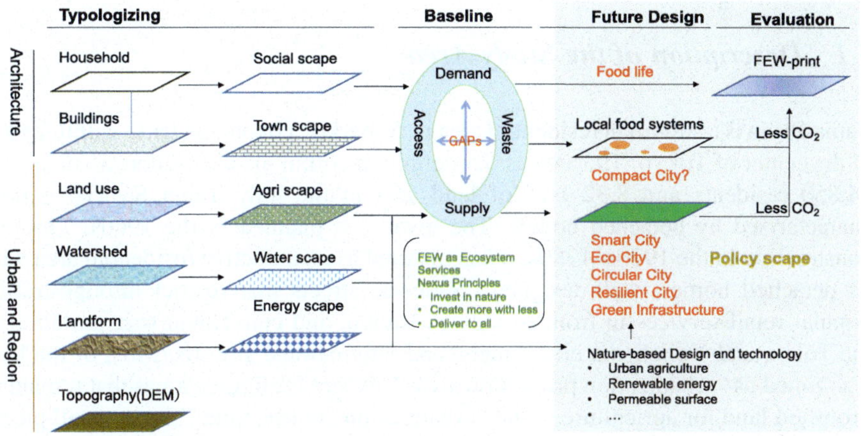

Fig. 4 Implementing the M-NEX method in urban design

the household or building unit level and then scaled to the neighborhood, city, or regional level. Present FEW demand is used as a baseline to begin analysis using FEWprint.

The FEW supplies in the equation include the natural provisioning and services, otherwise known as ecosystem services. Demand is satisfied when there is enough supply. However, the ability to produce sufficient food, energy, and water in cities is constrained by environment and technology, including parameters as diverse as topography, transportation systems, and the coverage of FEW services. A reduced ability to meet direct needs within an area implies poor services, which not incidentally may include a lack of access and mobility. There are always gaps between supply and demand. Shortfalls may be a matter of quality, or of quantity, and both may be affected by the uneven spatial distribution of urban services. The purpose of M-NEX is to identify these gaps and to propose solutions that might bridge them, through the mobilization of social-ecological resources and governmental policy. These typically would take the form of broad concepts in the political landscape such as the compact city, smart cities, eco-cities, resilient cities, zero-carbon cities, and so on.

5 Examples of M-NEX Tokyo Project Outputs

Considering its large residential area and the potential for low rise housing to achieve decarbonization, the M-NEX Tokyo team worked with the WISE Living Lab in the study area of Tama-Plaza, a suburban neighborhood around the Tama-Plaza train station in Aoba-Ku, Yokohama City, near Tokyo.

5.1 Description of the Study Area

Tama-Plaza is a suburban residential area in Aoba-ku, Yokohama City, within 25 km of the center of Tokyo. The area as our definition within the old Ishikawa village has 84,850 residents and 8.32 km^2 of land area in the hilly Tama Kyuryo region, characterized by detached homes. The town was planned in the 1960s. Through construction in the 1970s–1980s, it has matured as an attractive residential area with its detached homes, well-designed cul-de-sac streets that restrict through-traffic, popular retail services in front of railway station, and convenient transportation to the Tokyo and Yokohama city centers, and international airports. Most of the land was zoned as "urbanization promotion areas," except for limited areas that remain as protected land for agriculture in the "urbanization control zone." Some small pieces of land in the urbanization promotion zone were protected for food production based on "productive land for food" legislation where property taxes could be waived for a 30-year period. Most of the land in the study area was developed as low detached houses, while midrise buildings and retail are concentrated near railway stations. Residents of the study area are mostly commuters to the city centers of Tokyo and Yokohama, traveling from the nearest station at Tama-Plaza. As a result of these conditions, there is a big population gap between daytime and night-time.

5.2 Food–Energy–Water Nexus and Local Food Systems

Local farmland produces a very limited supply of vegetables, so most vegetables consumed in the study area are transported in from other prefectures in the Tokyo metropolitan region, from more distant Hokkaido in northern Japan, or from overseas. Residents purchase vegetables and other food from an abundant selection of local options:

- Tokyu Department Store food court, selling high-quality food for families
- Tokyu Store supermarket, targeting middle- and upper-class families
- Ito-Yokado, the most popular discount supermarket
- Hamakko farmers' market, with fresh vegetables supplied direct from local farmers
- Always-open convenience stores, serving prepared food boxes
- Vegetable vending machines, selling vegetables and rice
- Vegetable stands beside farmland, open irregularly
- Vegetable markets and food markets, with scheduled times and locations, operated by local brokers
- Food truck services, on scheduled times and routes

While there is diversity of choice, shops tend to concentrate in front of the Tama-Plaza train station. Thus, the need to use a car for food shopping is often inevitable, although seniors who have relinquished their drivers' licenses will face particular challenges getting around on foot in this hilly area. This made the CO_2 emission at the locations away from the railway station or the neighborhood higher. It indicates the importance of transportation modals to grocery stores in the calculation of FEWprint (Fig. 5).

5.3 Local Design Workshops

The M-NEX Tokyo project promoted the nexus concept at WLL through a series of participatory communications, with stakeholders, program development, a design workshop, and follow-up projects. The first stage aimed to nudge stakeholders through a series of events. Information was prepared on the five types of capital, conditions of the study area, and an inventory of stakeholders. The second stage was to develop an action program to work with stakeholders. A cooperative team, functional datasets, and tools were prepared. Information about the five types of capital, the FEWprint indicator, and food accessibility were kept ready for use in emerging communications. Third was the design workshop. Communication events were conducted bi-monthly in diverse forms, including field tours, street interviews, and forum exchange projects. Participatory design workshops provided the platform to share information, identify problems, and work out solutions. Supportive datasheets, access maps, FEWprint indicators, etc., were frequently used for

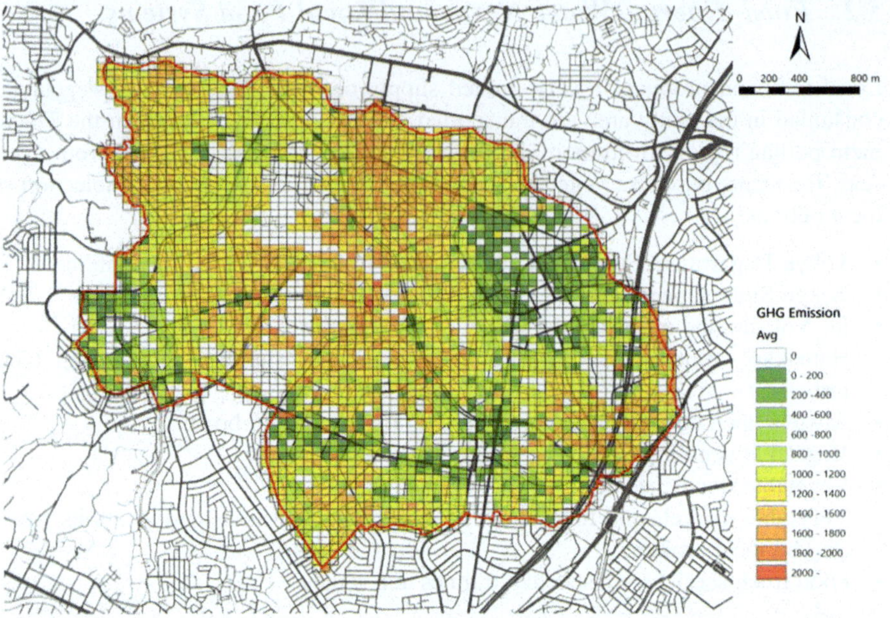

Fig. 5 Average CO_2 emissions (g-CO_2/cap-year) by 50 m grid for food access

communications. Fourth was follow-up. Swift feedback to participants on activities and workshops helped to keep the network active. Ideas and solutions were summarized on web pages, posters, and digital models for further discussion. Readers may refer to the details of the progress in Yan and Nakayama (2021).

5.4 *Building an Edible Garden City by the Home FEW Nexus Farm*

The M-NEX team held an online design workshop from 8 to 10 February 2021. Building upon the previous achievements of the project, the objective of this final design workshop of the event was to test the applicability of the M-NEX method in a suburban area of the metropolitan region.

The key issue in suburban Tokyo is to make the suburban community sustainable and responsive to the changing needs of society by revitalizing potential resources. Normally, land in built-up communities is fully covered by housing and paved roads and considered resource-poor in natural resources. Conversely, these areas are often rich in terms of rooftops, private gardens, and public services. Such a built-up neighborhood has advantages in the adoption of renewable energy (O'Brien et al. 2010). Residents could install solar photovoltaic or heating systems on rooftops and harvest rainwater for watering plants and vegetables in private gardens. Waste food

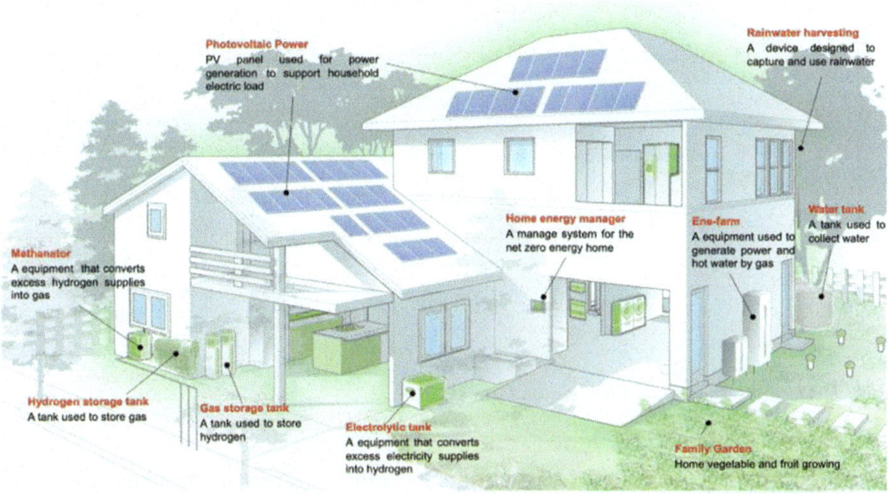

Fig. 6 Components of home FEW nexus farm. (Author: Ruiyi Zhang and EcoGIS lab)

can be composted for fertilizer. Surplus electricity could be sold to the grid or used to recharge electric vehicles, which could reduce peak demand on the grid or supply energy to houses in a power outage.

Applicable commercial products are already available, such as the ENE-FARM developed by Panasonic, which extracts hydrogen from LP gas and combines it with oxygen to generate electricity. This could be a transformative solution for suburban towns, where houses could produce and consume FEW independently.

The design could also include the collection of waste thermal energy to heat water. These concepts are capable of offering high resource-use efficiency and significantly reducing CO_2 emissions. Theoretically, about 60% of a household's electrical power demand could be met by this system, and the self-sufficiency ratio could be even higher with solar panels installed. A water tank could be added to harvest rainwater for gardening. We refer to this concept as the Home FEW Nexus Farm. As depicted in Fig. 6, this system has great potential to create a closed FEW nexus in suburban areas dominated by detached homes.

Figure 7 illustrates the cogenerative system flow of a nexus farm. Solar panels generate electricity, which can be consumed directly or stored in a battery. The system is controlled by a home energy management system (HEMS) that monitors and manages the generation and consumption of energy. Not only can the Ene-Farm system produce electricity using city gas, it could also support use of hydrogen. Hydrogen energy, distributed by hydrogen stations, could be produced using renewable energy such as solar and biomass. A family that owns a fuel cell vehicle (FCV) could connect it to the home system in times of disaster or power outage and use electricity produced by hydrogen fuel cells in the vehicle. This appears to be a promising concept that could make cities more resilient and adaptive to disasters.

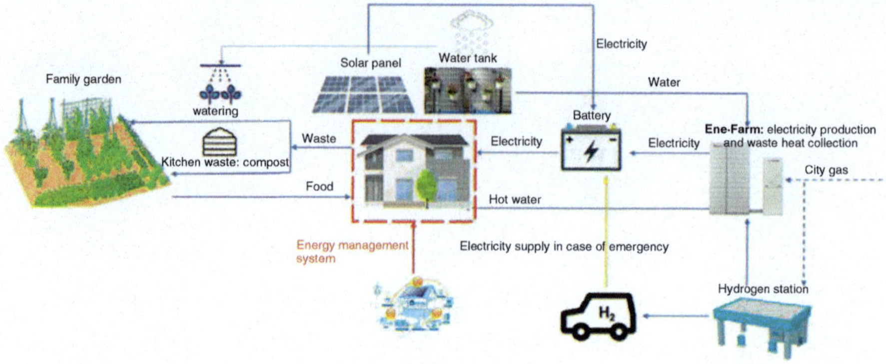

Fig. 7 Design of the home FEW nexus farm system

Rainwater can be collected for watering the family garden to produce food, and kitchen waste can be recycled through composting to fertilize the family garden.

By applying this home FEW nexus farm to a detached home community we could create a circular FEW system in the neighborhood. The study area, Tama-Plaza, contains many natural and social resources both upstream and downstream, including fruit and vegetable production, bakeries, fresh noodle makers, etc. There are farmlands, allotment gardens, and water tanks already in place. Areas along the Tama River could be restored and revitalized with permaculture, while both private and public gardens could be used for rainfall collection and to enhance green infrastructure. There is also a large wholesale market in Kawasaki City, neighboring this community. Social networks are active and rich, with various actors around the WISE Living Lab. By integrating all of these natural, social, and industrial resources, an "Edible City" was proposed through this project, with much improved self-sufficiency in terms of FEW supply and demand (Fig. 8).

Eventually, a donut-type system could be built as a new urban system based on the FEW nexus, as illustrated in Fig. 9. The white ring of the donut indicates the flow from consumption to market to production, in which waste is returned as fertilizer. This forms a cycle from production to consumption and disposal. The gray ring on the outside represents the current situation of a city where farmland, housing, and stores are connected to each other through direct physical access. The size of the rings is intended to express the physical scale of the movement of materials. The closer they are to each other, the lower the cost of access. The spatial balance between the three, in other words, the positioning of housing, stores, and farmland matters in terms of the circular city.

This donut-shaped urban FEW model redefines the relationships of production, distribution, and consumption. We consider the nexus at different scales in urban planning, from household to city. These multiple perspectives make design at the architectural level easier to scale for policy making at the city scale. Beyond production and consumption, the physical stocks of houses, farmland, and retail stores are all considered potential elements of the urban nexus. The donut would be a

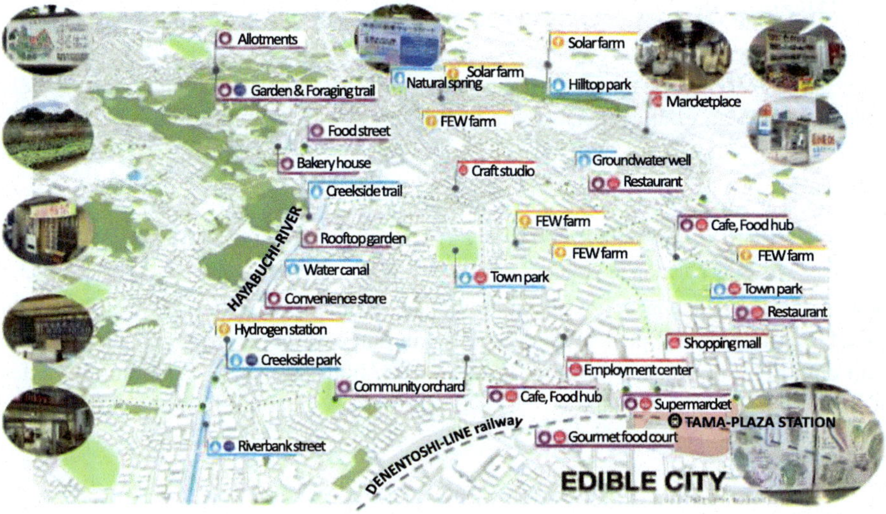

Fig. 8 Scaling up from home FEW nexus farm to an "Edible City"

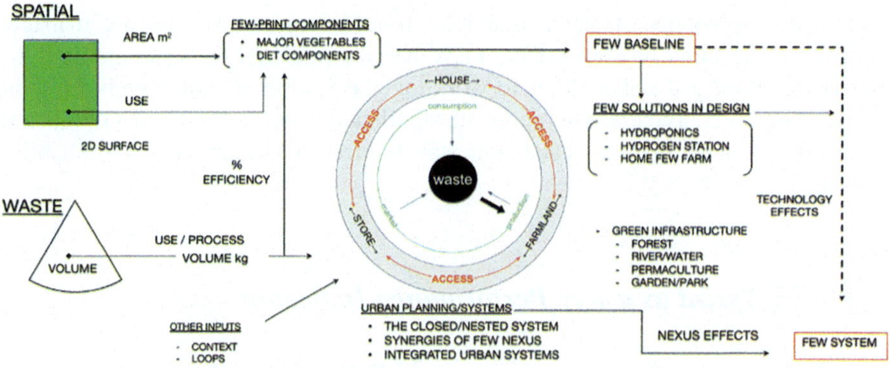

Fig. 9 An urban system featuring the FEW nexus

nested system across scales rather than a system of silos. The smallest element of the donut would be the home nexus farm. A network of home nexus farms would form the circulation system, with nature and social resources at the community or city level creating brilliant nexus effects. By applying these concepts and examining the dimensions of the circles in different scenarios, we can begin to see the vast potential of the FEW nexus effect. The FEWprint indicator can serve as a powerful evaluation tool when applied to cases, to compare before-and-after effects.

6 Discussion

Key issues of the design-led nexus approach for sustainable cities have been discussed in Yan et al. (2021). Some of the key points are highlighted below.

6.1 *Food as an Entry Point*

The food–energy–water nexus is a wicked problem in cities. To simplify the problem, the M-NEX project used food as an entry point for organizing workshops and eliciting solutions, with consideration from both the demand and supply sides. The choice to take advantage of the popularity of food helped facilitate our approaches with stakeholders.

6.2 *FEW as a Commons*

Generally, the supply of and demand for food, energy, and water are managed separately by actors in each sector. FEW has the potential to be a commons for otherwise silo-based stakeholders to work together and think collaboratively about services and infrastructure. This approach significantly supported the SUGI initiative (https://jpi-urbaneurope.eu/), catalyzing stakeholders to take action on sustainability issues.

6.3 *FEWprint as a Key Performance Indicator*

The challenge of how to deal with changes in FEW quantitively was a key issue in discussing "moveable" nexus thinking. The FEWprint indicator provides key performance indicators (KPI) to evaluate the effectiveness of design propositions under various scenarios, such as business as usual (BAU), incremental change, and transformational change. It can help to evaluate adaptation to climatic and demographic changes in comparison with current conditions as a baseline. With FEWprint, the M-NEX method can be used to (1) redesign the relationship between demand and supply, (2) reassess the costs and benefits of FEW resources and services, and (3) discover opportunities in cities for innovative FEW management in the future.

6.4 Technological Innovation

Technology plays a key role in solving problems toward sustainable urbanization by improving productivity, reducing carbon emissions, and creating jobs. No one solution fits all cities. Adoption of technologies should be based on an understanding of the political visions and fundamental factors of each study city. The Tokyo team demonstrated the innovation of a renewable energy and hydrogen-based home FEW nexus farm for a large number of detached homes in the suburbs of Tokyo.

6.5 Contextualization

Each city in this project had distinct concerns and priorities of problems to discuss, so the solutions were totally dependent on the relevant local socioeconomic and natural conditions. It is always important to consider how to make the design process context-specific. It is crucial to understand context, because context will directly impact the ability to design in any given situation. Having a common indicator like FEWprint or common language as in the sustainable development goals does help localize the moveable nexus approach appropriately in a distinct context.

7 Conclusions

This chapter reviewed the concept of nexus approach and its application for urban design. A new concept, *moveable nexus* was introduced in the context of urban dynamics and applied in the M-NEX project. In this project, a series of design charrette workshops was organized by the Living Labs of six cities, while knowledge and technology were harvested for the common M-NEX design support platform. The platform is also a strategic guideline for practicing nexus approaches in many situations. An easy-to-use indicator FEWprint was developed in the composition of the physical land area to meet the demand for food, energy, and water, and the virtual forest area to absorb the correspondent CO_2 emissions. This indicator can be used to quickly evaluate the status of decarbonization and the effect of design scenarios in policy making. Urban living labs are recommended for the engagement of stakeholders and the outreach of the nexus approach. Thanks to the WISE Living Lab, a partner of the M-NEX project from an early stage, making the transdisciplinary co-creative food-energy-water nexus activities possible. Although the solutions we have proposed at each project site may be context-sensitive, the method and procedure are prevalent for resilient and adaptive cities. These will contribute to the discussions on urban food, energy, water management in the following chapters.

Acknowledgments This chapter was based on the work of the M-NEX project, under a grant from the JST/Belmont Forum Collaborative Research Area: Sustainable Urbanisation Global Initiative (No. 11314551). Local governments, companies, and communities were involved in the activities of the national teams. We are grateful to JPI Europe Urban for initiating the Sustainable Urbanization Global Initiative—Food-Water-Energy Nexus and making the M-NEX project possible.

References

Agarwal P (2007) Walter Christaller: hierarchical patterns of urbanization. Centre of Spatially Integrated Social Science 36. http://www-personal.umich.edu/~copyrght/image/books/Spatial%20Synthesis2/CSISS%20Classics%20-%20Walter%20Christaller%20Hierarchical%20Patterns%20of%20Urbanization.htm

Allan JA (2003) Useful concept or misleading metaphor? Virtual water: a definition. Water Int 28(1):4–11

Arthur M, Liu G, Hao Y, Zhang L, Liang S, Asamoah EF, Lombardi GV (2019) Urban food-energy-water nexus indicators: a review. Resour Conserv Recycl 151:104481

Bazilian M, Rogner H, Howells M et al (2011) Considering the energy, water and food nexus: towards an integrated modelling approach. Energy Policy 39(12):7896–7906

Bhaduri A, Ringler C, Dombrowski I, Mohtar R, Scheumann W (2015) Sustainability in the water–energy–food nexus. Water Int 40(5–6):723–732

Brink E, Aalders T, Ádám D, Feller R, Henselek Y, Hoffmann A, Ibe K et al (2016) Cascades of green: a review of ecosystem-based adaptation in urban areas. Glob Environ Change 36:111–123

Cairns R, Krzywoszynska A (2016) Anatomy of a buzzword: the emergence of 'the water-energy-food nexus' in UK Natural Resource Debates. Environ Sci Policy 64:164–170

Carpenter SR, Booth EG, Gillon S, Kucharik CJ, Loheide S, Mase AS, Motew M et al (2015) Plausible futures of a social-ecological system: Yahara Watershed, Wisconsin, USA. Ecol Soc 20(2):10. https://doi.org/10.5751/ES-07433-200210

Cassandras CG (2016) Smart cities as cyber-physical social systems. Engineering 2(2):156–158

Covarrubias M (2019) The nexus between water, energy and food in cities: towards conceptualizing socio-material interconnections. Sustain Sci 2019(14):277–287

Covarrubias M, Spaargaren G, Boas I (2019) Network governance and the urban nexus of water, energy, and food: lessons from Amsterdam. Energy Sustain Soc 9(1):1–11

Daher BT, Mohtar RH (2015) Water–energy–food (WEF) Nexus Tool 2.0: Guiding integrative resource planning and decision-making. Water Int 40(5–6):748–771

Dhakal S (2010) GHG emissions from urbanization and opportunities for urban carbon mitigation. Curr Opin Environ Sustain 2(4):277–283

Endo A, Oh T (2018) The water-energy-food nexus: human-environmental security in the Asia-Pacific Ring of Fire. Springer, Singapore, 337p

Endo A, Tsurita I, Burnett K, Orencio PM (2014) A review of the current state of research on the water, energy, and food nexus. J Hydrol Reg Stud 11:20–30

Frank B, Delano D, Caniglia BS (2017) Urban systems: a socio-ecological system perspective. Sociol Int J 1(1):1–8

Gain AK, Giupponi C, Renaud FG, Vafeidis AT (2020) Sustainability of complex social-ecological systems: methods, tools, and approaches. Reg Environ Change 20(3):102

Gassmann O, Böhm J, Palmié M (2019) Smart cities: introducing digital innovation to cities. Emerald Group Publishing, Bingley

Grigg NS (1999) Integrated water resources management who should lead, who should pay? J Am Water Resour Assoc 35(3):527–534

Hoff H (2011) Understanding the nexus. background paper for the Bonn 2011 nexus conference. Stockholm Environment Institute, pp 1–52

Irena (2015) Renewable energy in the water, energy and food nexus. International Renewable Energy Agency, pp 1–125

Johansson J, Neset TSS, Linnér BO (2010) Evaluating climate visualization: an information visualization approach. In: Proceedings of the International Conference on Information Visualisation, pp 156–161

Kaushal SS, Belt KT (2012) The urban watershed continuum: evolving spatial and temporal dimensions. Urban Ecosyst 15(2):409–435

Khurian M, Turral H (2010) Information's role in adaptive groundwater management. In: Peri-urban water and sanitation services. Springer, Dordrecht, pp 171–191

Kumar P, Druckman A, Gallagher J, Gatersleben B, Allison S, Eisenman TS, Hoang U et al (2019) The nexus between air pollution, green infrastructure and human health. Environ Int 133 (pt A):105181

Kurian M, Ardakanian R (2015) Governing the nexus: water, soil and waste resources considering global change. Springer, Dordrecht, pp 1–230

Liu J, Yang H, Cudennec C, Gain AK, Hoff H, Lawford R, Qi J, de Strasser L, Yillia PT, Zheng C (2017) Challenges in operationalizing the water–energy–food nexus. Hydrol Sci J. https://doi.org/10.1080/02626667.2017.1353695

Martínez-Martínez L, Calvo J (2010) The growing problem of antibiotic resistance in clinically relevant gram-negative bacteria: current situation. Enferm Infecc Microbiol Clin 28(Suppl 2): 25–31

Masson-Delmotte V, Zhai P, Pirani A, Connors SL, Péan C, Berger S, Caud N et al (2021) IPCC, 2021: summary for policymakers. In: Masson-Delmotte V, Zhai P, Pirani A, Connors SL, Péan C, Berger S, Caud N et al (eds) Climate change 2021: the physical science basis. contribution of Working Group I to the Sixth Assessment Report of the Intergovernmental Panel on Climate Change. Cambridge University Press, Cambridge

McCalla A (1997) The water, food, and trade nexus. In: MENA-MED conference by the World Bank, Marrakesh

Merrett S (2003) Virtual water and the Kyoto Consensus. Water Int 28(4):540–542

Nature (2016) Buzzword off. 'nexus' is enjoying new-found popularity. but what does it actually mean? Nature 538:140

Newell JP, Goldstein B, Foster A (2019) A 40-year review of food–energy–water nexus literature and its application to the urban scale. Environ Res Lett 14(7):073003

O'Brien WT, Kennedy CA, Athienitis AK, Kesik TJ (2010) The relationship between net energy use and the urban density of solar buildings. Environ Plann B Plann Des 37(6):1002–1021

Ramaswami A (2020) Unpacking the urban infrastructure nexus with environment, health, livability, well-being, and equity. One Earth 2(2):120–124

Ramaswami A, Tong K, Canadell JG, Jackson RB, (Kellie) Stokes E, Dhakal S, Finch M et al (2021) Carbon analytics for net-zero emissions sustainable cities. Nat Sustain 4(6):460–463

Rockström J, Steffen W, Richardson K, Cornell SE, Fetzer I, Bennett EM, Biggs R et al (2015) Planetary boundaries: guiding human development on a changing planet. Science 347(6223): 1259855

Rogelj J, den Elzen M, Höhne N, Fransen T, Fekete H, Winkler H, Roberto Schaeffer F, Sha KR, Meinshausen M (2016) Paris agreement climate proposals need a boost to keep warming well below 2 °C. Nature 534:631

Roggema R (2021) TramsFEWmation. Springer, Netherlands

Roggema R, Yan W (eds) (2017) Tsunami and Fukushima disaster: design for reconstruction. Springer, Dordrecht

Roggema R, Tillie N, Keeffe G, Yan W (2021) Nature-based deployment strategies for multiple paces of change: the case of Oimachi, Japan. Urban Plan 6(2):143–161

Sachs I (1980) Developing in harmony with nature: consumption patterns, time and space uses, resources profiles and technological choices. Can J Dev Stud 1(1):154–175
Sachs I (1988) Work, food and energy in urban ecodevelopment. Econ Pol Wkly 23(9):425–427, 429–34.
Samuelson PA (1983) Thünen at two hundred. J Econ Liter 21(4):1468–1488
Scott M, Lennon M, Haase D, Kazmierczak A, Clabby G, Beatley T (2016) Nature-based solutions for the contemporary city/re-naturing the city/reflections on urban landscapes, ecosystems services and nature-based solutions in cities/multifunctional green infrastructure and climate change adaptation: brownfield greening as an adaptation strategy for vulnerable communities?/ Delivering green infrastructure through planning: insights from practice in Fingal, Ireland/ Planning for Biophilic cities: from theory to practice. Plan Theory Pract 17(2):267–300
Siddiqi A, Anandon LD (2011) The water-energy nexus in Middle East and North Africa. Energy Policy 39(8):4529–4540
Sperling JB, Berke PR (2017) Urban nexus science for future cities: focus on the energy–water–food–X nexus. Curr Sustain Renew Energy Rep 4(3):173–179
Urban Nexus (2013a) Health and quality of life in urban areas. Urban nexus WP3 synthesis report. http://www.urban-nexus.org.eu
Urban Nexus (2013b) Competing for urban land. Nexus synthesis report. http://www.urban-nexus.org.eu
Urban Nexus (2013c) Synthesis report: urban climate resilience. http://www.urban-nexus.org.eu
Varbanov PS (2014) Energy and water interactions: implications for industry. Curr Opin Chem Eng 5:15–21
Vogt C, Zimmermann M, Brekke K (2014)"Operationalizing the urban NEXUS: towards resource-efficient and integrated cities and metropolitan regions. giz and ICLEI. http://resilient-cities.iclei.org/fileadmin/sites/resilient-cities/files/Full_papers/Urban_NEXUS_Publication_ICLEI-GIZ_2014_web.pdf
Wackernagel M, Rees W (1998) Our ecological footprint: reducing human impact on the earth. New Society Publishers, Gabriola Island, 176p
Wester P, Warner J (2002) River basin management reconsidered. In: Hydropolitics in the developing world: a southern African perspective. African Water Issues Research Unit, Pretoria, pp 61–71
White DD, Wutich AY, Larson KL, Lant T (2015) Water management decision makers' evaluations of uncertainty in a decision support system: the case of WaterSim in the Decision Theater. J Environ Plann Manage 58(4):616–630
Yan W, Nakayama S (2021) Redesigning the urban food life through the participatory living lab platform: practices in suburban areas of the Tokyo Metropolitan Region. In: Roggema R (ed) TransFEWmation: towards design-led food-energy-water systems for future urbanization. Springer, Cham, pp 209–234
Yan W, Roggema R (2019) Developing a design-led approach for the food-energy-water nexus in cities. Urban Plann 4(1):123–138. https://doi.org/10.17645/up.v4i1.1739
Yan W, Sakairi T (2019) Geo CPS: spatial challenges and opportunities for CPS in the geographic dimension. J Urban Manage 8(3):331–341
Yang YJ, Goodrich JA (2014) Toward quantitative analysis of water-energy-urban-climate nexus for urban adaptation planning. Curr Opin Chem Eng 5:22–28
Yuan M-H, Chiueh P-T, Lo S-L (2021) Measuring urban food-energy-water nexus sustainability: finding solutions for cities. Sci Total Environ 752:141954
Zhang Y (2013) Urban Metabolism: A Review Of Research Methodologies. Environ Pollut 178: 463–473
Lehmann, Steffen. 2018. "2.10 - Conceptualizing the Urban Nexus Framework for a Circular Economy: Linking Energy, Water, Food, and Waste (EWFW) in Southeast-Asian Cities." In Urban Energy Transition (Second Edition), edited by Peter Droege, 371–98. Elsevier.

Yan, Wanglin, Andy van Delbsteen, Greg Keeffe, Geoffery Thün, and Rob Roggema. 2021. "Developing and Implementing the Design-Led Nexus Approach for Sustainable Urbanisation." In CITIES 20.50--Creating Habitats for the 3rd Millennium: Smart--Sustainable--Climate Neutral. Proceedings of REAL CORP 2021, 26th International Conference on Urban Development, Regional Planning and Information Society, 251–60. CORP--Competence Center of Urban and Regional Planning.

Climate Change in Global Cities

Yoshiaki Miyamoto

Abstract Scientific studies have found it to be unequivocal that human activities are causing climate change. Global warming is a result of processes in the atmosphere and oceans, for which the governing equations are relatively well known. Hence, by conducting climate simulations using supercomputers and cutting-edge prediction models, the magnitude of future warming can be quantitatively predicted with a certain degree of error. Such predictions are fundamentally important for many fields, including the future design of food-energy-water processes and political decisions. Predictions show that the future impact of global warming will differ in each city. This chapter introduces how predictions are conducted, limits to the use of data, and differences among cities in the impacts of global warming.

Keywords Climate change · Cities · Urban planning · Food-energy-water nexus

1 Climate Change

Scientific evidence shows that the global temperature at the Earth's surface has increased over time as a result of human activity (IPCC 2014). The rate of increase of mean temperatures at the surface was approximately 0.6 °C/100 years during the twentieth century. Although the rate may not seem so large, it is large compared with the past. The crux of the global warming issue is that rising temperatures result not from natural variability, but mainly due to anthropogenic emissions of carbon dioxide, which enhances the greenhouse effect. The amount of carbon dioxide in the atmosphere is rising rapidly. Nature, or in other words the Earth system, including the atmosphere and many other natural systems, can adjust to changes as long as they are not too rapid. For example, a typhoon's strong winds will mix the upper ocean layers, damage trees and other vegetation on land, and also affect atmospheric motion such as weather fronts. However, changes caused by the

---

Y. Miyamoto (✉)
Faculty of Environment and Information Studies, Keio University, Fujisawa, Kanagawa, Japan
e-mail: ymiya@sfc.keio.ac.jp

W. Yan et al. (eds.), *Resilient and Adaptive Tokyo*,
https://doi.org/10.1007/978-981-99-3834-6_3

passage of a typhoon will return to the original values after a certain time period. In other words, typhoons can be regarded as a matter of natural variability, which does not greatly change the Earth system. However, since the current rate of global warming is very rapid compared to natural variability, if carbon dioxide concentrations continue increasing, the Earth's atmosphere will change to a completely different state in the future.

2 Model Simulation

In response to increasing demand for projections of the Earth's future atmosphere under global warming scenarios, numerous studies have been conducted using long-term simulations of the global atmosphere (e.g., Tokioka et al. 1984; Rind et al. 1988; Gates 1992, 1995; Gates et al. 1998; Schmidt et al. 2006). To improve the accuracy of projections, some studies have conducted simulations for both the atmosphere and ocean (e.g., Tokioka et al. 1995; Meehl et al. 1997, 2000, 2007, 2009, 2014; Watanabe et al. 2010; Taylor et al. 2012; Schmidt et al. 2014; Eyring et al. 2016). Whereas the "coupling" of atmospheric and ocean models increases both uncertainty and computational costs, coupled simulations are essential in order to improve future projections.

Climate simulations for global warming are generally conducted by making assumptions about CO_2 concentrations. Natural processes such as air motion, clouds, and radiation are solved in the simulations, which result in quantitative predictions of meteorological quantities such as temperature on the ground. The Earth's atmosphere can be regarded as a fluid of air that is a mixture of nitrogen (78%), oxygen (21%), and other elements, with percentages being largely consistent up to an altitude of approximately 80 km. Since the equations of fluid motion are already well-established, the technology of numerical simulations for atmosphere and ocean is also well-developed.

Simulations for atmosphere can be conducted by "discretizing" the atmosphere into a finite number of grid points. This is a popular methodology used to numerically solve fluid motions that are governed by a given set of equations. The equations describe temporal changes in velocity or temperature, based on various processes such as advection and diffusion at the current state. The equations are discretized into a finite number of grid points in space and time, and then quantities such as velocity or temperature at the next time step are obtained at each spatial grid point based on values of the quantities at the current time step. The set of discretized equations is called a numerical model, and in particular, a set of equations for the atmosphere is referred to as a weather forecast model (Fig. 1).

The grid-point method inevitably causes errors as a result of discretization, reducing the accuracy of numerical integration. Even if all available observation data are collected, the initial condition will include some errors. This means that perfect information for the entire atmosphere cannot be obtained. Errors can be amplified when nonlinear effects are considered, known as chaos. The results of

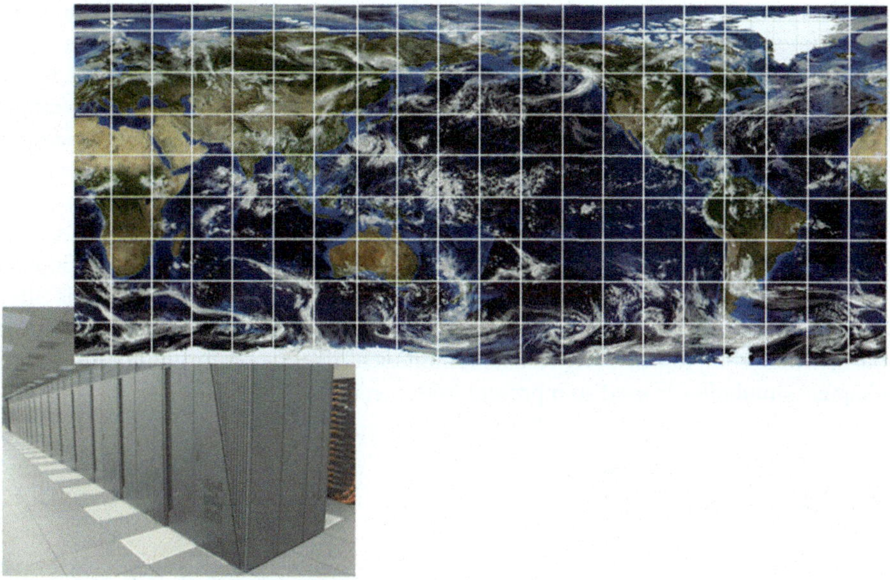

Fig. 1 Schematic of global simulation conducted by supercomputer

numerical integration of equations become more accurate as grid distance is shortened and initial and boundary conditions become more accurate. Nevertheless, for climate simulation, in which time scales are very long, boundary conditions play a more important role than initial conditions in determining atmospheric state.

Simulations are carried out by discretizing the world into a finite number of grid points. Meteorological quantities such as wind speed, temperature, and humidity are allocated at each grid point, as representative values around the grid points. Processes that have spatial scales smaller than the grid distance are "parameterized" by formulas obtained theoretically and empirically, if they play an important role in determining the grid-scale quantities.

3 Parameterization

A major difference from typical simulations for fluid dynamics is that atmospheric simulations need to consider physical processes such as solar and Earth radiation and cloud processes. Radiation and cloud processes both change the temperature of air, which affects air motion, which in turn changes the spatial field of temperature. The interaction between physical processes changing temperature and air motion is essential for predicting a future state of the atmosphere. Because the physical processes for radiation and cloud physics are complex and their spatial scales are small compared with large-scale atmospheric motion, it is computationally prohibitively expensive to solve for both processes and atmospheric motion at the same time.

Hence, the physical processes introduced above and other processes for which spatial scale is less than grid scale, which cannot be neglected in the governing equations, are *parameterized* in numerical models, whereas large-scale atmospheric motion is explicitly solved. To conduct numerical calculations for weather forecasts or simulations, various kinds of parameterization need to be solved in a numerical model. For instance, radiation, cumulus convection, cloud physics, sub-grid scale turbulence, boundary-layer turbulence, and surface processes are the most popular parameterizations for weather simulations. Note that the term "simulation" is often used to refer to a "hindcast" (a numerical integration for governing equations for the past), whereas a "forecast" represents a numerical integration for the future. In this chapter, simulation is used to represent both meanings.

3.1 Radiation

The parameterization of radiation solves for processes associated with solar radiation, and radiation from the air and Earth surface. The former is referred to as short-wave and the latter as long-wave radiation, based on the dominant wavelengths of electromagnetic waves. Solar radiation energy is mainly in the ultra-violet and visible wavelengths, while radiation from the air and Earth surface is mainly in infrared wavelengths. Wavelengths are determined by the temperature of the source that emits the electromagnetic waves. Short-wave radiation originating from the sun accelerates the molecular motion of air, which means that temperature of the air increases. Long-wave radiation is a result of the emission of energy from air molecules or the Earth's surface and hence tends to decrease their temperature. Temperature changes due to radiation are calculated by the divergence of radiation flux in the vertical direction, by assuming that radiation fluxes are constant in horizontal directions.

3.2 Cloud Microphysics

Another process that is essential for atmospheric simulations is associated with clouds, referred to as cloud microphysics. Clouds consist of a huge number of water droplets and/or ice particles. When water droplets or ice particles form, or water vapor condenses, heat is released to the surrounding air, and air temperature increases. In contrast, when the water droplets or ice particles evaporate, the temperature of surrounding air decreases. Temperature changes due to the phase change of water are not negligible. Temperature affects air motion, which in turn affects the spatiotemporal distribution of clouds. Since there is always some cloud in the atmosphere, cloud effects must be considered for accurate simulations. However, as in radiation, the spatial scale of cloud processes is almost 10^6 m smaller than that for large-scale air motion. It is unrealistic to explicitly solve cloud processes in atmospheric simulations, so the effects of microphysics need to be parameterized. Instead of solving temporal changes in mass or location of each cloud particle, many parameterizations solve the temporal change in mass, or mixing ratio (mass ratio of cloud to dry air), by assuming size/number distribution at a grid point. Water droplets are often classified into two categories: cloud droplets that do not fall because of small size, and rain droplets that fall due to gravity. Meanwhile, ice particles are classified into three or more categories: ice particles that do not fall, small-sized snow that falls, and graupel (large-sized snow that falls). To solve for cloud processes, the temporal changes in the mass or mixing ratio of all the water categories need to be solved, which increases the number of prognostic equations.

3.3 Cumulus Convection

In the troposphere, moist convection (or cumulus convection) plays a vital role in transporting mass and momentum. Especially in the convectively active tropics, isolated convection is very important for large-scale circulation or other convective phenomena such as typhoons. Hence, to realistically simulate the atmosphere, it is essential to incorporate the effects of cumulus convection. However, in typical climate simulations, the spatial resolution is not fine enough to explicitly resolve convection (Miyamoto et al. 2013, 2015, 2016). Because of the importance of cumulous convection, the effects need to be parameterized. Most climate simulations that have been conducted so far have included cumulus parameterization.

3.4 Sub-grid Scale Turbulence

Flow motion on a spatial scale smaller than or comparable to the grid distance cannot be explicitly solved. Strictly speaking, flow motion on a spatial scale larger than four to six times the grid distance can only be solved in simulations.

However, small-scale motion or sub-grid scale motion likely affects grid-scale motion. Specifically, sub-grid scale motion dissipates kinetic energy, which is in turn converted into thermodynamic energy. Hence, the effects of sub-grid scale motion are represented as diffusivity for grid-scale motion. The effects of sub-grid scale turbulence are also parameterized in atmospheric simulations.

While we have seen typical examples of major processes that are parameterized in climate simulations, other parameterizations also solved in simulations include surface processes and gravity wave drag. Since a number of processes affect the climate, various effects need to be incorporated into a climate simulation.

4 Model Errors

Even if the governing equations of atmosphere are solved numerically with a set of parameterizations, results of simulations include errors. The errors are due mainly to inaccurate boundary data, discretization of time and space, formulation of parameterizations, and many experimental parameters. Models often have biases for quantities in some regions. Hence, model intercomparisons are sometimes carried out to reduce the uncertainty of individual models. The Climate Model Intercomparison Project (CMIP) is an example of efforts to combine the outputs of climate simulations conducted by several research institutes around the world in order to obtain one of the most reliable datasets for future projections.

At present, a climate prediction for the near future (several decades) is conducted by numerically solving the equations of climate models using supercomputers. As introduced above, a number of parameterizations need to be solved as well as the air motion at grid points, so climate simulations require massive computational resources. Ensemble simulations are also conducted to reduce uncertainty. Since significant computer and human resources are required, climate simulations are conducted mostly by national agencies or groups of researchers. To examine future changes in meteorological quantities such as temperature, one should consider using datasets produced by national agencies, because that is the most reliable data which can be obtained at present.

5 Data Analyzed in the Present Study

Here, we use the data from the "database for Policy Decision making for Future climate change" (d4PDF) (Mizuta et al. 2017; Imada et al. 2017; http://www.miroc-gcm.jp/~pub/d4PDF/index_en.html) for the analysis of six target cities in the FEW nexus project (Fig. 2 and Table 1). The features of d4PDF data can be summarized as:

- Probabilistic future projections of low-frequency local-scale events.
- Large ensemble of climate simulations with various climate models.
- Intended to be utilized for impact assessment studies and adaptation planning for global warming, such as disasters, agriculture, water resources, ecosystems, human health, etc.

The dataset includes a historical climate simulation from 1951 to 2010 with 100 members, a non-warming simulation from 1951 to 2010 with 100 members, +2K future climate simulation from 2031 to 2090 with 54 members, and +4K future climate simulation from 2051 to 2110 with 90 members.

For this chapter we analyzed temperature and precipitation around the six FEW nexus project cities in 3 different years, representing the present and future states: 1980, 2010, and 2080. Two kinds of dataset were generated: time series of

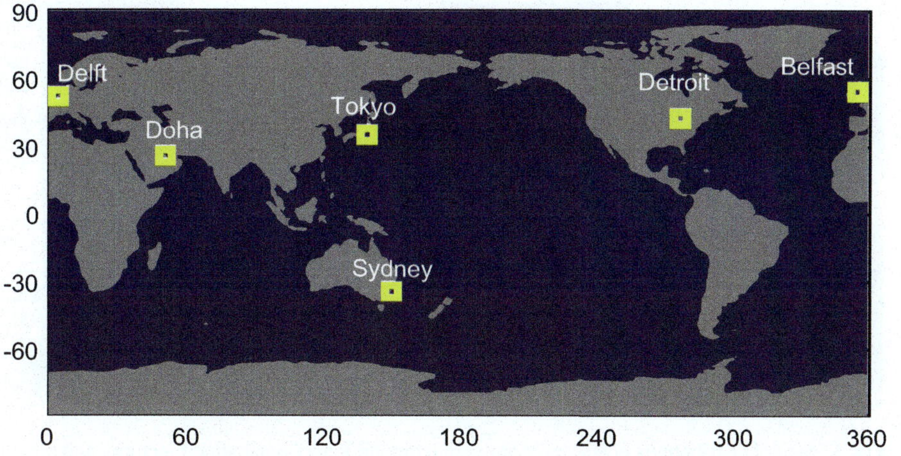

Fig. 2 Six cities featured in the FEW nexus project

Table 1 Geographical location of the six cities

City	Country	Longitude	Latitude
Belfast	UK	W 5° 55′	N 54° 35′
Detroit	USA	W 83° 2′	N 42° 19′
Tokyo	Japan	E 139° 41′	N 35° 41′
Doha	Qatar	E 51° 31′	N 25° 17′
Delft	Netherlands	E 4° 21′	N 52° 0′
Sydney	Australia	E 151° 12′	S 33° 52′

daily-averaged quantities for each year, and horizontal distributions for monthly-averaged quantities in area 3.0° × 3.0° centered at the city for each month.

Figure 3 displays the spatial distributions of sea-level pressure, precipitation, and horizontal wind in the simulation on July 20 in 1980 and 2080. On both days, a tropical cyclone is observed close to Japan, which accompanies a heavy precipitation area on the northern side of cyclone center. Around the center of the cyclone which can be detected as the local minimum of the sea-level pressure field, the wind direction rotates counterclockwise. The distributions indicate that the results of simulation appear to be reasonable.

Figure 4 shows the histogram of relative frequency of precipitation in Tokyo in 1980 and 2080. Whereas the overall number of frequencies are similar to each other, there are a couple of peaks of very strong precipitation. The largest precipitation in 2080 is over 260 mm/day, which is more than twice the largest precipitation in 1980 (120 mm/day). The second largest precipitation in 2080 is 130 mm/day, also larger than the maximum in 1980.

Figure 5 shows the time series of temperature in the six cities in 2010 and 2080. In the cities in the northern hemisphere (all except Sydney), the temperature is highest in the northern summer, specifically around the 220th day (mid-August) and lowest in the northern winter. Meanwhile, the temperature in Sydney is highest in the

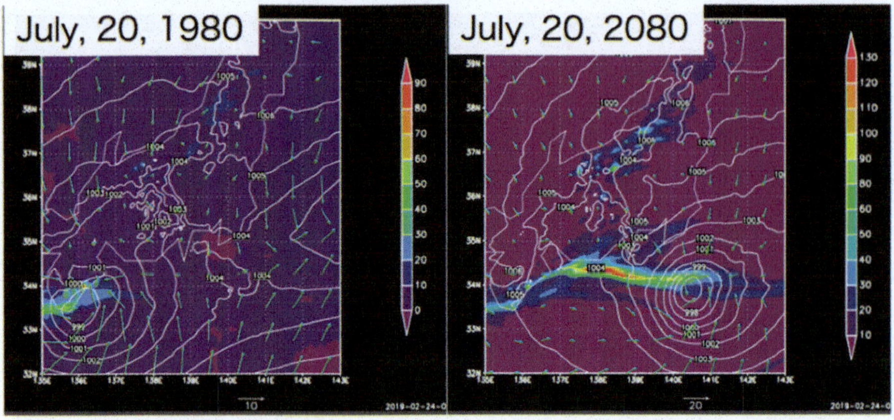

Fig. 3 Spatial distributions of sea-level pressure (contours), horizontal wind (vectors), and precipitation (shading) on July 20 in 1980 and 2080

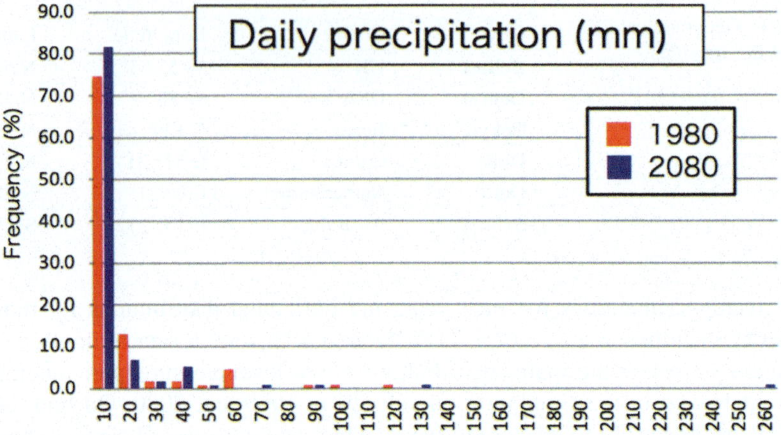

Fig. 4 Histograms of relative frequency of daily precipitation in Tokyo in 1980 (red) and 2080 (blue)

southern summer and lowest in the southern winter. In all cities, the temperature appears to be higher in 2080 than in 2010. In fact, both the maximum and average temperatures are higher in 2080.

Differences among cities in temperature between 2010 and 2080 merit attention. In Tokyo and Detroit, the temperature difference is especially large in early summer. In Doha and Delft, the temperature difference is notable in late summer. In northern hemisphere cities, highest temperatures in 2080 are in September, in contrast to July or August in 2010.

Figure 6 shows the time series of precipitation in the six cities in both 2010 and 2080. In contrast to temperature, precipitation is not projected to increase in every

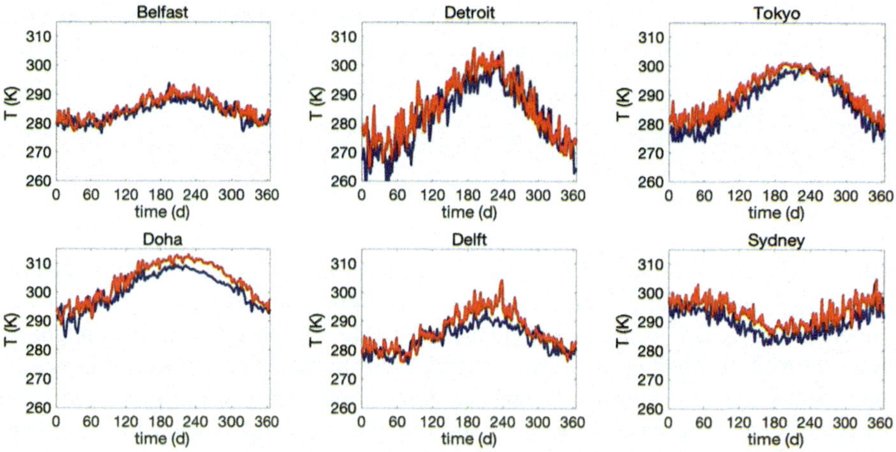

Fig. 5 Time series of temperature in six cities in 2010 (blue) and 2080 (red)

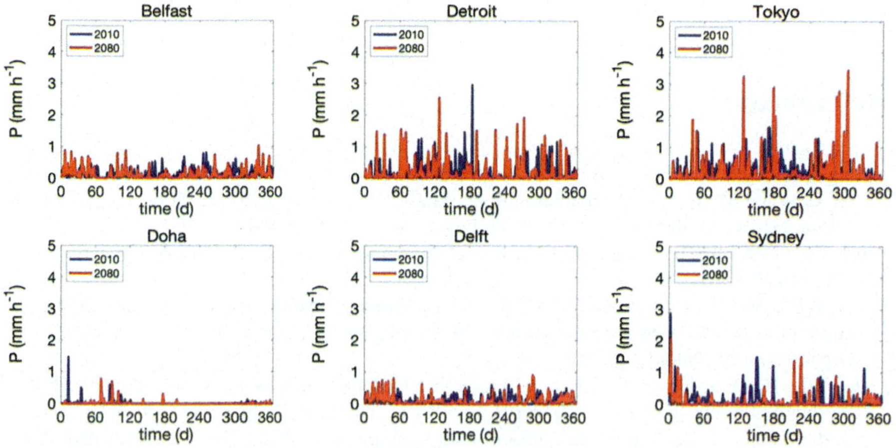

Fig. 6 Time series of precipitation in six cities in 2010 (blue) and 2080 (red)

one of the cities in 2080. In that year, precipitation averages are high in Detroit and Tokyo and low in Sydney, while differences in precipitation are negligible in the other three cities.

The above analyses for temperature and precipitation indicate that the expected impacts of global warming will differ among the selected six cities, which implies that future changes in climate will vary by location. This has demonstrated that simulations enable one to conduct various types of analyses for future conditions.

Nevertheless, as explained above, computer simulations for global warming include errors due to discretization and parameterization, and they are currently not sophisticated enough to resolve meteorological quantity in each city. Limitations of simulation data are important to recognize. Since climate is affected also by biological and chemical processes, models incorporating these effects, known as Earth system models, are being designed to better predict the future (e.g., Hibbard et al. 2007).

6 Conclusions

Climate change is a result of processes in the atmosphere and oceans, for which the governing equations are relatively well known. Climate simulations performed using supercomputers and cutting-edge prediction models can predict the magnitude of future warming, albeit with a certain degree of error. Among a myriad of uses, such predictions can also help us design our urban food-energy-water processes to be more resilient. The case study here showed that different cities will experience different impacts of global warming. Simulation results can be useful, but it is also important to be aware of their limitations. Improvements in climate simulations will help improve future simulations and predictions.

References

Eyring V, Bony S, Meehl GA, Senior CA, Stevens B, Stouffer RJ, Taylor KE (2016) Overview of the Coupled Model Intercomparison Project Phase 6 (CMIP6) experimental design and organization. Geosci Model Dev 9:1937–1958. https://doi.org/10.5194/gmd-9-1937-2016

Gates WL (1992) AMIP: the atmospheric model intercomparison project. Bull Am Meteor Soc 73: 1962–1970

Gates WL (1995) An overview of AMIP and preliminary results. In: Proceedings of the first international AMIP scientific conference, WCRP-92, WMO TD-no. 732. World Meteorological Organization, Monterey, pp 1–8

Gates WL, Boyle J, Covey C, Dease C, Doutriaux C, Drach R, Fiorino M, Gleckler P, Hnilo J, Marlais S, Phillips T, Potter G, Santer B, Sperber K, Taylor K, Williams D (1998) An overview of the results of the Atmospheric Model Intercomparison Project (AMIP I). Bull Am Meteor Soc 73:1962–1970

Hibbard KA, Meehl GA, Cox P, Friedlingstein P (2007) A strategy for climate change stabilization experiments. Eos Trans Am Geophys Union 88:217. https://doi.org/10.1029/2007EO200002

Imada Y et al (2017) Recent enhanced seasonal temperature contrast in Japan from large ensemble high-resolution climate simulations. Atmosphere 8(3):57

IPCC (2014) Climate change 2014: synthesis report. In: Core Writing Team, Pachauri RK, Meyer LA (eds) Contribution of working groups I, II and III to the fifth assessment report of the Intergovernmental Panel on Climate Change. IPCC, Geneva, Switzerland, 151 pp

Meehl GA, Boer GJ, Covey C, Latif M, Stouffer RJ (1997) Intercomparison makes for a better climate model. EOS Trans Am Geophys Union 78:445–451

Meehl GA, Boer GJ, Covey C, Latif M, Stouffer RJ (2000) The Coupled Model Intercomparison Project (CMIP). Bull Am Meteorol Soc 81:313–318

Meehl GA, Covey C, Taylor KE, Delworth T, Stouffer RJ, Latif M, McAvaney B, Mitchell JFB (2007) THE WCRP CMIP3 multimodel dataset: a new era in climate change research. Bull Am Meteorol Soc 88:1383–1394

Meehl GA et al (2009) Decadal prediction. Bull Am Meteor Soc 90:1467–1485. https://doi.org/10.1175/2009BAMS2778.1

Meehl GA, Moss R, Taylor KE, Eyring V, Stouffer RJ, Bony S, Stevens B (2014) Climate model intercomparisons: preparing for the next phase. Eos Trans AGU 95(9):77

Miyamoto Y, Kajikawa Y, Yoshida R, Yamaura T, Yashiro H, Tomita H (2013) Deep moist atmospheric convection in a sub-kilometer global simulation. Geophys Res Lett 40:4922–4926

Miyamoto Y, Yoshida R, Yamaura T, Yashiro H, Tomita H, Kajikawa Y (2015) Does convection vary in different cloudy disturbances? Atmos Sci Lett 16:305. https://doi.org/10.1002/asl2.558

Miyamoto Y, Yamaura T, Yoshida R, Yashiro H, Tomita H, Kajikawa Y (2016) Precursors of deep atmospheric convection in a subkilometer global simulation. J Geophys Res Atmos 121:12080–12088

Mizuta R et al (2017) Over 5000 years of ensemble future climate simulations by 60 km global and 20 km regional atmospheric models. Bull Am Meteor Soc 98:1383–1398. https://doi.org/10.1175/BAMS-D-16-0099.1

Rind D, Suozzo R, Balachandran NK, Lacis A, Russell G (1988) The GISS global climate-middle atmosphere model. Part I: model structure and climatology. J Atmos Sci 45:329–370. https://doi.org/10.1175/1520-0469(1988)045<0329:TGGCMA>2.0.CO;2

Schmidt GA, Ruedy R, Hansen JE, Aleinov I, Bell N, Bauer M, Bauer S, Cairns B, Canuto V, Cheng Y, Del Genio A, Faluvegi G, Friend AD, Hall TM, Hu Y, Kelley M, Kiang NY, Koch D, Lacis AA, Lerner J, Lo KK, Miller RL, Nazarenko L, Oinas V, Perlwitz JP, Perlwitz J, Rind D, Romanou A, Russell GL, Sato M, Shindell DT, Stone PH, Sun S, Tausnev N, Thresher D, Yao M-S (2006) Present day atmospheric simulations using GISS ModelE: comparison to in-situ, satellite and reanalysis data. J Clim 19:153–192. https://doi.org/10.1175/JCLI3612.1

Schmidt GA, Kelley M, Nazarenko L, Ruedy R, Russell GL, Aleinov I, Bauer M, Bauer SE, Bhat MK, Bleck R, Canuto V, Chen Y-H, Cheng Y, Clune TL, Del Genio A, de Fainchtein R, Faluvegi G, Hansen JE, Healy RJ, Kiang NY, Koch D, Lacis AA, LeGrande AN, Lerner J, Lo KK, Matthews EE, Menon S, Miller RL, Oinas V, Oloso AO, Perlwitz JP, Puma MJ, Putman WM, Rind D, Romanou A, Sato M, Shindell DT, Sun S, Syed RA, Tausnev N, Tsigaridis K, Unger N, Voulgarakis A, Yao M-S, Zhang J (2014) Configuration and assessment of the GISS ModelE2 contributions to the CMIP5 archive. J Adv Model Earth Syst 6(1):141–184. https://doi.org/10.1002/2013MS000265

Taylor KE, Stouffer RJ, Meehl GA (2012) An overview of CMIP5 and the experiment design. Bull Am Meteor Soc 93:485–498. https://doi.org/10.1175/BAMS-D-11-00094.1

Tokioka T, Yamazaki K, Yagai I, Kitoh A (1984) A description of the Meteorological Research Institute atmospheric general circulation model (MRI GCM-1). Technical Report of the Meteorological Research Institute, 13, 249pp

Tokioka T, Noda A, Kitoh A, Nikaidou Y, Nakagawa S, Motoi T, Yukimoto S, Takata K (1995) A transient CO_2 experiment with the MRI CGCM—quick report. J Meteor Soc Japan 74:817–827

Watanabe M et al (2010) Improved climate simulation by MIROC5: mean states, variability, and climate sensitivity. J Clim 23:6312–6335. https://doi.org/10.1175/2010JCLI3679.1

Scaling the Food–Energy–Water Concept in Tokyo

William Galloway

Abstract This chapter is about the FEW Nexus in Tokyo. The nexus is by its nature complex and complicated, with multiple intertwined parts. To manage complexity, it is essential to place local processes and customs at the center of any design or development. With that in mind this chapter aims to bridge the ideal goals of the FEW Nexus with the specific reality of Tokyo by combining two narratives. The first is a historical overview of the shift of urban planning from equilibrium to dynamic resilience, making use of several ideas from the realm of ecology. The second describes the nature of change in the Japanese city, supported with a simple case study of a residential block in Tokyo. The importance of the sanitary movement is specially recognized for its scale of impact, and the impossible prototype that it became for nearly all grand urban planning theories thereafter. The FEW Nexus is close in spirit to that precedent because it too has the ambition of changing the infrastructure of cities in their entirety, much like the sanitary movement did in its time. The existing Japanese planning system offers several entry points for the overlay of the FEW Nexus. It is therefore proposed that a bottom-up approach to design can be used to transform large parts of Tokyo through the fast pace of replacement of buildings in the city, as well as through partnership with building typologies that function inside logistical networks.

Keywords Scaling theory · Logistical urbanism · Panarchy · Resilience theory

1 Introduction

This chapter examines the subject of the FEW Nexus in Tokyo; however, the lessons taken from the subject matter can be generalized and are presented with that intention. To be clear, Tokyo is not representative of every Japanese city. In the same way, observations made in one part of Tokyo do not perfectly represent the

W. Galloway (✉)
Department of Architectural Science, Toronto Metropolitan University, Toronto, ON, Canada
e-mail: will.galloway@torontomu.ca

W. Yan et al. (eds.), *Resilient and Adaptive Tokyo*,
https://doi.org/10.1007/978-981-99-3834-6_4

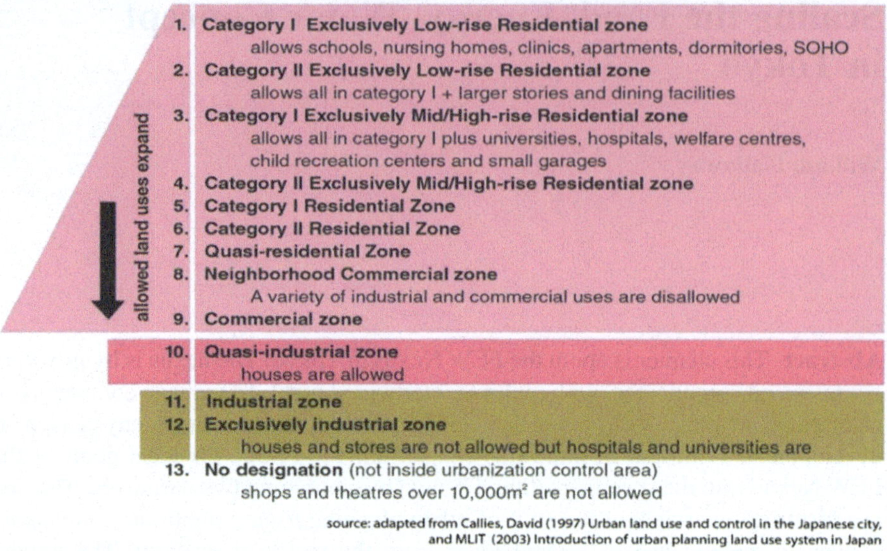

Fig. 1 Zoning categories in Japan

entirety of its metropolitan area. Even so, because zoning regulations in Japan are primarily governed through the national building code, and they control the size and allocation of all but the largest developments, it is possible to make a number of generalizations based on the study of even a small area.

In practice, Japan's urban landscapes are formed by a series of simple, if not generic, regulations. They are somewhat unique in comparison to other parts of the world in that they are parametric and, with notable exceptions, not normally open to negotiation or challenged by an esthetic review. To be more precise, zoning regulations work in partnership with the building code through the use of mathematical proofs; that a building meets strict requirements with regard to quantifiable characteristics, including the size of a building, its proper ventilation, access to sunshine for both the building in question as well as for neighbors, and a number of life safety issues (Galloway 2009). Because the regulations are parametric there are few absolutes. Instead, there is a range of solutions for how a building might be placed on a site and how it looks. A prediction of forms that could be built in a neighborhood based on zoning regulations has more in common with a quantum diagram, filled with multiplied potential, than to anything approaching certainty. Zoning creates tendencies but does not impose strict outcomes. The same approach governs the rules regarding the zones themselves. While Japan has only 13 official land use zones, which are applied locally over the entire nation, they are not siloed. Instead, they are organized in a kind of pyramidal relationship (Fig. 1), where land uses and building sizes are more and more inclusive as they step down the pyramid of zones (Callies 1997; Sorensen 2005). In application, a typical Japanese city makes full use of that freedom. When walking on a street it can be difficult to recognize the zoning of an area only by observing the building types and sizes.

Taken together, the regulations and zoning maps are behind the creation of a mixed-use urban landscape. Such landscapes can be found in dense and iconic Japanese areas like Shinjuku, but also in the suburban neighborhoods that grow around every town and city across Japan. One important outcome of this approach is that suburbs are potentially as walkable—and as well served—as central urban areas when it comes to services, from medical care to restaurants. The key word here is potential, and it is important to acknowledge that potential does not always translate to reality. Many suburbs are mixed-use, and homes sit side-by-side with shops, clinics, and other functions (Galloway 2009).

Conversely, there are also many suburbs where the car is a de facto requirement. However, that outcome is a matter of local activity and not the result of legal restrictions. Interestingly, the lack of exclusivity runs in both directions, and in central Tokyo, where intense land use is the norm, it is possible to find farmland and other productive green spaces, although they only become truly common in the inner suburbs and outwards.

Though it is satisfying to marvel at the potential diversity of the Japanese city, the important point here is that neither end of the gradient is restrictive, and as noted later in this text, it is possible to imagine that even large areas can change their characteristics dramatically over time because of the high turnover of buildings. Before proceeding further, we should acknowledge that there are outliers to all of these norms. There are areas planned with quite different rules, and these are important to recognize as exceptions. However, they are not the subject matter of this chapter. Instead, we will focus on the lessons of the apparently (if not actually) unplanned city that makes up most of Tokyo's urban landscape.

In addition to its particular zoning regulations, growth and change in Japan's cities are governed by a characteristically constant demolition and reconstruction—and for decades, an investment in larger and denser building types. That is a pattern that is not mirrored in most cities of the developed world (Gleeson 2018; Holling 2017). Ownership and the use of land are both constitutionally protected in Japan, meaning there is little recourse for those who wish to block development that is otherwise permitted on a site. Even if we only consider recent history, it is entirely correct to note that the city has been rebuilt several times over. In the process it avoided a housing crisis, even though Tokyo's population continues to grow (as of 2021). Unlike other large world cities—from London, to Toronto, to San Francisco—Japan built its way out of a housing crisis beginning in the 1970s. Faced with a crisis, it increased its housing stock year after year, beginning in 1973 (Gleeson 2018). The same cannot be said of any other major city in the world. Tokyo is, in this respect, unique.

These characteristics are relatively well documented, and familiar to residents of Japan. The cities that emerge from the patterns summarized above are not necessarily the most beautiful in the world and often chaotic in appearance. However, there is a high degree of flexibility built into the processes and systems that maintain Japan's cities. That flexibility is central to any project aimed at changing the way cities function. Observing Tokyo's urban characteristics is useful because they indicate an ability to change quickly and to absorb unexpected perturbations. More importantly, they indicate a capacity to respond to changing needs without changing the regulatory system in a significant way. Compared to western cities that is a rarity, and worth closer examination. For us it is important because it represents a different approach to thinking about global concepts such as the FEW Nexus. For instance, roads do not change very easily in Japan, but buildings can be swapped in comparatively short time. The opposite was true in cities in Europe and North America, where the urban planning response to the COVID-19 pandemic (as of July 2022) was dominated by changing street use (Jefferies et al. 2020; O'Sullivan 2020). If one of the purposes of the FEW Nexus is to change urban form, it is important first of all to acknowledge the specific nature of the city and to propose FEW concepts at a scale and manner that aligns with existing processes.

With that goal in mind, Tokyo is considered in this chapter with a group of theories that combine ecology with urban planning and design. Considering the FEW Nexus, which includes ecological services as well as man-made infrastructure, an appropriate framework for analysis should include both ecology and the built environment. While ecological theories have been applied before to the study of urban planning, this chapter critiques and proposes a variation grounded in the reality of Japanese urbanism.

2 Theoretical Ecology of Change in Cities

2.1 A Brief History of Change

The recognition that cities are constantly changing is central to any realistic description of urbanity. Yet when cities are theorized, and plans made for the future, the goal is almost always totalizing, transformational, and climactic. As the modern city emerged from the Industrial Revolution, the planning theories that attempted to contain its excess energy and its negative outcomes tended to presume some kind of future but final resolved state. From the Garden City of the early twentieth century to the 15-min city of the twenty-first century, the presumption is that the correct outcome can finally be obtained once the idea is carried out everywhere, all at once, and at scale.

That generally is not what happens. Instead, we relearn an old lesson, namely that urban form is unstable, a process rather than an object. Built fragments of ideas and theories can be found in small doses, scattered across cities around the world, but a single vision of what a perfected urban life might look like is a mirage. A partial explanation for why cities resist correction in this way is because the rate of urban change is unevenly distributed, and every city marches to its own drum when it comes to change—growing, shrinking, abandoning, and re-inhabiting neighborhoods as the "animal spirits" (i.e., the collective will of a society) guide them (Akerlof and Shiller 2011). Why is this a problem? If planning only needed to be local it would not be. However, when we are forced to respond to large scale problems like climate change, global migration, and the economics of regions and nations, it is necessary to develop policies and tools that can work at that level.

There are precedents. Arguably, the total transformation of modern urban form has happened twice. First, at the end of the nineteenth century with the introduction of the sewers and infrastructure needed to make cities livable (Snowden 2019; Black 2008). Then again throughout the twentieth century, when cities were reformed around automobiles and suburbia (Callahan and Ikeda 2004; Hélie 2019; Sies 1987). These changes were transformational because they were supported by radically entwined financial, governmental, and social systems. Together they could overcome the inertia of their time. The fact that they were constructivist (i.e., totalizing) reforms sets them apart as historical planning movements. To give an example, consider the remaking of New York by Robert Moses in the early and mid-twentieth century (Callahan and Ikeda 2004), or the previous renovation of London by engineers such as Bazelgette in the nineteenth century. It is argued here that all other urban planning theories are a reaction to, or a correction of, those two radical transformations. No plan since has acted at the same scale, and none has yet reached the same degree of integration.

The ability to adapt has become more important as external pressures placed on urban form multiply and take on greater significance. Climate change, radical shifts in population, economic change, and interactions between these pressures force urban planners to prepare for a significant degree of uncertainty. Reveling in this reality in the 1990s, Rem Koolhaas blithely declared architecture impotent to change anything as wild as a city, and that cities themselves were unmanageable in any case. As an antidote, he concludes that urbanism "...will no longer be concerned with the arrangement of more or less permanent objects but with the irrigation of territories with potential; it will no longer aim for stable configurations but for the creation of enabling fields that accommodate processes that refuse to be crystallized into definitive form; it will no longer be about meticulous definition, the imposition of limits, but about expanding notions, denying boundaries, not about separating and identifying entities, but about discovering unnameable hybrids..." (Koolhaas 1995, p. 3).

Given this history, we might hypothesize that expanding the impact of the FEW Nexus may hinge on the placement of procedural change at the center of the project. Alternatively, the concept will need to be perfectly integrated with popular social movements and economics (Galloway et al. 2021). In any case, the approach will need to be local, even while the concept is intended to have global impact. With that combination in mind, we should consider the last time a transformation on that scale took place.

2.2 First Steps

The sanitary movement, which began in the 1830s and continued to define urban theory well into the 1900s, was a necessary requirement of modern urban planning—indeed it was the central transformation needed to make cities function at all (Black 2008; Morley 2007). Before that, cities were disease-laden and deathly, growing only by massive in-migration. Unmanaged density and poor sanitation led directly to sickness and death. It was not a sustainable model and constrained the size of cities and their populations just as the Industrial Revolution was provoking change in social and economic structures around the world (The Economist 2020). Cities eventually underwent what McKeown called a "Mortality Revolution," by which "...industrial cities in the developed world... [became] salubrious places with low rates of death and high life expectancy" (Snowden 2019, p. 184). Without repeating the history of the sanitary movement here, and also recognizing that it was not without flaws, we can take away several key lessons.

Chief among these is that the scale of the problem was answered by an equally scaled response. Over a period of decades cities were transformed from the underground up, through the introduction of sewerage systems and a host of attendant technological and social infrastructure. Arguably the scale of response was only possible because it was planned and carried out by multiple actors, each working on different aspects of both social and physical systems. The actors in question included urban planners, lawyers, politicians, engineers, architects, medical scientists, social reformers, and economists. The movement brought together professions and ideological groups, including those who otherwise opposed each other on different topics (Snowden 2019). While this was an amazing achievement, for us the changes brought about by the sanitary movement are most relevant because they were global. Though the movement took decades to take hold at a global scale, the idea moved quickly enough from its roots in France and England, to the rest of Europe, the USA, and eventually to all of the industrial world (Snowden 2019). This last group includes Japan (Sorensen 2005).

Combined with later variations, such as the City Beautiful movement (briefly popular before the advent of the First World War), the way cities were conceived and sustained changed fundamentally, and in the process, created what we can now recognize as the clearly formed modern city. A city where infrastructure and networks are not only built into the urban fabric, but where their absence becomes a measure of poverty and failure. In London the sewerage system of the engineer John Bazalgette, for instance, has taken on a near mythical stature. And yet the physical transformation of cities is not even the most radical change that took place. That distinction goes to the new expectation that sewerage and other infrastructure would be installed and maintained by a centralized government—a radical idea at the time, now taken for granted. It is hard to overestimate the impact of that shift. It represents the first acceptance of something like our contemporary information society, where governments know the location and consumption patterns of residents with unheard-of intimacy (Snowden 2019).

Reliance on centralized authority to secure a safe living environment had implications for the way cities were theorized and planned from that time onward. A top-down approach, where the state re-ordered urban form and daily life in a simplified way was normalized. It is possible that later changes coming from the primacy of automobiles were so overwhelming because of the large top-down systems put in place during the sanitary movement. This pattern is as true of Japan as of other parts of the world, even though it was not a direct copy in its application (Sorensen 2005; Hein 2010). Cities around the world were transformed from top to bottom. The change was so transformational that it took a global pandemic as severe as the COVID-19 outbreak to question their viability (Badger 2021). As successful as the sanitary movement was, its reliance on centralized authority and its role in creating data-driven planning can be understood as the pre-requisites of what would become high-modernist planning in the mid-twentieth century.

As a movement, high modernism was defined by top-down decrees, including large scale rearrangements of communities. Like the sanitary movement before it, it was attached to a large investment in massive infrastructural projects. This time it was not just the pipes and machines that made sewerage systems work, but also the roads that carried automobiles, and the electrical grids that powered a more sophisticated urban life. We would be remiss if we did not pause for a moment and make clear that the changes described here in such broad terms were not always a good thing. The negative social effect of building highways through precarious neighborhoods is undeniable (Callahan and Ikeda 2004) and there are currently some attempts underway to redress those mistakes (Fitzgerald and Agyeman 2021; Popovich et al. 2021). At the same time, as energy production begins to take place at the scale of individuals or small groups, we can see that the infrastructure created for one-way delivery of energy is no longer adequate. Instead, infrastructure that allows energy to move in both directions is changing the way the infrastructure itself is viewed. We are poised to reconsider the basic concepts that created our cities, and as a result the infrastructure that supported their real-world realization. In the case of Japan this question takes on some poignancy as it is additionally connected to the 2011 Fukushima nuclear disaster and the on/off discussion about whether to restart the nuclear power industry (Stapczynski 2021). That question is in the end asking whether or not Japan is capable of running a decentralized infrastructure. If the FEW Nexus is about bringing production to the community level, that answer will need to be addressed.

Before we consider what needs to change, it is useful to understand the particular failure we are considering. High modernism, simply stated, was the ultimate outcome of a philosophy bent toward efficiency, both of production and consumption, reaching its peak in the 1960s and 1970s. Like the sanitary movement it was a global phenomenon, though it took different forms in Asia than in the West (Marcotullio et al. 2003; Hein 2010). James Scott describes its development as a shift toward legibility, taken for the simple and direct purpose of making it easier for states (and corporations) to measure, and hence to manage, the problems they were responsible for. Legibility under those terms did not mean complete understanding. To the contrary, it meant focusing on a select set of information at the expense of local knowledge and complexity, bringing with it a host of consequences. In his book, Seeing Like a State, Scott uses the subject of scientific forestry, developed in eighteenth century Prussia, as a parable for the failure of modernist thinking. He describes how forests were measured and then remade to include only a small selection of species. Natural ecosystems were literally destroyed and replaced to maximize economic value and to make management easier. The outcome was a forest remade as a mono-culture plantation (Scott 1998). Ecological networks were reduced to a limited number of variables, creating a predictable and constant future. As it turned out, the lack of diversity eventually became a serious problem and the Germans needed to invent a word for the death of forests that resulted.

By now this critique is well discussed and recognized, and we view cities with a more open eye. Contemporary cities are understood to be complex, adaptable, and constantly changing. Yet the ideal of planning cities from the top down, using a completist approach, has never lost favor. It can be seen in recent examples such as Masdar (Abu Dhabi) and other contemporary smart cities, where the ideal is to control inputs and outputs in a highly refined and deterministic way (Cugurullo 2016). Even smaller, self-contained examples, such as Re-Gen village (a planned village as yet unbuilt in the Netherlands) share this potential liability (Galloway et al. 2021). While Re-gen Village is technically sophisticated, it also relies on a singular form, a singular closed-loop infrastructure, and a very specific and controlled esthetic. For all of its ecological credentials, its reliance on certainty and order is too reminiscent of modernist city plans such as Brasilia, which Scott convincingly argued was ordered for the sake of legibility at the expense of its inhabitants (Scott 1998). Ideologically there is a wide gap between the intentions and execution of these projects; however, they all presume (and enforce) a limited range of variation and diversity that is risky in times of stress and change (Gunderson 2000). We will return to this point later when we consider the Japanese case, which, as we saw above, is substantially different.

2.3 Ecological Resilience and Engineered Resilience

Even if cities are understood as changeable, the tools of urban planning are not generally attuned to that reality, often relying on heuristics or technocratic moves. They also tend toward ideal outcomes that are expected to remain fixed in place once achieved. There is some movement on this point. In recent decades, we have witnessed a shift toward human-centered design, for example, as exemplified by the work of Jan Gehl. In his approach, human activity guides urban form directly. This could be considered an antidote to the technocratic approach and leads to interesting outcomes. For the purpose of this chapter however we will set that direction aside and focus on the physical aspects of the city.

It is useful to linger briefly on the subject of resilience and its implications for how cities are conceived. The key point is to recognize the difference between *engineered resilience* and *ecological resilience*, a distinction first raised by the theoretical ecologist C.S. Holling as a way to explain processes of change in ecosystems (Holling 2017). In the case of the former, resilience is a measurement of how long a system needs before it returns to a state of equilibrium after a disturbance. The key to this view of resilience is that there is only a single viable stable state, and all efforts and costs are thus directed toward preserving that state. Ecological resilience, however, refers to a multi-equilibrium system, where resilience does not always mean returning to a previous mode. Instead, there are several possible states of equilibrium, each as valid as the other. Resilience in that case is measured by the amount of disturbance that can be absorbed before a system changes to a new state (Gunderson 2000). Engineers and system designers tend to prefer the single state definition and will employ safeguards to ensure that outcome. From their point of view, change and disturbance are recognized as natural processes but are designed out of all systems as much as possible. As a general approach it can work very well, as long as a system sits within a predictable and constant environment. As we have seen with the extremes of high-modernist planning outlined above, one way to create that kind of certainty is to simplify the systems themselves and to remove unwanted variation.

Those who allow the idea that ecological resilience is possible in urbanism will look to other means of control and recognize more than one future as viable. From the point of view of urban planning the implication is that a design will be able to function with several distinct states, and that a disturbance will set the direction toward one state or another. By way of an example, a city designed for a large population could also be designed in such a way that it functions just as well with a population reduced to a fraction of its former size. The physical attributes of a city designed to work with that amount of uncertainty would necessarily be different from a city designed for a single optimal form. From the point of view of urban planning theory, this type of city necessarily would recognize that plans and regulations intended to enforce constancy are friable in periods of change. More open-ended governance is likely to be messier, but also more capable of managing unexpected forces. In the case of the Japanese city, uncertainty at the small scale is entirely acceptable, even built into the system. With that in mind, the FEW Nexus in Tokyo may be better considered from the perspective of ecological resilience.

As it happens, the concept has precedent in urban planning, though not without some controversy. It is most transparently on display in its application by the New Urbanist concept of the Transect (Han 2021; Hélie 2019). Based on the concept of ecological succession, the model defines a range of nine urban zones, shifting gradually between natural, rural, and urban landscapes. Each of the zones is distinct and metaphorically maintains a resemblance with the ecological succession we can see in real landscapes. Conceptually the idea is a good fit for a description of urban cycles of change. As Hélie (2019) reminds us, the chief idea behind it is that ecosystems become more complex as simple forms die out and, in their dying, create the substrate in which more complex and "higher-order" organisms can grow. Failure of one system is a necessary requirement for the birth of the next. In urban terms this pattern is familiar. Run-down neighborhoods are often host to artists and outliers who change the character of a place and then are themselves pushed out as it becomes economically or culturally attractive for an entirely different community. Without their first occupation of the area the next step would not happen. The lesson to take away from this observation, however, is not that there are many kinds of communities in a city, but rather that change is not always in a single direction, i.e., toward richness and quality. It is just as possible that a community can change from a place with rich social ground, deep roots, and strong connections, into a place where connections are broken and vitality is erased altogether. Such examples were common near the end of the last century in many North American cities. Perhaps the most extreme example is Detroit, which has become famous for its devastating loss of population and removal of vast swathes of homes, leaving behind large areas that have no certain future. What is left represents a new urban typology in search of new uses. Japan similarly faces a shrinking population, with the same uncertainty of how to address its empty homes and spaces left behind.

By contrast, the New Urbanist vision is based firmly in a model of growth. Its adoption of ecological succession as a guiding principle is proposed as a way to ensure multiple scales and types of community. Notably it aims to escape from a "linear" or reductive approach to planning by adding mixed land use and spatial variation to the system. However, the ideology also ignores the possibility of change and reversal. The passage of time is entirely excluded as a possibility. To the contrary, New Urbanism often works hard to deny change through extremely strict regulation. As Hélie (2019) describes it, "What is missing from the New Urbanist transect is its "geological" dimension—how does one layer arise out of the previous one? We move across zones in space, but never in time. Without removing the structure of the previous zone and starting over, we cannot "upgrade" a zone. We must assume that a zone comes into existence fully realized and functional." This critique captures a common issue with the conversion of abstract theory to concrete practice in urban planning. While New Urbanism captured the idea of accumulating multiple typologies through zoning, there is no room in their plans for a zone to change from one type to another, never mind to develop new forms as they are needed. In this regard the transect is not sufficiently different from the high-modernist ideal that distributes tightly controlled and siloed forms within a city. Perhaps this is why the concept has become yet another utopian ideal that is built only sporadically in cities—convincing enough to be tested, yet not enticing enough to be adopted by more than a few communities or governments.

2.4 *Panarchy and Hyperobjects*

To avoid a similar fate with the FEW Nexus, it is here proposed that we might make better use of ecological theory by avoiding its application as a metaphor, and instead consider it as a description of urban processes. The focus of an urban master plan is then defined as much by process as by form. The distinction presumes cities are not machine-like and should therefore not be governed by the rules of engineered resilience. To be fair, this is an approach that is in apparent contrast to the massive urban transformations of the past. However, it has the benefit of suggesting a model where cities can prepare for multiple futures without divesting from existing processes.

A useful framework for this ambition comes from the concept of panarchy. After establishing the field of resilience theory in the 1970s C.S. Holling continued to develop his concept, often in partnership with others. He eventually coined the term "panarchy" (Holling 2001), which is more directly applicable to urban forms, as it is bound up with the idea that connections between multiple scales of activity can influence one another. To be fair, the connection with urban planning is a somewhat uneasy fit. This is clear even with sympathetic ideas such as sustainable design. The issue is not so much with the connection between the concepts themselves, but rather with the way urban planning tends to be conceived. Specifically, there is a noticeable tendency toward the search for an ideal form when urban plans are developed. If we use the terms of resilience theory, it is a kind of search for a climax state, or a state of perfect equilibrium. On the face of it this is not problematic. However, the issue becomes clear even from Holling's earliest insights on the subject. Writing in 1973, he concludes that "…an equilibrium centered view is essentially static and provides little insight into the transient behavior of systems that are not near the equilibrium" (Holling 2017, p. 2). While he was writing of ecosystems in general, a later critique by Ahern addresses urban planning directly. Ahern writes that the tendency of urban plans toward an ideal climax state "…denies the inherent spatial and functional dynamics of complex, self-organizing socio-economic systems, like cities—that are subject to frequent, regular disturbances as well as stochastic disturbances occurring at highly irregular intervals and durations of time and affecting multiple spatial scales" (Ahern 2013, p. 1204).

Panarchy offers a possible antidote to that tendency. As it is well explained and analyzed elsewhere (Gotts 2007; Gunderson 2000; Holling 2004, 2017; Berkes and Ross 2016) it will not be described in detail here. To summarize, the idea was created to describe the fact that adaptation to disturbances takes place at different speeds depending on the scale under consideration. Though it was not developed to explain architecture or urban planning, if we were to apply the concept within that realm, it could be described most simply by saying that buildings change quickly, infrastructure changes slowly, and cities and regions change even more slowly. But they are all changing over time.

For urban planning the idea of panarchy is useful because it proposes that adaptation at one scale can affect adaptation at another scale, both larger and smaller. This is different from a more "normal" understanding of hierarchies; it recognizes that not all adaptations work from the top down and that significant change and impact can take place from the bottom up. If we consider the critique of ideal urban planning, panarchy is additionally useful because it recognizes that as changes take place, there are also processes of conservation at play—and these are the locations around which adaptations will coalesce. We should note that the concept is not without criticism. One of the main concerns is that it works better as an abstract idea than a world-tested proof (Gotts 2007; Allen et al. 2014). For the purposes of this chapter, that critique is less of an issue, because the kinds of change that are measured in Tokyo fit the expectations that a panarchy-based model predict.

Before returning fully to the Japanese city, it is again useful to examine a more recent addition to the subject of hierarchy and scale, this time in the form of hyperobjects. The term "hyperobject" was coined by the popular philosopher Timothy Morton as a way to describe the enormity of the world we exist in and the ideas that describe it. A leading thinker of the Anthropocene, he paints a picture where humans are not only part of the world we live in, but absolutely integrated and inseparable from it (Morton 2013). Conceptually it is similar to the way that the parts of a living cell form a larger organism and cannot stand outside of it—an important point, simply because it is difficult to measure something when we are inside of the very thing that we are attempting to measure. A hyperobject is a thing so large that it cannot be understood all at once, and in many ways is not possible to grasp, even though it has concrete impact on our daily lives. Following this idea, climate change is a hyperobject, as is the internet. All of these are as real as hammers, yet frustratingly intangible and hard to interact with directly, even though we are entangled with them (Blasdel 2017). We could almost imagine being able to measure something so complicated and large. We have after all managed to measure and understand things once considered impossible, from X-rays to distortions in space-time. However, a hyperobject is additionally complicated by its equally long timescale. Because it is so large and exists for such a long time it is difficult for humans to imagine, never mind measure. Simply stated, the human timescale is much shorter than the kinds of timescales that a hyperobject exists within. That disconnect between both time and spatial scales is what makes hyperobjects so difficult to see and to engage with purposefully.

The idea of hyperobjects is also tied to communication around climate change and ecology. This topic is useful to consider because it is one way that people might engage with such large subject matter in a direct way. In particular, there is some concern over how to convey the urgency and importance of the message that change is needed before the Anthropocene ends in disaster (Boulton 2016; Nisbet 2009). Morton proposes that we need to stop making arguments over climate change (a hyperobject) from a top-down rational approach and to instead turn toward feelings, making use of art that can reach to the subconscious. Otherwise, we will fail to see the problem as relevant to us, simply because it is too large (Morton 2013). This particular way of seeing the world is reminiscent of the Animal Spirits of Keynes, which describe the collective actions of a kind of public mind in the field of economics. Animal Spirits are difficult to predict and are often irrational and emergent (Akerlof and Shiller 2011). As is often pointed out, humans quite often discount future problems in the face of the problems that need to be addressed in the present. Both of these ideas explain the weirdness of the world we live in and the difficulties in taking action at a large scale to achieve important goals, such as avoiding climate change or economic collapse. Interestingly, they both share the innate understanding that people act collectively, but that communication needs to be engaging enough that individuals will respond. If the FEW Nexus concept can be scaled, it will need to follow a similar mantra.

From the point of view of panarchy, hyperobjects are interesting because they describe the highest point of a scale of hierarchies. From our point of view, that is a useful connection because panarchy describes a way to see the connection to large things from the point of view of small or even human scales (both spatial and chronological). With a little imagination we can predict that the hyperobject of climate change can be addressed directly through something as small as architecture and urban planning, and that we can engage with long timespans even through the short-term activities of daily human life. The implication of the ideas expressed by Morton is that the answer to massive problems is to become intertwined with our world as much as possible. However, his description also implies a danger, because many of our systems are problematic. It will require some care if we do not want to become caught up in the unexamined certainties that placed us so firmly within the Anthropocene.

3 Tokyo and the Passage of Time

Returning now to the Japanese city, in light of the discussion above and the various imagined ways that cities can be comprehended and acted upon, the FEW Nexus faces two challenges.

First, we are faced with the fact that technical systems are generally governed by centralized organizations and authorities, and this is justified appropriately by the need to escape the negative consequences of humanity gathered in something as dense and compact as a city. The FEW Nexus implies a disturbance to that system in that it will attempt to decentralize parts of a network, some of which approach the scale of hyperobjects. Food production, electrical production, and water use are all very large systems that are already interconnected in complicated ways. If there is an intent to reimagine those connections, then they will need to be reconsidered at that large scale, even if a design focuses on the local manifestation.

Second, if a building design or urban plan is considered with the framework of panarchy, the most logical way forward will be to develop a method where the designs will act at multiple scales, both physical and temporal. Taking on projects in that way requires some understanding of both the systems in place but also the rhythm of change of the environment.

In the Japanese city the pace of change is relatively fast. More importantly, as we have already discussed above, change is not controlled by placing the parts of a city into discrete zones, but rather is allowed to take place from the bottom up, with individuals choosing and building their environment much more directly than is common in the West. The ideas of the transect and the compact city are misplaced in a Japanese city because they both respond to spatial and social segregation that is not normal in Japan (Galloway 2009).

From the point of view of resilience theory, the Japanese city is more a model of ecological resilience than engineered resilience, at least at the small scale of buildings and neighborhoods. Infrastructure in the form of roads and sewerage is similar

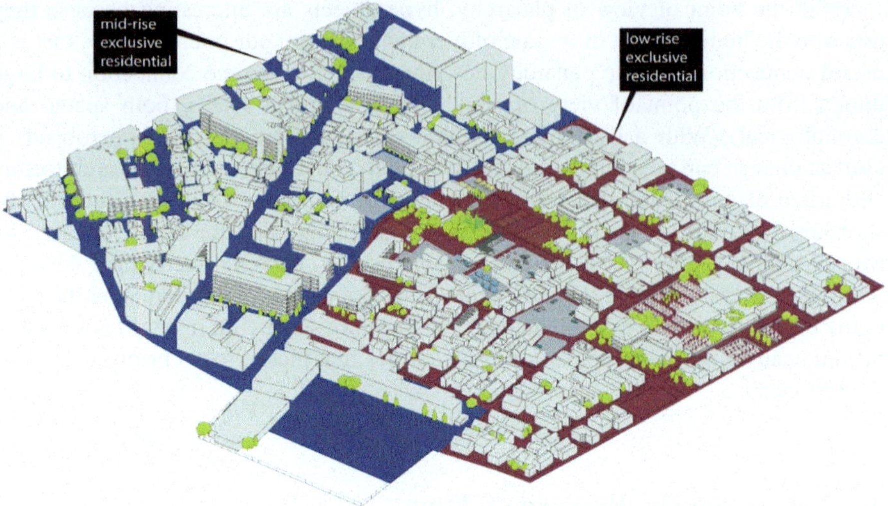

Fig. 2 Case study in Tsurumaki, Setagaya Ward, Tokyo: Zoning as described by regulation in a typical suburban neighborhood in Tokyo

to the western model in that they are controlled from above and slow to change. Yet even then the adoption of new roads is not imposed so much as planned for and built when it becomes convenient to landowners.

A simple study of a typical Tokyo suburban block illustrates both the degree of variety that can be found in a suburb (Figs. 2 and 3) as well as the pace of change that is normal for the city (Fig. 4). The study area is in Tsurumaki, Setagaya ward, located about 5 km from Shibuya, one of Tokyo's more recognizable urban centers. It is predominantly filled with single-family homes, but as a result of the parametric zoning regulations, land use is well mixed and the area is walkable. Indeed, the degree of walkability and variation of building types and scales within even this single sample would be considered a remarkable success if measured from the ambitions of the transect or the compact city model. While this is a typical Tokyo neighborhood it is nonetheless important to recognize that the degree of variation and walkability both decrease the further one moves from Tokyo's center. The furthest suburbs will require a car to be viable.

The substantial variety in land use is important because it is normal and not an outlier of Japanese urban form. It is possible to understand, for instance, that a typical community in Tokyo will not usually be defined by a single social class, as low-income studio apartments are often located within areas otherwise dedicated to single-family homes owned by the middle class and the truly wealthy. Similarly, it is not uncommon to find small businesses, restaurants, and even workshops within a quiet residential area. Areas zoned for taller buildings are similarly filled with low-rise houses and shops, meaning that a diversity of land use and residents is normal. Urban farms are less common in suburban areas especially toward the city center, but they are not so unexpected that they warrant special attention for their

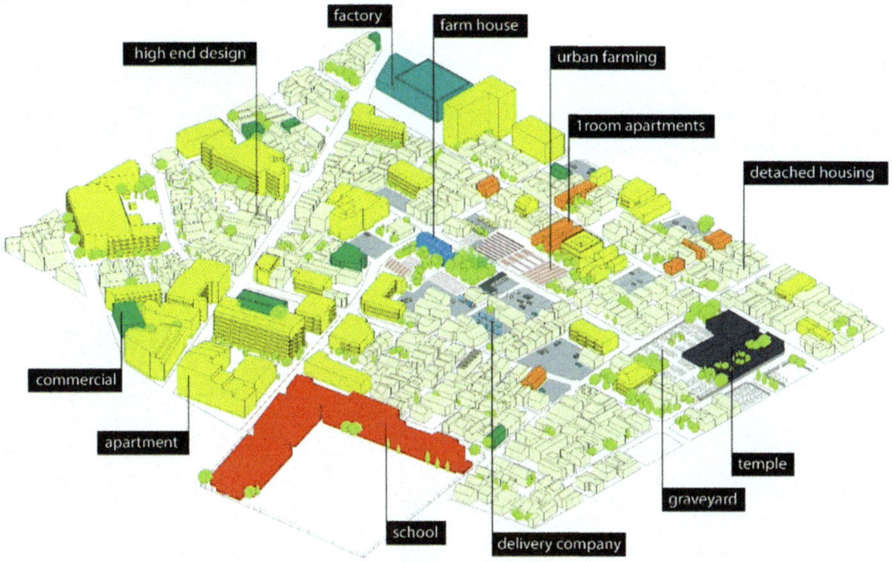

Fig. 3 Case study in Tsurumaki, Setagaya Ward, Tokyo: Actual land use as of 2019

existence. The fact that the urban planning regulations allow for that degree of variation is telling. Diagrammatically, the variation describes a range of potential. In actuality it creates a certain amount of livability that may not otherwise be possible in such a dense urban setting.

Potential is often considered just that. Something that could possibly happen, but does not. In the case of the Japanese city, potential is tested and acted upon, not only when there is a decision by urban planners acting on high, but every day of every year by residents as they build and use the land they live on in constantly changing ways, and urban planners have nothing to do with it. Landowners are required to follow strict regulations that govern the general size and shape of the buildings they place on their land; however, within that envelope land use is quite free, as we see in Fig. 3. The particular economics of development and land ownership in Japan tend to encourage landowners to tear down and replace older buildings over a period of about 30 years for houses (Fig. 4) and a bit longer for larger buildings (Gleeson 2018; Harding 2016). A kind of pulse of change results, sweeping constantly through the entire city as new ideas about housing, quality of life, and construction are translated by everyday people. This can be very adaptive. As populations age, for example, it is not uncommon to see day homes for the elderly begin to appear in older suburban areas. Small clinics are also very common, giving residents a great deal of choice when it comes to medical care in their community. In the case study area over a period of 28 years approximately 30% of the buildings were replaced. Changes in land use are not shown.

In a parallel manner, convenience stores in Japanese communities are equally changeable and adjust their offerings over time in response to the accumulation of

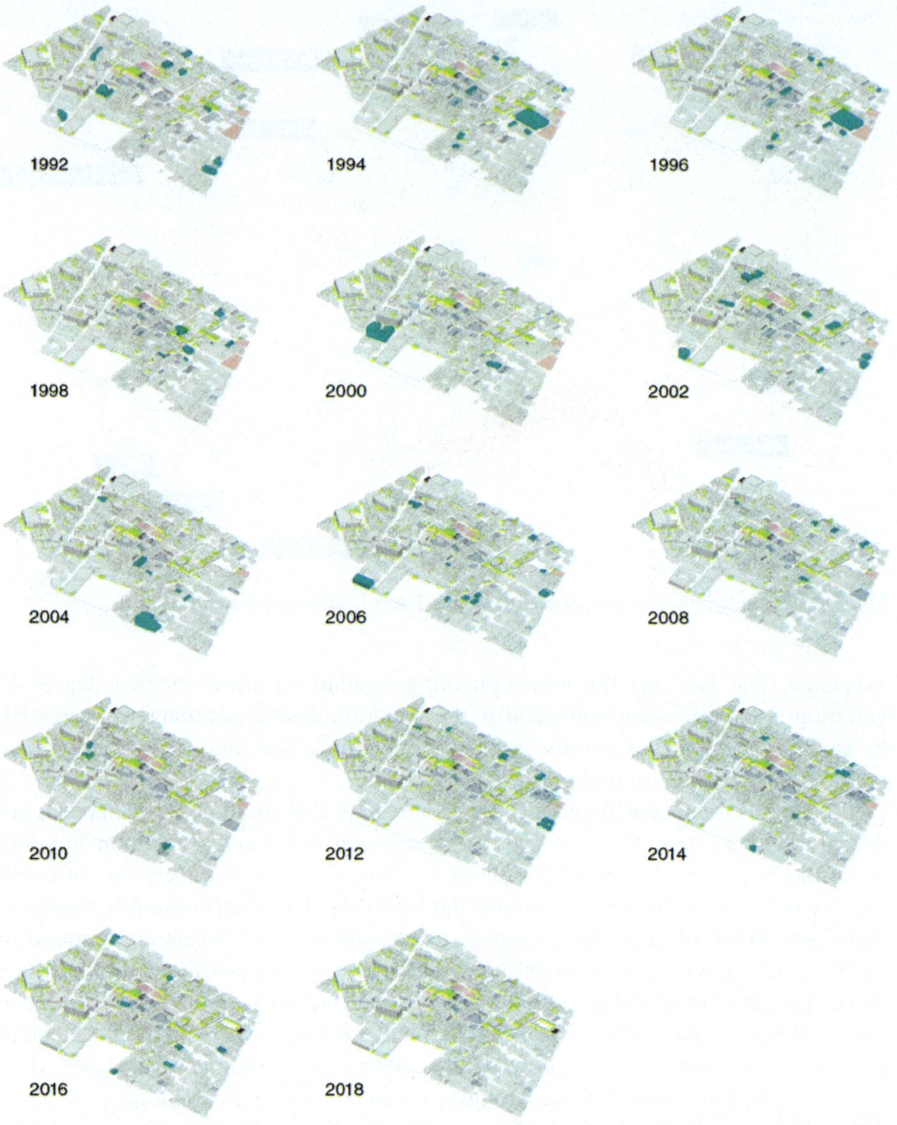

Fig. 4 Case study in Tsurumaki, Setagaya Ward, Tokyo: Pace of change. Blue indicates new buildings added each year between 1990 and 2018. Over a period of 28 years approximately 30% of the buildings were replaced. The process is continuous and continuing. It can be considered a proxy for the typical pace of change in Tokyo

local behavior. Small changes are to be expected; however, in recent years the typology has begun to take on more typical community functions, from farmer's markets, to package delivery, and even disaster recovery centers (Creasy and Galloway 2017). For better or worse, the convenience store, which is ubiquitous in

Japan, has taken on outsized importance. As we look further from the center of Tokyo or if we step away from train stations the number of convenience stores decreases (Nakayama and Yan 2019), which is problematic. However, from the point of view of the FEW Nexus model, the fact that there is an existing small scale typology that functions according to rapid and constant redefinition is important. Even more important is that the convenience store is part of a sophisticated logistical and digital network that can be adapted to extend the scale of otherwise individual tests of sustainable design, including the FEW Nexus. Considered from the point of view of panarchy they represent the conservative nature of adaptation and can be used to propel ideas further than they might otherwise reach.

4 Conclusion: A Proposal

Given the breadth and depth of ideas outlined above, it is not surprising we have largely forgotten how much of the modern city is a recent construct. That our cities are still so new gives us hope that our choices so far are not permanent, and that circumstances can be improved as we face larger problems. In the face of climate change and shifting economics and demographics, how we see the world matters.

Arguably the problem we face with climate change is not that it is met with skepticism by too many people, but rather that it is not yet considered relevant to our daily lives. James Scott and Timothy Morton offer a framework to understand our collective perspective, and Morton especially is popular for placing the subject matter into a slightly more palatable perspective. Yet, while Scott and Morton both offer cogent critiques of the current situation, they also share the distinction of not offering a way to change direction. If the FEW Nexus is intended to be a probable (and not just possible) model for the future, where food, energy, and water are combined synergistically in a single place, then it will require a transformation equal to the shift that was brought about by the sanitary movement of the nineteenth century, or the shift to a car-centered world in the twentieth century. And yet the powerful pushes and pulls that created those changes do not exist today. And so we need more subtle strategies—methods and models that allow us to affect the large scale from the bottom up. Additionally, we need to develop ways to transform cities at the infrastructural scale without replacing them entirely.

The Japanese city offers a special case because it is attuned already to bottom-up action, with massive replacement of its parts taking place as a matter of course. With the use of panarchy as a framework, the best way to introduce the FEW Nexus (and any similar concepts) is to piggyback on existing processes of change in the city, leveraging them to create a direction of change even as the details are left fluid. With this in mind, a plan for the FEW Nexus in Tokyo should consider simultaneous actions that work at multiple timescales and multiple sizes. That implies designing several layers within masterplans, each with a different rate of change and scale, and each designed to intentionally influence outcomes at higher and lower orders of both physical space and time. That the Japanese city is already shifting constantly offers a way to change both land use and the physical buildings in relatively short order, affecting a city-scale transformation without relying on large-scaled plans to determine the city itself. Buildings can be replaced with new typologies—designed to produce energy, to conserve water, and to grow food. It is not as easy to change infrastructure in the city. However, we can consider building typologies and networks that can be modified and combined to create an infrastructure-scaled transformation to go along with the shifting of individual buildings. Further study is needed; however, it is not out of place to consider railways, supermarkets, and convenience stores as essential typologies within the FEW Nexus. Notably each typology fits within its own pace of change within the city, from long-lasting rail stations to short-term convenience stores. The fact that they are all semi-public community buildings, with strong logistical and network effects, means they can have a high impact on a community and become powerful multipliers of change. They should be targeted as partners for any serious FEW Nexus plan in Japan. Meaningfully, none of these typologies requires governmental action.

If the government were to become involved, policy might be created to nudge behavior and outcomes and further amplify processes. Naturally every city, town, or region would need to be treated differently to match local patterns and inputs. Even so, the basic idea of the FEW Nexus is applicable as a general approach. Making use of both process and policy, Japan could become much more resilient and prepared for the changes that are already coming into play. This proposal is based on ecological resilience and would be open to multiple sustainable futures. As Japan continues to develop we can aim for several robust futures and avoid the trap of making another singular utopia that can only be partially realized.

References

Ahern J (2013) Urban landscape sustainability and resilience: the promise and challenges of integrating ecology with urban planning and design. Landsc Ecol 28(6):1203–1212

Akerlof GA, Shiller RJ (2011) Animal spirits: how human psychology drives the economy, and why it matters for global capitalism. Princeton University Press, Princeton

Allen CR, Angeler DG, Garmestani AS, Gunderson LH, Holling CS, Sveriges lantbruksuniversitet. (2014) Panarchy: theory and application. Ecosystems 17(4):578–589

Badger E (2021) Covid didn't kill cities. Why was that prophecy so alluring? International New York Times

Berkes F, Ross H (2016) Panarchy and community resilience: sustainability science and policy implications. Environ Sci Pol 61:185–193

Black M (2008) The experience of the first sanitary revolution: are there lessons for today's global sanitation crisis? Waterlines 27(1):62–70

Blasdel A (2017) 'A reckoning for our species': the philosopher prophet of the Anthropocene. The Guardian. https://www.theguardian.com/world/2017/jun/15/timothy-morton-anthropocene-philosopher. Accessed 24 Jul 2021

Boulton E (2016) Climate change as a 'Hyperobject': a critical review of Timothy Morton's reframing narrative. Wiley Interdiscip Rev Clim Change 7(5):772–785

Callahan G, Ikeda S (2004) The career of Robert Moses: city planning as a microcosm of socialism. Independent Rev 9(2):253–261

Callies D (1997) Urban land use and control in the Japanese City: a case study of Hiroshima, Osaka, and Kyoto. In: Karan P, Stapleton K (eds) The Japanese city. University Press of Kentucky, Lexington

Creasy M, Galloway W (2017) A convenient future. Planphase 6:38–47

Cugurullo F (2016) Urban eco-modernisation and the policy context of new eco-city projects: where Masdar city fails and why. Urban Stud 53(11):2417–2433

Fitzgerald J, Agyeman J (2021) Removing urban highways can improve neighborhoods blighted by decades of racist policies. The Conversation. https://theconversation.com/removing-urban-highways-can-improve-neighborhoods-blighted-by-decades-of-racist-policies-166220. Accessed 15 Sep 2021

Galloway W (2009) Of suburbs and cities. Town and country planning, pp 322–327

Galloway W, Logan K, Yan W (2021) Proposal for a database of food-energy-water-nexus projects. In: Roggema R (ed) TransFEWmation: towards design-led food-energy-water systems for future urbanization. Contemporary urban design thinking. Springer, Amsterdam, pp 281–312

Gleeson J (2018) How Tokyo built its way to abundant housing. https://jamesjgleeson.wordpress.com/2018/02/19/how-tokyo-built-its-way-to-abundant-housing/

Gotts NM (2007) Resilience, panarchy, and world-systems analysis. Ecol Soc 12(1):24

Gunderson LH (2000) Ecological resilience - in theory and application. Annu Rev Ecol Syst 31(1): 425–439

Han S (2021) The use of transects for resilient design: core theories and contemporary projects. Landsc Ecol 36(5):1567–1582

Harding R (2016) Why Tokyo is the land of rising home construction but not prices. FT.com

Hein C (2010) Shaping Tokyo: land development and planning practice in the early modern Japanese metropolis. J Urban Hist 36(4):447–484

Hélie M (2019) How urban change is like ecological succession. Strong Towns. https://www.strongtowns.org/journal/2019/7/25/how-urban-change-is-like-ecological-succession

Holling CS (2001) Understanding the complexity of economic, ecological, and social systems. Ecosystems 4:390–405

Holling CS (2004) From complex regions to complex worlds. Ecol Soc 9(1):11

Holling CS (2017) Resilience and stability of ecological systems (1973). In: Robin L, Sörlin S, Warde P (eds) The future of nature. Yale University Press, New Haven, pp 245–260

Jefferies T, Cheng J, Coucill L (2020) Lockdown urbanism: COVID-19 lifestyles and liveable futures opportunities in Wuhan and Manchester. Cities Health 5:S155

Koolhaas R (1995) Whatever happened to urbanism? Des Q 164:28–31

Marcotullio PJ, Rothenberg S, Nakahara M (2003) Globalization and urban environmental transitions: comparison of New York's and Tokyo's experiences. Ann Reg Sci 37(3):369–390

Morley I (2007) City chaos, contagion, Chadwick, and social justice. Yale J Biol Med 80(2):61–72

Morton T (2013) Hyperobjects: philosophy and ecology after the end of the world. University of Minnesota Press, Minneapolis

Nakayama S, Yan W (2019) The package redelivery problem, convenience store solution, and the delivery desert: case study in Aoba Ward, Yokohama. J Urban Manag 8(3):355–363

Nisbet MC (2009) Communicating climate change: why frames matter for public engagement. Environ Sci Policy Sustain Dev 51(2):12–23

O'Sullivan F (2020). Barcelona will supersize its car-free 'superblocks.' Bloombergcom. https://www.bloomberg.com/news/articles/2020-11-11/barcelona-s-new-car-free-superblock-will-be-big. Accessed 21 Jul 2021

Popovich N, Williams J, Lu D (2021 Can removing highways fix America's cities? International New York Times

Scott JC (1998) Seeing like a state: how certain schemes to improve the human condition have failed. Yale University Press, New Haven

Sies MC (1987) The city transformed: nature, technology, and the suburban ideal, 1877-1917. J Urban Hist 14(1):81–111

Snowden FM (2019) Epidemics and society: from the black death to the present. Yale University Press, New Haven

Sorensen A (2005) The making of urban Japan: cities and planning from Edo to the twenty-first century. Routledge, New York

Stapczynski S (2021) Japan restarts first nuclear reactor since 2018 amid hurdles. Bloomberg.com. https://www.bloomberg.com/news/articles/2021-06-23/japan-restarts-first-nuclear-reactor-since-2018-amid-hurdles. Retrieved 28 Aug 2021

The Economist (London) (2020) Cleanliness is next to growth 436(9205):69–71

Land Use Planning and Conservation Policy in the Tokyo Metropolitan Area

Ruiyi Zhang, Wanglin Yan, and Rajib Shaw

Abstract Tokyo has witnessed rapid growth and dramatic socio-economic transformation in the last half-century. The most remarkable changes in land use have taken place in suburban municipalities, where agricultural and natural land continue to be converted to residential and industrial uses, often without comprehensive planning. This chapter qualitatively explores land use changes in the Tokyo Metropolitan Area (TMA) by analyzing policies, discussing the institutional features, including flexibility and possible loopholes, and socio-economic context behind changes of land use. And, this chapter calls for an evolution of land use planning and land resource conservation system geared more toward adaptive and sustainable urbanization.

Keywords Land use planning · Zoning · Urban transformation · Conservation and development

1 Introduction

Overall, since 2010, more people in the world are living in urban than rural areas, and by 2020, 56.2% of the world population was living in cities. In the coming decades, ongoing urbanization will be still the most significant socio-economic and spatial driving forces, especially in developing countries. Land use will keep changing in the internal parts of mature metropolitan areas of developed countries while

R. Zhang (✉) · R. Shaw
Graduate School of Media and Governance, Keio University, Fujisawa, Kanagawa, Japan
e-mail: zry1030@sfc.keio.ac.jp; shaw@sfc.keio.ac.jp

W. Yan
Faculty of Environmental Information Studies, Keio University, Fujisawa, Kanagawa, Japan
e-mail: yan@sfc.keio.ac.jp

W. Yan et al. (eds.), *Resilient and Adaptive Tokyo*,
https://doi.org/10.1007/978-981-99-3834-6_5

more extensively outward in emerging metropolitan areas of developing countries. In either case, land use changes are likely to be dramatic, with magnitudes and impacts far greater than any other region on Earth. A variety of land use planning and conservation systems exist in different metropolitan areas, all intent on efficiently utilizing regional resources, developing sufficiently and orderly toward environmental sustainability.

There are many socio-economic factors and temporal uncertainties intertwined with competing land use relationships in metropolitan areas, between residential, industrial, and cultivated lands. How cities in metropolitan areas should manage land use systems and policies remains challengeable. The answers are often ambiguous.

The objective of this chapter is to sort out the institutional origins and changes in land use planning and land conservation systems in the TMA and through basic investigation of industrial and farmland land use change. The question is whether land use changes in the TMA have responded to the characteristics of regional planning in different time periods, and whether they are in conflict with institutional intentions at the regional scale. The resilience and flexibility of the land use systems are discussed from the perspective of its adaptability and planning.

Tokyo Metropolitan Area (TMA) has experienced drastic land use change in last 50 years. For instance, the manufacturing shipping value and numbers of employees continued to decrease since the 1980s, while the industrial land stocks remained unchanged in some places and increased at a slower pace in others (Aiba 2021). Up to now, metropolitan peripheries have typically not been a main focus of research on sustainable land use planning and conservation systems besides context-specific and solution-oriented case studies (Geneletti et al. 2017).

In-depth study on the interrelationships between land use policy and land use change in metropolitan areas could lead us to a better understanding of the missing piece for sustainable urbanization. It could again nurture the scientific integrity of regional development and local administration and provide lessons about land use change from a regional perspective.

2 System of Land Use Planning and Conservation in Tokyo

2.1 Land Use Planning in Japan

2.1.1 The Urban Land Use Planning System

Among all policies regulating the relations between land uses, the most notable ones are regulatory and economic measures taken under the planning law (Tsubota 2006). Planning systems, including development guidance and land resource protection, vary from country to country, because of differences in socio-economic contexts and institutional characteristics, such as governance and ownership structures. Meanwhile, systems in one country or region can evolve to adapt to the different contexts of different stages as developing, so they could not remain static. Urbanization and

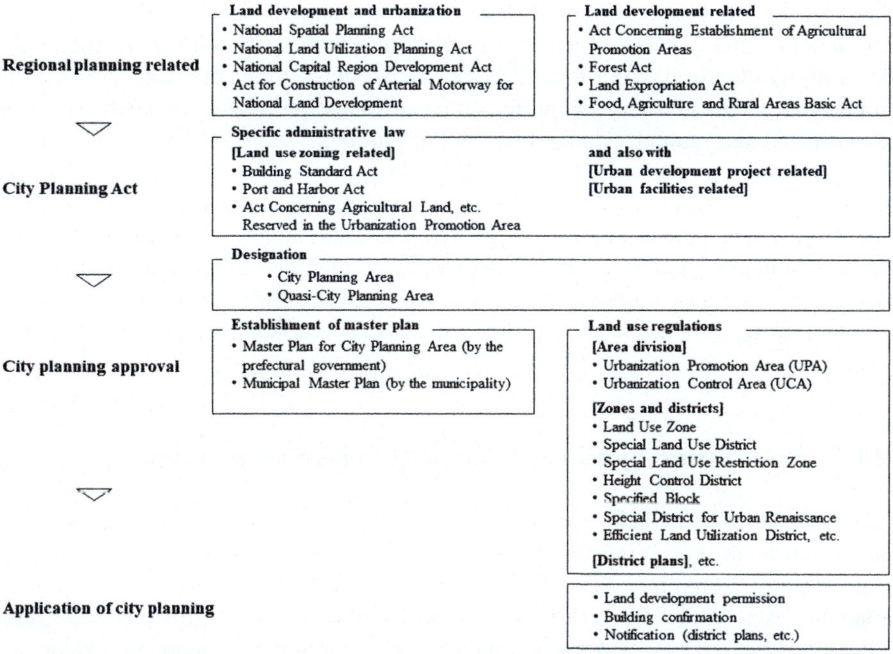

Fig. 1 Land use planning and conservation system within the City Planning Act and relevant institutions. (Based on report by MLIT 2003)

land use change in the TMA at the regional scale could be seen as a consequence of the interaction, or conflict, between two factors: socio-economic demands and the urban land use planning system.

Figure 1 shows the Japanese urban land use planning system in terms of both the development and conservation of land. The City Planning Area and delineation of other policy zones, as the meeting point of policy and implementation for the promotion and control of land development, are designated by the prefectural governments. A Quasi-City Planning Area is an area outside the City Planning Area, commonly seen in areas next to expressway interchanges, where orderly land use or preservation of the environment is required. These lands can be designated as a Land Use Zone, Special Land use District, Special Land use Restriction Zone, etc., while excluded are a District Plan, Urban Facilities, and Urban Development Projects. Outside the City Planning Area, there are special plans for other land uses related to the national land use plan, such as the agricultural promotion area plan.

Prefectures issue the Master Plan for the City Planning Area and municipalities issue their Municipal Master Plan. The land use plan for the City Planning Area includes guidelines for the assignment of each urban function and the formation of urban centers at all levels. The planning system at the national and regional levels in Japan has gradually evolved from being "state-led" and "development-oriented" from the early post-war period onward, toward a system of "state-local cooperation" for national land development and conservation. Planning systems at the municipal level are gradually moving toward "local autonomy without breaking the existing framework." The ideal is for prefectural and municipal plans for land use to be coordinated and mutually compatible, but not unified.

Specifically, the municipalities decide on the areas for land development and land conservation, respectively, based largely on local socio-economic factors such as industrial development, population change, and land supply and demand. Distinctions in land use efficiency are widely recognized in the group of Japanese planners and researchers. Local implementation in the context of differences in local understanding of regional policy has shaped the landscape of land resource utilization in Japan.

2.1.2 Area Division System and Farmland Conservation System

The basic context of land resource regulation and conservation in the process of urbanization in Japan is mainly determined by the City Planning Act and the Agricultural Promotion Act. A City Planning Area (CPA), based on the City Planning Act, is divided into Urbanization Promotion Area (UPA), Urbanization Control Area (UCA), and non-delineated area, aiming to encourage, guide, or control urban development. Meanwhile, the Agricultural Promotion Area (APA), based on the Agricultural Promotion Area Act, aims to encourage farmland conservation and agricultural development. The two areas overlap with each other in zoning (Fig. 2), while paddy field and other farmland mainly exist in the (1) UPA, (2) APA inside the UCA, (3) non-APA inside the UCA, (4) APA outside the CPA, and (5) non-APA outside the CPA. The priority of farmland conservation is in the areas of (2) and (4), which have the highest hierarchy in the policy and regulations, while the areas of (3) and (5) should be protected against land use conversion due to urbanization because of their location outside the UPA or even the CPA. However, controversies have occurred between administrative ministries in identifying the priority between urbanization promotion and farmland protection in Japan (Moreno-Penaranda 2011). In many cases, even the farmland inside the APA could be converted for urban development.

Institutionally, in the UCA, a development approval system including land development permission and building confirmation is applied; not that development is prohibited but that "in principle" all development needs to be permitted before construction (as opposed to "no construction without permits within the urbanization area and no permits outside the urbanization area"). There are also cases in which construction is not subject to the development approval system, depending on the scale, even though the types of permittable development and construction are restricted. Meanwhile in the APA, a system of permission for the conversion of agricultural land is adopted, in which only the conversion of agricultural land in special agricultural land areas is not permitted, and the conditions of agricultural land in general agricultural promotion areas range from "no permission in principle" for

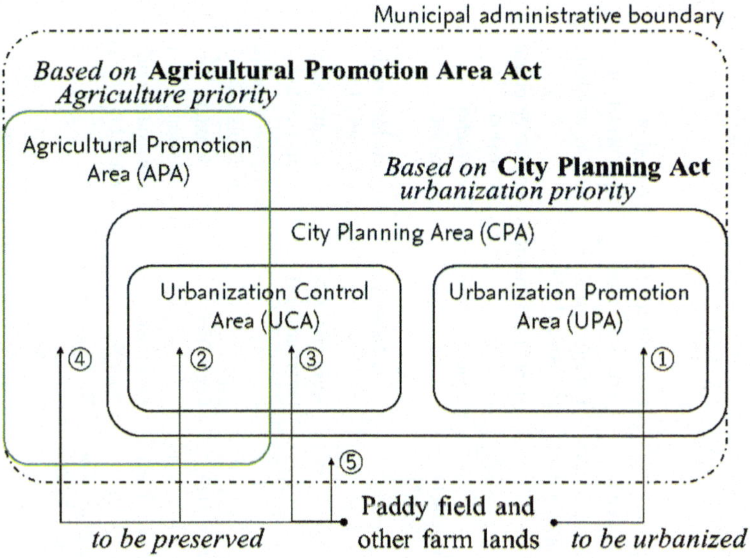

Fig. 2 Conflict between farmland conservation and urban development (APA can be furtherly divided into special agricultural land and general agricultural promotion areas)

large agricultural land with improvement plans, to "permittable" for small agricultural land without improvement plans.

2.1.3 Features of Japan's Conservation Policy from an International Perspective

In general, Japan's land use planning and conservation system has some commonalities with other developed countries, such as Germany and the USA. At the same time, Japan's system is seen as being relatively flexible due to cultural factors in Japan (Sorensen 2000). Table 1 shows an international comparison of land use regulation, from which we can recognize there is no obvious difference in land use planning and conservation systems between Japan and other developed countries, at least before designation and implementation. So, the questions have been framed as: (1) What is the result of land use policy in Japan under such a planning and conservation system? (2) Whether or not the land use change corresponded to the institutional intention, and why?

Table 1 Comparison of land use regulations across developed countries

Type	Land use regulation			Conservation area (urbanized area)	Method of implementation
	General regulation	Control over use	Project implementation		
Japan	Zoning	Common criteria all over Japan	Redevelopment promotion district plan	Urbanization Control Area (UCA)	Building confirmation
	District plan			Area of agriculture land	Land development permission
France	Plan Local d'Urbanisme (PLU)	Different criteria	Zone d'Aménagement Concerté (ZAC)	Zone Agricole	Construction permission
		Common criteria if located in joint city planning area		Zone Naturelle	Review for consistency with PLU
Germany	Bebauungs-plan (B-plan)	Different criteria. Construction possible in an area without B-plan if located in connected built-up area	–	AuBenbereich (external area)	Baugenehmigung (building code)
					Review for constancy with B-plan
England	No clear zoning	Development individually reviewed within the framework of the plan approval system	Action area plan	Green Belt	Development permission
USA	Zoning	Different criteria	Planned unit development (PUD)	Subdivision control	Subdivision regulation and building code
			Transfer of development rights (TDR)	Building control	

Source: JICA (2007)

2.2 *Basic Land Use Policy in the TMA*

Meanwhile, this land use planning and conservation system has an additional layer of patchwork in major metropolitan areas like Tokyo and adjacent cities in response to specific demands, such as housing, industrial development, and natural resource protection. Due to the heterogeneity within different parts of the metropolitan area, principles for land use planning are basically determined by the distance from the metropolitan center, which is further summarized as a regional governance agenda:

Main intensively urbanized core of the TMA From the perspective of strengthening the international competitiveness of the city, effective and advanced land use is promoted, as well as improvement of the urban environment and safety.

Suburbs of the TMA There are municipalities facing population decline and increasing cost of urban infrastructure. Urban functions such as administration, medical and nursing care, welfare, and commerce, as well as residences, are to be centralized in fewer municipal central areas around railway station. Urban sprawl in suburban areas is to be controlled.

Periphery and hinterlands of the TMA With regard to agricultural and forest land use, prime farmland is legislatively protected in this area, which is indispensable for the stable supply of food and other services. The occurrence of underused farmland is prevented, and land use efficiency is to be enhanced through promoting the accumulation of the lands of farmers. As well, the development and conservation of forests will be promoted, which play an important role in preserving the land resource and recharging water sources.

In the 1950s, the deterioration of the environment in existing urban areas and the problem of overcrowding became a serious issue in the TMA, due to the concentration of industries and the accompanying rapid population influx from rural areas during the post-war period of high economic growth. For this reason, the Tokyo Metropolitan Area Development Act was enacted in 1956 in order to prevent excessive concentration of industries and population in existing urban areas of the TMA, to prevent disorderly sprawl, and to promote planned infrastructure development. According to the Act, the areas that have already been urbanized are considered part of the Existing Urban Area (further EUA, Fig. 3). It is an area where the infrastructure necessary for the capital is maintained or promoted while suppressing extreme concentration of industry and population.

The Suburban Development Zone (further SDZ, Fig. 3) is an area in the suburbs of the EUA where green spaces are preserved, and planned urban development is promoted. The SDZ was to preserve the environment in urban areas and to function as a buffer zone between existing urban areas and the natural environment (Akita 2017). The area outside the SDZ is partially planned for urban development outside the TMA as Urban Development Area (further UDA, area colored beige in Fig. 3). It is designated as an area that eases the concentration of industry and population and promotes development for the purpose of proper allocation of industry and

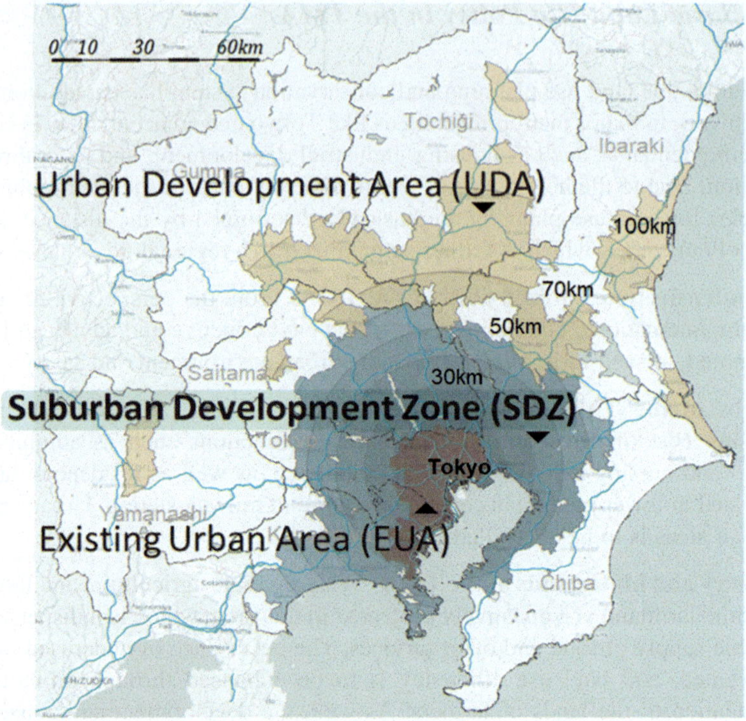

Fig. 3 Three governance areas based on the TMA Development Act. (Based on: data from MLIT 1956)

population throughout the metropolitan. Since the UDA is above 60 km away from the center of the TMA, and it is identified as a destination of industrial transfer from the national perspective, in this chapter, we survey land use change in both the SDZ and the EUA.

2.3 History and Trends of Land Use Policy in the Tokyo Metropolitan Area (TMA)

Based on the National Capital Region Development Act, enacted in 1956, the "1st Basic Plan for the Capital Region" was established in 1958 for comprehensive development and resource management in the TMA. Followed by four updates afterward until 2000s, an institutional reform for Capital Region Development Plan in 2006 and an important amendment in 2016, the policy and socio-economics have continuously interacted each other and formed the structure of TMA. Those are "recorded" in the urban landscape and land use, chronicling the mutual effects of different backgrounds and local contexts. Thus, the spatial features of land resource development and conservation vary through time periods and geographic locations in particular by distance from the city center.

Table 2 shows the policy and environmental changes of the TMA after the Second World War. Demographic conditions and land use development were the key factors in the changes. Analysis reveals the flexibility of regional policies that keep updated to adapt to the social transition.

During the rapid economic growth accompanied with fast urbanization in the 1960s, it was difficult for city planners and policy makers to predict the frontier boundaries in advance (Usui 2019). The 1st Basic Plan to some extent pictured a blueprint of a "dart board" structure across the whole TMA. That was rejected by public especially urban developers and municipalities soon. The 2nd Basic Plan accepted the reality and opened door to urban sprawl. To avoid the confusions caused by the second version, the 3rd Basic Plan demarcated the Existing Urban Area (EUA) as the center of the metropolitan, the Suburban Development Zone, and the Urban Development Area with specific development visions. Policy goals were set to foster polycentric urban complex and to reduce dependency on the metropolitan center by thriving social and cultural functions in peripheries.

When entering to the 1980s, natural population growth became shrinking, and social growth diminished to a low level. The 4th Basic Plan lowered the projected population in year 2000 to 40.9 million, and the 5th Basic Plan kept to a lower scenario of urban growth at 41.9 million for 2011 and 41.8 million for 2015. Instead of figureprint structure along commuting railways, the "great ring linkage axis of the TMA" in the northern, eastern, and western inland areas of the Kanto region was emphasized then and preserving orderly land use and promoting regional cooperation in a circular direction were highlighted.

In the beginning of this century, the New National Capital Region Development Plan was issued with some institutional integration and legislative amendment for a new era of depopulation and aging. A megalopolitan structure was proposed, containing a unipolar dependency structure with multiple functions concentrated in the metropolitan center, clusters of self-sufficient suburban cities, and regional business core cities in the metropolitan fringe. The current version of the Plan was issued in 2016, based on amendment of the Plan 2006. The population was expected to peak off at 43.6 million in 2015 and then decline to 42.4 million in 2025. Three basic consensual principles were proposed for future development of the TMA:

- To selectively disperse the functions of the EUA while consolidating and improving local functions.
- To compact the built-up areas and preserve green areas in the SDZ.
- To promote clustering of various functions in the UDA and nurture regional business cores.

Table 2 Plans for national capital region development

Capital region plan	Planned period	Background	Expected population size	Main emphases
1st Basic Plan	1958–1975	Concentration of population and industry	26.6 million	Green belt
		Needs for a metropolitan area with a functional center		Satellite cities
				Curbing expansion
2nd Basic Plan	1968–1975	Railway network and highway development	33.1 million	Governance Area (Fig. 3)
		Urban sprawl		Harmonious coexistence between built-up areas and green space in the SDZ
		Green belt action failed		Satellite town promotion in the UDA
3rd Basic Plan	1976–1985	Oil shock in 1973	38.0 million	Multi-polar metropolitan area
		Growth of peripheral areas		Regional business core city
				Self-sufficient suburban cities
4th Basic Plan	1986–2000	Low stationary phase of demographic transition	40.9 million	Reverse unipolar dependence on the metropolitan center
		Emergence of aging issue		Multi-core and multi-ring regional structure
		Technology innovation and information society		Self-sufficient suburban cities
5th Basic Plan	1999–2011	From age of growth to age of maturity	41.9 million	Redevelopment of living environment in the EUA
		Safe metropolitan area with symbiotic environment		"Ring of hub cities" in the SDZ
				"Great ring linkage axis of the TMA" in fringes of the metropolitan area
New National Capital Region Development Plan	2006–2015	Value of diversity	41.8 million	Global city of Tokyo in the EUA
		Saturation phase of urbanization		Cluster of suburban cities to share functions of the center
		Stabilized regional urban system		Regional business core city

(continued)

Table 2 (continued)

Capital region plan	Planned period	Background	Expected population size	Main emphases
Plan amendment (current version)	2016–2025	Aging population issue speeding up	42.4 million	Urban renaissance in the EUA
		Global competition		Consolidate built-up areas and preserve green space in the SDZ
		Risk of huge disaster		Promote clustering of multi-functions in the UDA
		Deteriorating infrastructure		
		Global environment issue		

Source: Based on MLIT report, 2016. https://www.mlit.go.jp/common/001116833.pdf and https://www.mlit.go.jp/common/001128802.pdf

Table 3 Population changes in the Tokyo Metropolitan Area (TMA)

	Population size (millions)			Proportion of TMA (%)	
Time period	Nationwide	Capital region	TMA	To nationwide	To capital region
1950	83.20	19.05	13.05	15.7	68.5
1960	93.42	23.79	17.86	19.1	75.1
1970	103.72	30.26	24.11	23.3	79.7
1980	117.06	35.70	28.70	24.5	80.4
1990	123.61	39.40	31.80	25.7	80.7
2000	126.93	41.32	33.42	26.3	80.9
2010	128.06	43.47	35.62	27.8	81.9
2020	126.15	44.34	36.91	29.3	83.2

Source: based on data from e-Stat, Japan, 1950–2020. (The statistics range for TMA is decided as Tokyo prefecture, Saitama prefecture, Chiba prefecture, and Kanagawa prefecture)

3 Problems Addressed, and Problems Left Behind Rapid Urbanization Period

The policy of the TMA has been reviewed and issued approximately every 10 years since the 1950s, as it has adapted to rapid growth in the region. Table 3 shows the growth of population size in the TMA, the Capital Region, and Japan nationwide. The national population kept growing until 2010 while the growth rate has been low since the 1990s. The population of the Capital Region and the TMA saw nearly the same rapid and slow growth profile as the nation did, but it outran the nation after the 1980s and has not yet met its peak. All incremental growth in the Capital Region went to the TMA after the 2000s, which means the start of a decline in the rest. Still, the TMA is growing slower and slower, and it is believed that it will meet its peak around 2030. In accordance with these waves of growth and shrinkage, the intense of land use between development and conservation has been kept challenging.

3.1 Possible Shortcomings in Land Management and Conservation

Figure 4 illustrates the transition of land use in TMA from the late 1970s to the 1990s. At the end of the rapid growth period, a conversion of 400 sq km of non-urban land out of the approximately 6600 sq km area of the Suburban Development Zone in the TMA went to urban development, of which 200 sq km were carved out of forest or "wasteland" and another 200 were agricultural land, including paddy fields and other agricultural lands. According to the matrix of land use change (Table 4), a huge amount of natural and cultivated lands were converted to urban land use. While residential land was more often converted from agricultural land other than paddy fields, the vast majority of urban land was converted from forest and wasteland. Paddy fields represent the smallest type of land converted to urban use compared to other natural and cultivated lands. In addition, vacant land accounts for about one-third of all urban land conversions from natural and cultivated land, which exceeds new residential land. This shows the characteristics of land conversion in the suburbs of the TMA, and there is the issue of less efficiency in land use, which laterally supports the widely seen problem of urban sprawl. This also indicates possible shortcomings in the management of land use and the conservation of land resources in the TMA during the late part of the last century.

Therefore, as of 2016, the urbanization intensity of land use within 30 km of the TMA had peaked, while beyond 50 km there still existed a large amount of natural and cultivated land. Figure 5 shows the land use classification of the current land use status in 2016 according to the spatial distance to the metropolitan center. There is an obvious difference of land use structure in different spheres. In the TMA, more than 75% of the land in the 10 km radius is built-up land. Of that, 70% of the land in the 30 km radius is built-up land, with only a small amount of farmland and forests.

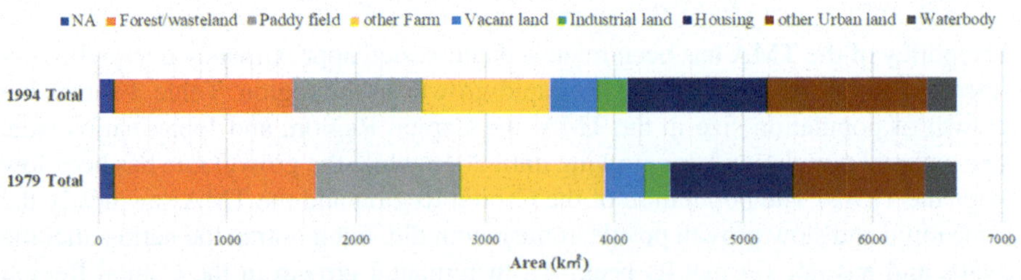

Fig. 4 Land use changes in the SDZ of the TMA (1979–1994)

Table 4 Matrix of land use changes in the SDZ of the TMA (1979–1994) unit: ha

Land use area (ha)		1994								
		A	B	C	D	E	F	G	H	1979 Total
1979	NA									12,106
	A	136,488	215	1240	5844	856	2825	8879	210	156,557
	B	488	97,839	1475	4384	556	1736	5463	359	112,298
	C	737	819	93,751	4965	822	4920	6016	139	112,170
	D	477	611	1318	15,810	1490	5245	6000	156	31,107
	E	45	19	43	502	17,620	400	1087	5	19,720
	F	90	48	281	1993	385	88,840	2312	33	93,982
	G	460	333	479	4164	942	3212	92,472	178	102,241
	H	80	108	355	588	98	60	1384	22,078	24,750
1994 Total		138,865	99,991	98,942	38,249	22,768	107,239	123,611	23,158	664,930

Notes: (1) A: Forest and wasteland; B: Paddy field; C: Other farmland; D: Vacant land; E: Industrial land; F: Housing; G: Other urban land use; H: Waterbody.
(2) NA: Data not available (12,106 ha) in both years.

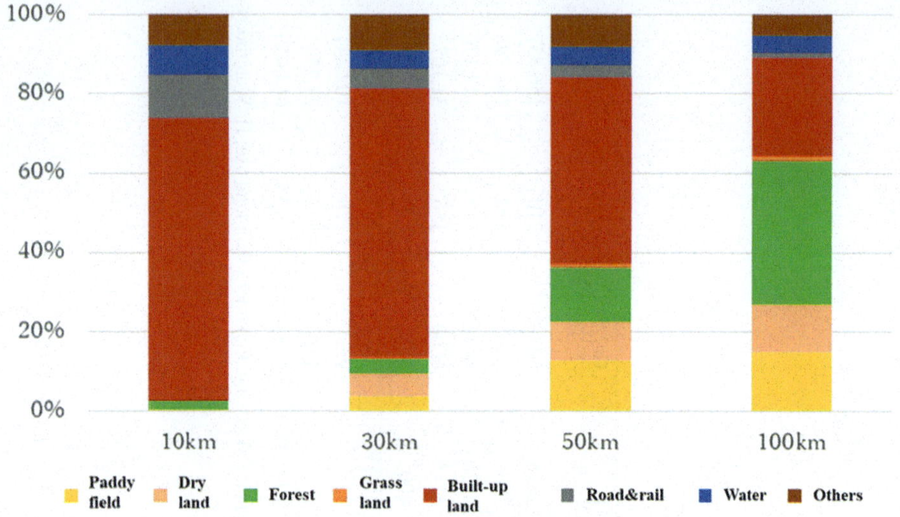

Fig. 5 Land use composition of circle areas in Tokyo, 2016. (Made by ECO-GIS Lab, Keio University)

While 50% of the land in the 50 km radius is built-up land and 50% is natural land, 20% of the land in the 100 km radius is built-up land.

The land use conversion and urbanization that occurs in the context of rapid regional growth is justified, but evaluation of whether the relevant land management system and planning have worked, and whether they have achieved their original institutional intent needs to be considered in the context of specific land use changes. Here, we use changes in industrial land use and changes in farmland d to explain the gains and losses as a result of land management policies on development regulation and resource conservation in the TMA.

3.2 *Industrial Land Use Change*

Industrial land transitions from the metropolitan center to the north and east part of the TMA. Figure 6a, b shows the spatial feature of macro-industrial land transition in both absolute quantity and relative proportion. Specifically, industrial land has been significantly reduced in the core of the metropolitan area, the EUA, except for the areas from the eastern part of Kanagawa Prefecture: Yokohama and Kawasaki and other coastal industrial zones. Tokyo's overall industrial land area is decreasing and in the central part of Kanagawa Prefecture and the southern part of Saitama Prefecture near the metropolitan core in a 20 km area, industrial land has also been significantly reduced. Conversely, other areas in the TMA have seen a significant increase in industrial land. Specifically, there has been a significant increase in industrial land in the northern area along the Kenodo Expressway in a 40- to

a.

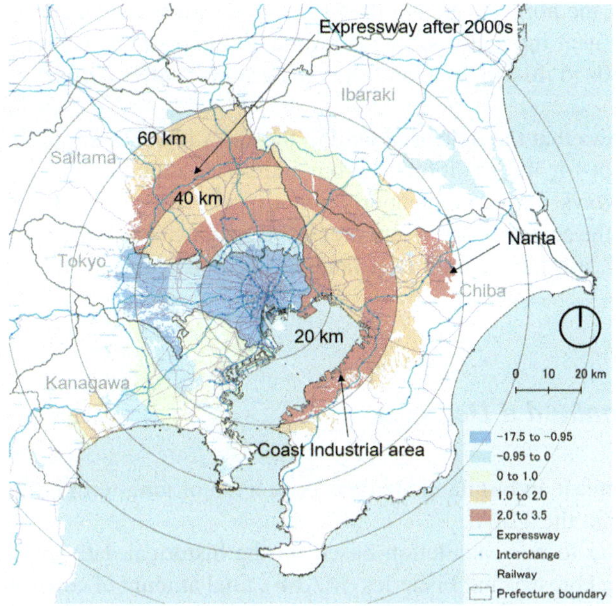

Area of industrial land change to Area of land slope below 11 degree
(classified by equal count, unit: %)

b.

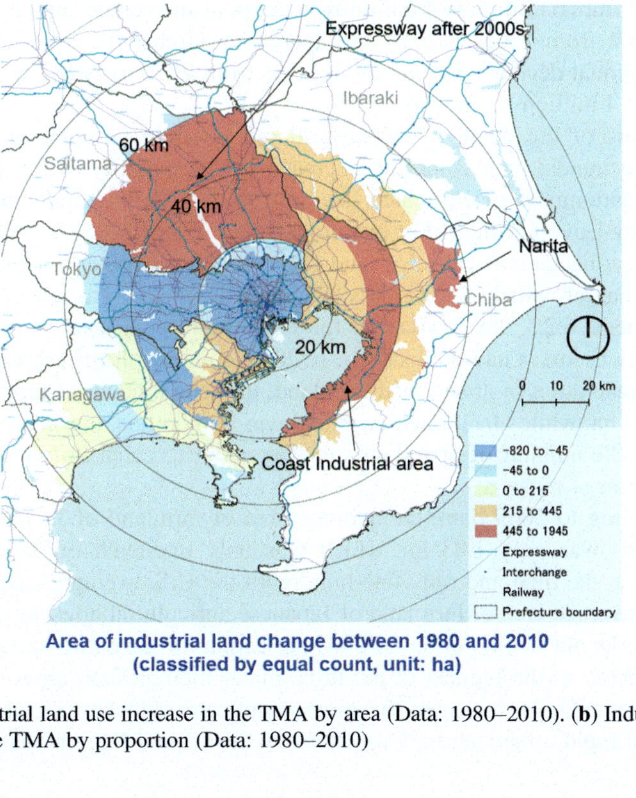

Area of industrial land change between 1980 and 2010
(classified by equal count, unit: ha)

Fig. 6 (**a**) Industrial land use increase in the TMA by area (Data: 1980–2010). (**b**) Industrial land use change in the TMA by proportion (Data: 1980–2010)

50-km area, the northern area of the Gaikando Expressway in 20- to 30-km area, the eastern part near the metropolitan core in a 20-km area, the Chiba Harbor industrial zone in a 30- to 40-km area, and the area around Narita Airport in a 50- to 60-km area.

This shows that the industrial land in the EUA is shifting into the SDZ and other areas, except for the traditional harbor industrial zone. The main destinations are along highways, important seaports and airports. However, the suburban areas of Tokyo and the central and western areas of Kanagawa Prefecture, which are also part of the SDZ, are facing industrial land use decrease during the regional industrial transition.

3.3 Farmland d Decrease

A rapid decrease in farmland area has been a major long-term issue nationwide and even worse in the TMA.

According to our calculation based on the historical data from the Ministry of Agriculture, Forestry and Fisheries (MAFF), total amount of cultivated land in Japan has decreased every year since the late 1960s, and even after the national population peaked in 2008, the tendency is not changing. The country's cultivated land area decreased from 6 million to 4.6 million hectares at an average annual decrease rate of about 0.6% from 1965 to 2010 and decreased another 0.2 million hectares at an average annual decrease rate of about 0.5% from 2010 to 2018, resulting in a total stock of 4.4 million hectares by 2019.

In terms of the detail of the overall reduction in farmland d, there are two competing trends: increasing and decreasing. The majority of the decrease is due to the abandonment of cultivated land and the "de-agriculturalization" of farmers (when they leave farming). For example, from 1993 to 2015, farmland area gained 90,000 new hectares because of land regeneration and other reasons, but also lost 760,000 ha, of which 45.6% was abandoned, 34.9% was converted for urban development, 6.2% was converted for infrastructure such as roads and railways, and 4.1% was lost in natural disasters. In 2019, counting the increase in farmland due to the regeneration of abandoned farmland, the total amount of farmland decreased by 23,000 ha, while abandoned farmland was 280,000 ha, the same as the previous year. So, the amount of abandonment was about to catch up with the amount of regeneration or reuse.

According to 2009 data, the average area of farmland of Japanese agricultural households was only 1.83 ha, which was only one-tenth of the average in EU countries at the time and only one-hundredth the U.S. average. According to 2015 data, the average area of farmland of Japanese agricultural households was 2.09 ha, which works out to only about 350 m^2 per capita for the entire nation.

The TMA, as the biggest of the three major metropolitan areas in Japan, along with its hinterlands, has seen the most significant reduction in farmland d in the context of rapid urbanization. Farmland d in the Capital Region has decreased from about 1.16 million hectares in 1965 to 750,000 ha in 2010, an average annual decrease of about 1.0%, and then a further decrease to 715,000 ha in 2018, an average annual decrease of 0.6%. The speed of farmland loss in the region has been consistently greater than the national average; the proportion of farmland in this region decreased from 19.3% in 1965 to approximately 16% by 2019. With 29.3% of the national population, the TMA had only about 5.4% of the nation's farmland , and it decreased faster than the average of the whole Capital Region.

4 Adaptive Planning in Response to Socio-economic Change

4.1 Historical Adaptability of the Land Use Planning and Conservation System

An evaluation of the adaptability of TMA land management to socio-economic change cannot be separated from an observation of the adaptation process of Japan's land use planning and conservation system. Table 5 shows the evolution of regional and urban land use policy.

4.1.1 During the Rapid Growth Era

The City Planning Act, enacted in 1968, introduced the current zoning and development permit systems, as key instruments in the development and conservation of land resources in Japan (Nakai 1988). In 1980, the district planning system was established, mainly for urbanized areas. In 1982, Temporary Inverse Area Division System were established within the UCA, which allows changes in delineation of the Urbanization Promotion Area (UPA) in accordance with the maturity of Land Readjustment projects without having to wait for the periodic review of zoning every five years. These institutional adjustments to the UCA responded to the rapid urbanization demands of the time, but also made it easier to convert land resources that had been protected.

The end of the 1980s brought another series of changes, with the support of the redevelopment district planning system, the deregulation of specific town areas, the introduction of comprehensive design, and the designation of intensive use districts, etc.

4.1.2 After the Collapse of Bubble Economy

During the bubble economy, the farmland inside the UPA in specified cities of the three major metropolitan areas, including the TMA, was encouraged to convert for

Table 5 Evolution of regional and urban land use policy (based on Asano et al. 2017)

Year	Event in urbanization	Land use planning related legislation	Institutional change
1968		New City Planning Act	UPA/UCA Area Division, development permission
1970	Osaka Expo	Building Standards Act, amendment	8 Land Use Zones, FAR regulation
1971	Junior baby boom		
1972	Club of Rome report, Limits to Growth		
1974	Total fertility rate falls below the replacement fertility level (2.14). First oil shock	City Planning Act, amendment	Deregulation in UCA (urbanized area)
	Compact city by G. Dantzig and T. L. Saaty in 1973	Productive Green Space Act	
1977	The Third National Comprehensive Development Plan (Settlement Area Concept)		
1979	Second oil shock		
1980		City Planning Act, Building Standards Act, amendments	District planning
			Inverse Area Division in the UPA
1982		Productive Green Space Act, amendment	Temporary Inverse Area Division System in the UCA
			Tax rate of farmland inside the UPA of specified cities within the three major metropolitan areas at the same level as residential land use in the neighborhoods
1987	Forth National Comprehensive Development Plan (multipolar distributed land use)		Expected 40 people per hectare for UPA designation
1989	Total fertility rate falls to 1.57		
1992		City Planning Act, Building Standards Act, amendments	Municipal Master Plan
			12 Land Use Zones, and District Plan in the UCA
1998	Fifth National Comprehensive Development Plan (twenty-first century grand design for national land)	City Planning Act, amendment	Regulatory Liberalization for Special Land Use District
		Revitalization of City Center Act	Revitalization of City Center Basic Plan
			Compact City into Municipal Master Plan

(continued)

Table 5 (continued)

Year	Event in urbanization	Land use planning related legislation	Institutional change
2000	Decentralization of planning authority	City Planning Act amendment	City Planning Area Master Plan
			Selectivity of Area Division
			Quasi-City Planning Area
			Land Development Permission
			Specified Land Use Restriction Zone
2002		Act on Special Measures concerning Urban Reconstruction	Urban Reconstruction Urgent Development Area
2003	Basic Act for Birthrate Declining Society		
2005	Population growth rate was negative for the first time. Total fertility rate falls to 1.26		
2006		City Planning Act, amendment	Land Development Permission on public facilities
		Revitalization of City Center Act, amendment	Designation liberalization for Quasi-City Planning Area
			Restriction on large-scale tourist facilities
			Revitalization of City Center Basic Plan Certification
2011	Great East Japan Earthquake		
	Era of depopulation		
2013	2020 Tokyo Olympics decided		
2014	Act on Creation of City, People and Jobs	Act on Special Measures Concerning Urban Reconstruction, amendment	Location Optimization Plan
2015	Basic Act on Promotion of Urban Agriculture	Act on Special Measures Concerning Vacant Houses	Vacant House Measures Plan
			Specified vacant house
2017		Productive Green Space Act, amendment	
2018		City Planning Act, Building Standards Act, amendments	13 Land Use Zones (Garden City Residential Zone)

urbanization. The tax rate of those farmland was implemented on a par with the residential land use in the neighborhoods in order to keep the supply of buildable land in the metropolitan areas.

In 1992, the master plan for city planning system was introduced at municipal level. It was a palliative measure to the socio-economic downturn after the bursting of the bubble economy. It gave local governments and developers more flexibility in urban development, while, as to the central government and the prefectural administration, it weakened their coordination responsibility and constraints on land management and conservation for long-term regional development and land use planning.

4.1.3 Era of Urban Shrinkage

In 2011, right after the national population peaked, authority in the three major metropolitan areas to decide on area division of UPA/UCA and urban facilities was transferred to local municipalities, with the exception of urban planning decided by prefectures. Due to the changes, local planning does not require the consent of the national government unless it has a significant bearing on national interests, and urban planning decided by municipalities does not require the consent of prefectures. This is one of the institutional measures to cope with the declining population and the aging issue occurring in varying degrees in each municipality. In 2014, the Act on Special Measures concerning Urban Reconstruction was amended to promote compact urban development in order to more seriously address the declining birthrate and aging issue. It does not deal much, however, with land use strategies for areas outside the promotion areas, such as the areas where exactly urban sprawl had occurred.

4.2 New Opportunities for the Land Use Planning and Conservation System?

At the end of the 2010s and in current times, regarding land use planning and conservation, the most important adaptation to the socio-economic situation is the Location Optimization Plan (LOP), based on the Act on Special Measures concerning Urban Reconstruction.

This program aims to concentrate urban services and residential functions around high-capacity public transportation nodes such as railway stations in the suburbs of the TMA. At the local scale, centralization-promotion areas are designated in each suburban municipality, and these are then divided into urban service function-promotion areas and residential function-promotion areas. In these areas, the city will be revitalized and land use will be improved more efficient through the effective use of underused land and vacant houses. Conversely, outside of these areas, public services will be readjusted in response to the increasing low density, and new land uses, such as the development of parks, farmland, and forests, and the restoration of the natural environment will be taken into consideration. In cases where it is difficult for a single region to provide sufficient services, efforts will be made to provide the necessary functions by linking the regions in a network.

Therefore, the concept of LOP follows the traditional urban structure theory of "decentralized concentration." It could be a new opportunity to rebuild the land use planning and conservation system in the peripheries of the metropolitan.

5 Discussion

5.1 At the Regional Scale: Institutional Resilience and Flexibility in the TMA

The evolution of land use policy and regional planning in the TMA shows that, in general, the institutional authority of land use planning and the conservation system has gradually evolved from state control to local decentralization, the emphases of the system have gradually evolved from regional integration and cooperation to local autonomy including deregulation of private investments as well as embracing diversity, and the value orientation of the system has gradually evolved from expansive growth guidance to inclusive stock enhancement.

During this process, Japan and the TMA have experienced explosive population growth, rapid industrialization and urbanization, suburbanization and the widespread use of private cars, slowing population growth, economic bubbles, and aging, among other things, as it evolved from a youthful to a mature society.

Among the several regional plans for the TMA, except for the first one that misjudged the boundary control of the metropolitan area development, several subsequent plans have adhered to some key topics regarding land use planning and conservation over decades, such as suburban new town development, land use guidance for circling of the metropolitan area, industrial and population evacuation, and polycentric urban structure.

In addition, in order to build disaster-resistant and flexible national land use, the government can comprehensively enhance the safety of the national lands from the perspective of a land use planning and conservation system through efforts at each stage, from the regional to the local level; for example, by preventing urban sprawl, securing open spaces for recovery and reconstruction, such as temporary storage areas, conserving and managing farmland , and improving the land conservation functions of forests and other ecosystems.

5.2 *Institutional Loophole? Avoiding Excessive Flexibility in the System*

From the perspective of agricultural land conversion management, there is a possible loophole in the system, which might be a consequence of interaction between the special institutional structure (see Fig. 2 in "Land Use Planning in Japan" section of this chapter) with its permission system and the land ownership structure in the TMA.

As mentioned above, with the evidence of it seen in the amount of abandoned, formerly cultivated land that has increased at an alarming rate, from 150,000 ha a year in 1985 to 400,000 ha a year in 2010. Comparing the total amount of farmland, it is conceivable that a large amount of farmland in Japan is being regenerated and restored to cultivation every year, while at the same time a large amount of farmland is being abandoned. The central government and local administration, however, have little control over the abandonment of farmland caused by multiple factors, such as aging of farmers, shortage in labor, and difficulties in land recovery. Although operations are guided by institutional and economic measures, such as subsidies, tax reduction, support for agricultural labor training, and encouragement of market-oriented intensive agricultural production, it has been always easy for large amount of abandoned farmland to be converted to urban development.

One of the most important reasons for the lack of strict regulations to control the loss of farmland could be the possible institutional loophole. The conditions of use of agricultural land in general agricultural promotion areas range from "no permission in principle" for large agricultural land with improvement plans to "permission" for small agricultural land without improvement plans. In addition, due to the context of "local municipal autonomy" in Japan, permission for development in the UCA and permission for farmland conversion in the APA is granted by the governor of the prefecture and the mayor of the ordinance-designated city (20 cities nationwide, e.g., Yokohama City, Kawasaki City, and Saitama City). Thus, in terms of implementation and management, development in the UCA and the conversion of non-urban land use in the APA have put a lot of pressure on farmland, even some of the areas that are given the highest protection status in policy and regulation.

One of the fundamental contexts for the widespread abandonment of farming is probably the structural characteristics of land ownership in Japan. The proportion of privately owned land is especially high in the three major metropolitan areas, with 60% of the land in the TMA being privately owned, including urban development and farmland. The high proportion of privately owned land in the TMA has largely determined land use in the suburbs of Tokyo, where agricultural production is dominated by small farmers and market-oriented operations. This makes it difficult to achieve strategic control and guidance in agricultural policies and mandatory regulations in agricultural planning, which have negative impacts on land conservation. Although there is the need to supply infrastructure and other municipal services in the villages or in the urban area inside the UPA according to legislation, uncertainties about landowners' intention for land use transition (Sorensen 2001) therefore make the delineation of zoning a challenge over time. And inappropriate zoning could lead to inappropriate institutional implementation, which harms the land use planning and conservation system itself.

Abandonment of farmland is backing the unplanned land use conversion to urban land use, while the loss of agricultural labor and aging of farmers are contributing to the abandonment. Meanwhile, unregulated urbanization due to flexible local implementation of zoning could be a key cause for land use transition and commonly occurs in the area zoned as UCA (Saizen et al. 2006). So, the TMA can and must do more on land resource conservation, and intensive or large-scale farmland land use should be a long-term priority for consideration of planning.

5.3 *At the Local Scale: Adapting to Demographic Transformation*

According to an urbanization conceptual model of Mulligan in 2013, the TMA today is in its saturation or terminal stage of urbanization with an urbanization degree above 90%. And it will be facing a population peaking in the near future and continuing to decline, as shown in the projections in Fig. 7. However, according to the national 5-year survey in 2020, the actual population of the TMA in 2020 was 36.91 million, which is 1.54% higher than projections. This may have been a message for policy makers fearing depopulation, but the down-sizing demographic transformation remains a long-term challenge.

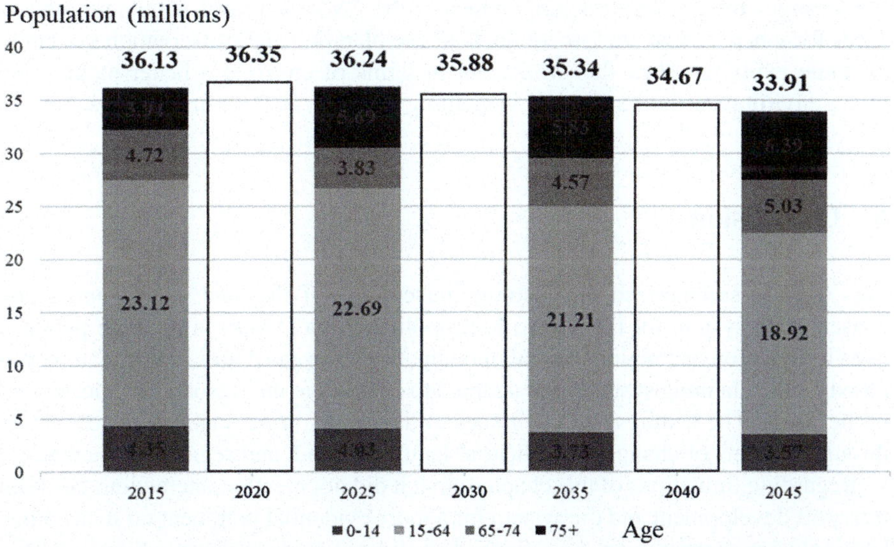

Fig. 7 Population projection in the TMA. (Source: based on data from NIP&SSR, 2018)

In the TMA, the further away from the metropolitan center or local central station, the faster the population is decreasing. According to estimates of the National Institute of Population and Social Security Research (NIPSSR), the population of the TMA will decline by approximately 6% until 2045, compared to the 16.3% decline for Japan nationwide. Since the Existing Urban Area (EUA) of the TMA is expected to increase by nearly 4.6%, the Suburban Development Zone (SDZ) and other periphery areas of the TMA can be expected to decrease by 10–15% or more.

Within the SDZ, the degree of population decline is likely to be severe in some areas but insignificant in others. This may bring many new challenges for the land use planning and conservation system from a regional scale down to a local one. First, there could be a dilution of urban area. Although the number of residents in some areas decreases, the already built-up environment, including housing, cannot be easily changed back to non-urban land use in response to the population decrease. Second, urban public services might have to be re-allocated. The issue of vacant facilities and community center relocation are expected to be more significant. Because of these two concerns, the land use planning and conservation system will have to evolve in response to a new era.

Some latest statistics and forecasts show that the population is returning to the metropolitan center. The residents who moved to the metropolitan suburbs during the period of rapid economic development gradually aged and now rely on well-supplied urban services. Conversely, the urban renewal projects in the metropolitan center encourage the mixing of commercial and residential land use, which increases housing supply. Therefore, the elderly and high-income groups have gave up their real estate in the suburbs and moved back to the center of the TMA. Therefore, the demand and supply of land in the SDZ or peripheries of the TMA could change more dramatic than before, in terms of both quality and quantity of land use. As we can see, changes in the overall demographics of the TMA and local population preferences for where to live are forcing the land use planning and conservation system to continue evolving. How this works out in terms of ensuring a better natural and living environment in the TMA is something to watch in the coming years.

6 Conclusion

This chapter summarized the institutional context of the land use planning and conservation system for the Tokyo Metropolitan Area (TMA) in multiple perspectives from nation and region to local municipality. It revealed how regional strategies based on the planning system have adapted to changes in the socio-economic context of the TMA. The features and key issues of the system were qualitatively discussed through the lens of changes in industrial land, farmland, and regional population.

Regarding limitations of this chapter, first, it did not mention interactions between transport development and the system for a more evidential judgment on such a huge and inclusive metropolitan area. Second, it did not provide direct evidence of the quantity and efficiency of land use changes corresponding to every policy change in the TMA. Those limitations are to be addressed in some follow-up studies.

As a summary, the resilience and adaptation of the land use planning and conservation system were highlighted by examining its historical changes while also noting its excessive flexibility. In an era of slowing and even negative population growth and an aging society, the system is expected to play a crucial role in ensuring a better natural and living environment for residents of the TMA in the future.

Acknowledgments This chapter is part of the M-NEX (m-nex.net) project, funded by a grant of the Sustainable Urban Global Initiative (SUGI): Food-Water-Energy Nexus by Belmont Forum and the Japan Science and Technology Agency (JST). This research was also supported by the Ushioda Grant 2020 from Keio University and Taikichiro Mori Memorial Research Grants 2020. Follow-up studies are being funded by a JST Doctoral Grant in 2022.

References

Aiba S (2021) Chapter 10: Mediating the conflict: urban planning for land use. In: Heisei urban planning history: what 30 years of transition left behind and what will be inherited, pp 294–343

Akita N (2017) Suburban Municipality in the Tokyo Metropolitan Area that is expanding and hollowing out. Land use planning in the era of urban shrinkage, pp 39–46

Asano J, et al (2017) 都市縮小時代の土地利用計画. 学芸出版社, pp 53–54

Geneletti D, La Rosa D, Spyra M, Cortinovis C (2017) A review of approaches and challenges for sustainable planning in urban peripheries. Landsc Urban Plan 165:231–243. https://doi.org/10.1016/j.landurbplan.2017.01.013

JICA (Japan International Cooperation Agency) and MLIT (2007) Urban land use planning system in Japan, 2nd edn. https://jica-net-library.jica.go.jp/library/jn325/UrbanLandUsePlanningSystem_all.pdf

MLIT (2003) Urban land use planning system in Japan. https://www.mlit.go.jp/crd/city/plan/tochiriyou/pdf/reaf_e.pdf

Moreno-Penaranda R (2011) Japan's urban agriculture: cultivating sustainability and well-being

Nakai N (1988) Urbanization promotion and control in metropolitan Japan. Plan Perspect 3(2): 197–216. https://doi.org/10.1080/02665438808725659

Saizen I, Mizuno K, Kobayashi S (2006) Effects of land use master plans in the metropolitan fringe of Japan. Landsc Urban Plan 78:411–421. https://doi.org/10.1016/j.landurbplan.2005.12.002

Sorensen A (2000) Land readjustment and metropolitan growth: an examination of suburban land development and urban sprawl in the Tokyo metropolitan area. Prog Plan 53(4):217–330. https://doi.org/10.1016/S0305-9006(00)00002-7

Sorensen A (2001) Building suburbs in Japan. Town Plan Rev 72(3):247–274. https://doi.org/10.3828/tpr.2001.72.3.247

Tsubota K (2006) Urban agriculture in Asia: lessons from Japanese experience. In: Conference proceedings of Food and Fertilizer Technology Center for the Asian and Pacific Region

Usui H (2019) A bottom-up approach for delineating urban areas minimizing the connection cost of built clusters: comparison with top-down-based densely inhabited districts. Comput Environ Urban Syst 77:101363. https://doi.org/10.1016/j.compenvurbsys.2019.101363

Green Infrastructure in Tokyo

Keidai Kishimoto and Wanglin Yan

Abstract The term "green infrastructure" refers to the aspects of land use and planning that utilize the diverse functions of natural ecosystems to bring social, ecological, and economic benefits to human society. People hike and jog in green spaces and obtain fresh vegetables grown on the land. Freshwater springs up from the ground. Whether or not people notice nature's services, natural ecosystems are one of the few natural resources available in urban areas. Without them, cities would just have gray infrastructure, concrete, and asphalt. It is important to be aware of the significant contributions of green infrastructure to urban systems. Does Tokyo have green infrastructure, does it need more, and what efforts are being made to create more? Which governmental plans and initiatives have introduced green infrastructure components? This chapter describes changes in Tokyo's natural ecosystems over time; the significance, definition, and strategies for green infrastructure in Tokyo and Japan; existing green infrastructure in Tokyo; and efforts to revitalize it.

Keywords Green infrastructure · Land-use change · Urban development

1 Natural Ecosystems Lost as Tokyo Metropolitan Area Urbanized

1.1 Green Space Loss

Tokyo Metropolitan Area began a long period of growth in 1603 when the shogun Tokugawa Ieyasu made Edo the principal capital of Japan. During the Edo period

K. Kishimoto (✉)
Graduate School of Media and Governance, Keio University, Fujisawa, Kanagawa, Japan
e-mail: keidai@keio.jp

W. Yan
Faculty of Environmental Information Studies, Keio University, Fujisawa, Kanagawa, Japan
e-mail: yan@sfc.keio.ac.jp

W. Yan et al. (eds.), *Resilient and Adaptive Tokyo*,
https://doi.org/10.1007/978-981-99-3834-6_6

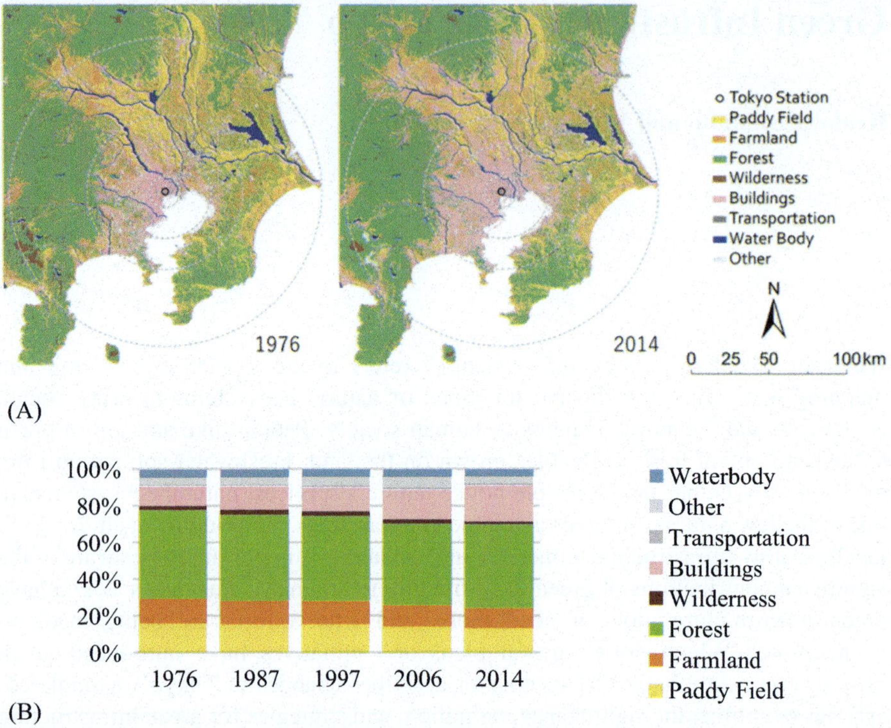

Fig. 1 Land-use changes over time in the Kanto region. (Source: Authors created using digital national land information from Ministry of Land, Infrastructure, Transport and Tourism). (**a**) Land use in 1976 and 2014. (**b**) Land-use change from 1976 to 2014

(1603–1867), it became one of the largest cities in the world. Once industrialization began in the nineteenth century, Tokyo Metropolitan Area faced several phases of urbanization: (1) urbanization with the construction of modern urban systems in cities (1860s to 1910s); (2) suburban development after the Great Kanto Earthquake and the Second World War (1910s to 1950s); (3) rapid urbanization and suburbanization with the building of urban infrastructure during a period of rapid economic growth (1950s to 1980s); and (4) urbanization and suburbanization with further enhancement of urban functions (1980s to 2010s) (Tokyo Metropolitan Government 2020). Tokyo Metropolitan Area is still today one of the world's largest cities, on par with New York and London.

Urbanization resulted in a significant reduction in green space in Tokyo Metropolitan Area. Figure 1a shows land-use maps in 1976 and 2014, and Fig. 1b shows changes in percentage of each land-use type. In the entire Kanto region, urban land use expanded greatly during this period, from a ratio of 10.9% to 19.6%. In Tokyo Metropolitan Government Area, built-up areas accounted for the majority of land use, rising from 34.1% to 45.9% during the period, with increases also in Kanagawa

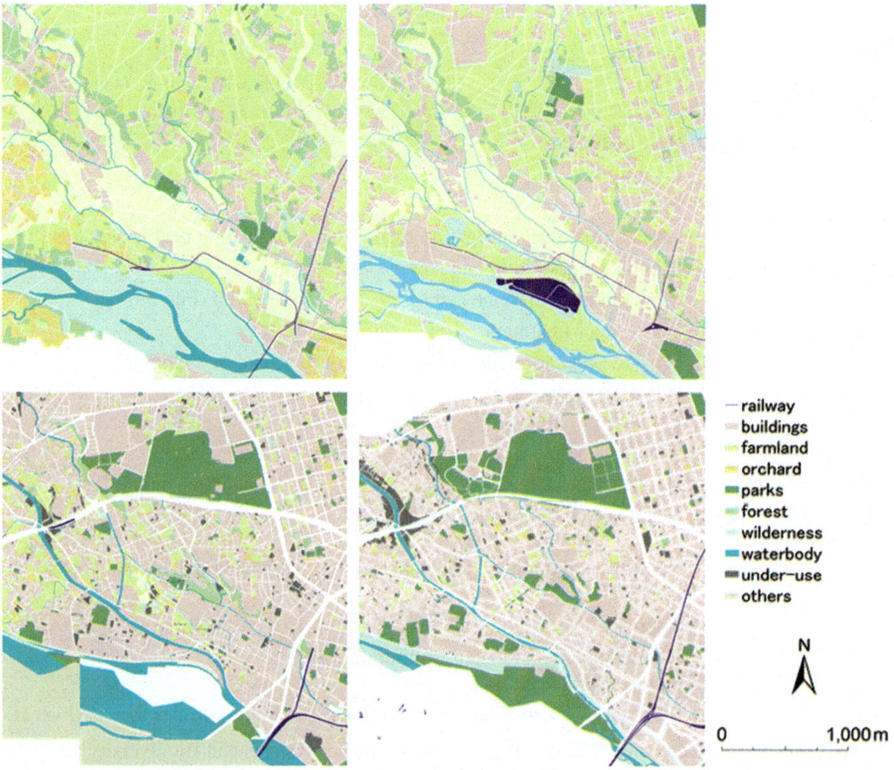

Fig. 2 Historic land-use changes in Setagaya City, Tokyo: *Above left*, 1929; *right*, 1955; *below left*, 1986; *right*, 2016. (Source: Authors created using the land-use data from the Bureau of Urban Development of Tokyo Metropolitan Government (1986 and 2016) and authors (1929 and 1955) based on topographic map)

Prefecture (from 23.2% to 36.4%), Saitama Prefecture (from 16.7% to 29.5%), and Chiba Prefecture (from 10.4% to 21.1%). The ratio of area covered by rice paddies decreased from 16.3% to 13.6% (a decrease of 16.5%), and by field from 13.3% to 10.8% (a decrease of 19.0%) in the Kanto region. In Tokyo Metropolitan Government Area, rice paddy area decreased by 83.2% during the period, and field by 60.3%, signs of the immense pressure of development on urban agriculture. In Kanagawa Prefecture as well, the situation was extreme, with a 48.3% decrease in rice paddy area and 33.7% decrease in field, while forest area decreased from 48.4% to 45.8%. The expansion of Tokyo Metropolitan Area occurred mainly at the expense of agriculture (especially urban agriculture). Correspondingly, the proportion of green space declined.

Figure 2 shows land-use changes around the Futako-Tamagawa Station in Setagaya, 15 km from Tokyo Station. In the 1929 map, rice paddies covered most of the lowland, while field covered most of the plateau, giving the area a rural, rustic

Table 1 Categories of ecosystem services and Tokyo example

Category	Services	Tokyo example
Provisioning services	Food	Urban agriculture (see chapter 9, Fig. 3a) or fishing
	Raw materials	
	Freshwater	Spring (Fig. 3b)
	Medicinal resources	
Regulating services	Local climate and air quality regulation	Mitigating heat island
	Carbon sequestration and storage	
	Moderation of extreme events	Dry well (Fig. 3c)
	Wastewater treatment	
	Erosion prevention and maintenance of soil fertility	Kokubunji Cliff
	Pollination	
	Biological control	
Habitat or supporting services	Habitats for species	Biodiversity (Fig. 3d)
	Maintenance of genetic diversity	
Cultural services	Recreation and mental and physical health	Recreation along riverbeds (Fig. 3e)
	Tourism	
	Esthetic appreciation and inspiration for culture, art, and design	
	Spiritual experience and sense of place	"Jinja" (shrine) forests (Fig. 3f)

Source: Authors created using TEEB—The Economics of Ecosystems and Biodiversity (2011)

appearance. The 1955 map shows the early stages of housing development, when fields around the station area was being converted to housing. Economic growth in the latter half of the 1950s led to rapid urbanization, and the city expanded significantly starting in 1986. As a result, many agricultural irrigation waterways were lost. By 2016, urbanization had expanded even further from the station, and as in the rest of the Kanto region, it progressed at the expense of agricultural land use.

The history of urban development described above has led to a reduction in the size and scale of intact green spaces, especially in agricultural land use, as well as a decline in their quality and continuity. The Tokyo Metropolitan Area suffered great losses of green spaces, as the next section explains.

1.2 Ecosystem Services Loss

Natural ecosystems, even in cities, provide various services locally, regionally, and globally, known as ecosystem services. These ecosystem services are beneficial for their many uses, for ethical and esthetic reasons, and for educational purposes (Bolund and Hunhammar 1999). Ecosystem services have been clustered into 4 major categories and 17 subcategories, as shown in Table 1 (The Economics of

Ecosystems and Biodiversity 2011). How have ecosystem services changed with urban development and reduced green space?

The Ministry of the Environment identified four factors driving an ecosystem crisis in Japan: (1) human activities such as urban development and overuse, (2) reduction of human–nature connections in *satoyama*—traditional Japanese landscapes with mosaic land-use including semi-natural ecosystems (Ichikawa et al. 2006; Takeuchi et al. 2003), (3) human impacts such as the introduction of invasive species and chemical substances, and (4) changes in the global environment such as climate change (Ministry of the Environment 2012, 2016).

Among the four drivers, urban development, as shown in "Green Space Loss" section, has had the most destructive effect on living things, and it continues to exert significant influence, even though urban development pressure has decreased in recent years (Ministry of the Environment 2016). Today, urban development is now easing in impact (Numata 2014), and flora and fauna appear to be thriving better in the city. Meanwhile, negative impacts of the second and third causes, under use of *satoyama* and invasive species, are severe (Numata 2014). *Satoyama* landscape, mosaic with rice paddies, secondary forests, and ponds, is a habitat for many plants and animals. Improvement of agricultural methods and depopulation of rural areas influence on use of *satoyama*. The conversion to an unmanaged and abandoned area results in the loss of biodiversity (Uchida et al. 2018; Iwachido et al. 2020) and a healthy natural environment. The magnitude of impact by climate change is now recognized scientifically (IPCC 2014; Ministry of the Environment 2016). Plants and animals are forced to change their phenology and habitats (Primack et al. 2009; Ibáñez et al. 2010).

As a result of these four factors and an ecosystem crisis, ecosystem services in Japan are considered to be worsened over the past 50 years (Ministry of the Environment 2016). Table 2 shows an assessment of changes in ecosystem services made by a committee set up by the Ministry of the Environment with the help of 120 researchers. The reports showed that among providing services, matsutake mushrooms were reduced to 1% of their peak level, while the marine fisheries were reduced to about 30% of their peak level, and groundwater recharge, one of the regulating services, was reduced by about 8% (Ministry of the Environment 2016). Conversely, interest in natural environments has increased these days.

Thus, the supply of ecosystem services is declining, though it benefits our lives and well-being. How should the remaining green spaces be positioned? What can cities do? The concept of *green infrastructure* has been developed to solve these questions. The following chapter describes "green infrastructure" concept in Tokyo.

Table 2 Changes in ecosystem services in 50 years. ++: up, +: slightly up, 0: not changed, –: slightly down, ——: down,?: insufficiently evaluated

		Past 50–20 years	Past 20 years to today	Use level
Provisioning services	Agriproducts	——	–	Under use
	Forestry	+	–	Under use
	Fishery	+	–	Overuse
	Freshwater		0	Overuse
	Wood	–	0	Under use
	Material	–	–	Under use
Regulating services	Climate regulation		–	
	Air control		0	
	Water flow		–?	
	Soil control	0		
	Disaster moderation	+?	0?	
	Biological control		–?	
Cultural services	Religious festivals	——	–	
	Education	–	0	
	Landscape		–	
	Traditional entertainment, arts	–	–	
	Tourism, recreation	+	–	
Disservice	Animal damage		+	

Source: Authors cited and arranged using the Ministry of the Environment (2016)

2 Significance of Green Infrastructure and Strategies

2.1 What Is Green Infrastructure?

The term *green infrastructure* refers to land use and planning/management that can be used to achieve positive social, ecological, and economic outcomes for human society, utilizing available diverse ecosystem services. The concept of green infrastructure varies regionally in definition and perception due to social, ecological, and economic differences. Research in the USA emphasizes improving urban stormwater management, for example, while in Europe the focus is more on improving ecosystems and ecological networks. Japan has its own definition of green infrastructure, a combination of the European and US approaches, a result of reevaluating traditionally recognized, multifaceted functions of green infrastructure, and a broader recognition of diverse services provided by nature, including disaster prevention and mitigation (Iwasa and Nishida 2017). In particular, the 2013 Great East Japan Earthquake was a major event that added impetus to the discussion of green infrastructure (Iwasa and Nishida 2017).

The evaluation and creation of green infrastructure should first recognize the magnitude of ecosystem services provided and the diversity of those services and include a comprehensive understanding of local ecosystems. Green infrastructure supports as many diverse species, as much mitigation of heat islands, and as much carbon fixation as possible. A component of green infrastructure should be designed to provide more diverse services and catalyze synergies at the same time. In addition, policymakers should develop green infrastructure respecting and shaped to local ecosystems. It would be a disservice, for instance, to introduce exotic species or grow vegetables unsuitable to the local environment without prior assessment.

The significance of green infrastructure has been studied academically in Japan. This is evident in the various efforts, both formal and informal, to establish new groups within academic societies (e.g., the Japanese Institute of Landscape Architecture and the City Planning Institute of Japan), to create new networks among researchers and municipalities beyond academic communities (e.g., the Green Infrastructure Association), and to develop regional networks (e.g., the Green Infrastructure Study Group in Setagaya).

2.2 *Strategy for Green Infrastructure in Japan*

The focus on development of green infrastructure to maximize and support ecosystem services is being promoted mainly by the national government, local governments, and private companies.

The national government has over time positioned green infrastructure to be included in policy for various federal lands and social conditions. First, in 2014, the Fundamental Plan for National Resilience mentioned the disaster prevention and mitigation functions of natural ecosystems (Cabinet Secretariat 2014). In 2015, the Second National Spatial Planning initiative noted green infrastructure in the "appropriate management of national land," and the Fourth Priority Plan for Social Infrastructure Development promotes green infrastructure for "improvement of the quality of life" (Cabinet Secretariat 2015; Ministry of Land, Infrastructure, Transport and Tourism 2015, 2020). Furthermore, the Fifth Basic Environment Plan, adopted in 2018, recognized that green infrastructure development "improves resilience" (Ministry of the Environment 2018). In response to these directions, the Ministry of Land, Infrastructure, Transport and Tourism (MLIT) established a strategy to promote green infrastructure and activated nationwide efforts for stakeholder collaboration (Ministry of Land, Infrastructure, Transport and Tourism 2019).

Local governments have adopted policies to promote green infrastructure, and some are frontrunners in implementing various initiatives. Setagaya City, for example, set the goal in its 2018 Green Master Plan to increase the green coverage of the municipal area to 33% by 2032 (Setagaya City 2018). Another example is the

"Setagaya Dam Initiative," which aims to maintain and increase the amount of rainwater infiltration in the city by preserving rainwater infiltration surfaces, including agricultural land, and assisting in installing rainwater infiltration facilities. There are also many citizen science projects, including surveys of flora and fauna, and compiling the data for use in policymaking.

Private companies can also be important actors in creating and managing green infrastructure in cities. One example is Tokyu Corporation, which operates railways in the Tokyo area and has developed many retail and office complexes along its rail lines. Its Futako-Tamagawa Rise and Minami-Machida Grandberry Park are major developments that incorporate green infrastructure elements featuring biodiversity, walkability, and water infiltration. AEON Corporation is a major retailer in Japan that works with citizens to plant native trees, among many other initiatives.

3 Tokyo's Urban Green Infrastructure: A Treasure for Resilience and Adaptation

Above, we learned that Tokyo has had a long-term trend of losing green space and agricultural land to urbanization, and that it needs green infrastructure to more fully benefit from the ecosystem services nature provides. In this section, we summarize the state of green infrastructure in Tokyo.

3.1 Remnant and Conserved Green Infrastructure

Traditional green infrastructure includes many green spaces. Forested land is the most common land type that provides excellent ecosystem services, however, as shown in Fig. 1a, there is not much forest land left in most urban areas. In some places, though, such as Setagaya City, notable areas of forest land have been retained, as shown in Fig. 3b. Tokyo has a complex mix of plateaus and lowlands that contribute to the unique culture of the area.

Natural springs bubble up at the base of terraces or ridges and the end of an alluvial fan and have become the location of many historical parks and residences. In the past, people used the springs for refreshment of body and mind, while today they also continue to keep urban ecosystems healthy and landscapes beautiful. At the same time, however, the long-term reduction of permeable surfaces and the overuse of groundwater has caused groundwater levels to drop and spring water availability to decline. In Setagaya City, forests along the Kokubunji Cliff Line have been designated as scenic areas in order to preserve landscapes, due to the presence of historic houses, natural springs, and a diversity of species living in the rich habitat.

Many Shinto shrines and Buddhist temples are located at the edge of a plateau, where they have a good view and are less likely to be damaged by floods. The term

Fig. 3 Images of landscapes and nature providing ecosystem services in Tokyo (refer to Table 1 for descriptions of labels (**a**) to (**f**)). (Photo credits: Keidai Kishimoto)

chinju no mori refers to a forest or grove around a *jinja*, or Shinto shrine. Many shrines have buildings for worship, offices, an approach or promenade, and the sacred forest or grove. These treed areas play a key role as the setting for spaces for worship, festivals, and biodiversity. *Chinju-no-mori* in Tokyo contribute to urban greenery and create networks of green areas (Fujita 2007). Buddhist temples as well often are often surrounded by groves of ancient trees. The lands of the Imperial Palace, residence of the imperial family, is a vast green space covering an area of 2.3 sq. km. Even though it is in the center of Tokyo, it is rich with diverse species such as the raccoon dog (Kuramochi et al. 2014) and helps to mitigate the heat-island effect (Ministry of the Environment 2006). Of the mitigation of heat-island impacts, the greenery of the Shinjuku-Gyoen park, or the former Imperial Park, in summer cools surrounding temperatures by 2–3 °C (Narita et al. 2004).

Urban agriculture, in particular, still supplies vegetables, fruits, and other food despite the intense pressure of urban development. While food provisioning in Tokyo itself has declined over the long term with the loss of farms, in recent years urban agriculture has become more recognized as a land use that society values and belongs in cities, and efforts are being made to conserve and promote it (Ministry of Agriculture, Forestry and Fisheries and Ministry of Land, Infrastructure, Transport and Tourism 2016; Tokyo Metropolitan Government 2017). Detailed evaluation and changes are explained in chapter "Urban Agriculture as a Tool for Adapting Cities for the Future" of this volume.

Besides these areas of greenery, trees and home gardens can be widely found on private land. As one of the significant areas supporting urban habitats (Rudd et al. 2002), they function as a place for household food production and as a place for rainwater infiltration (Yokota and Niwa 2020).

It is desirable that these be left untouched by urban development and continue to be conserved through proper land management, by considering the functions of green infrastructure that can improve the quality of life and contribute to resilience.

3.2 Built Green Infrastructure

A significant amount of new green infrastructure has been created in cities as a result of technological improvements and greening policies. Rooftop greening is a major greening strategy being employed, and green rooftops in Tokyo currently amount to as much as 2,508,000 sq. m (Tokyo Metropolitan Government, 2022), with many large commercial facilities and offices greening their rooftops. Rooftop greenery is used by birds (Motegi and Yanai 2004) and as a place for allotment gardens. In Tokyo, new construction and expansion of buildings of a certain size are required to be greened according to ordinances of the Tokyo metropolitan government and municipalities. Figure 4 shows the annual trends in ground and rooftop greening since the early 2000s, with 150,000–200,000 sq. m of greenery being created in most years.

Futako-Tamagawa Rise is a large retail and residential complex that opened in 2015, with its design including rooftop greening inspired by the original local environment. In addition to evoking the Tama River which runs nearby in the décor and design, the development used local plants, stones, and other materials from the Tama River basin. The complex's rooftop greening projects provide habitat to help conserve an endangered plant species known as kawara-no-giku (*Aster kantoensis Kitam.*). These efforts contribute to the local ecological network as a green stepping stone connecting the Tama River and Kokubunji Cliff Line. The rooftop greenery is popular for shoppers, residents, and office workers. For its contributions to the natural environment and urban development, Futako-Tamagawa Rise has received LEED for Neighborhood Development (LEED ND) Gold

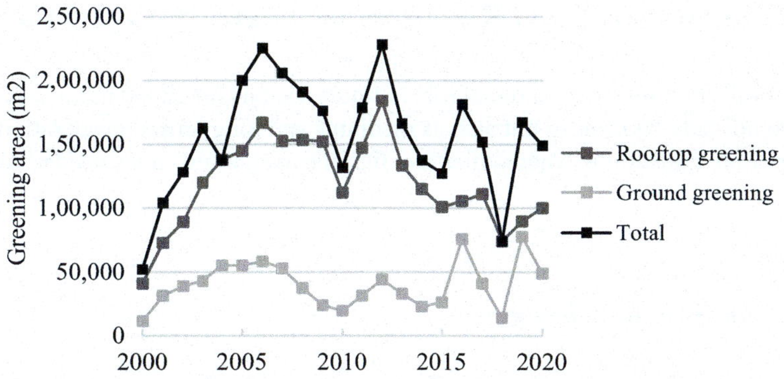

Fig. 4 Area of green rooftops on private developments of more than 1000 sq. m and on public developments more than 250 sq. m. (Source: Author created using Tokyo Metropolitan Government (2022))

certification, as well as AAA rating under the Japan Habitat Evaluation and Certification Program (JHEP), the highest ranking in this environmental certification system.

Permeable pavement and rain gardens are also attracting attention in Japan, and their use is gradually spreading. Japan faces typhoons and heavy rainfall every year and should work to prevent inland flooding in cities. Setagaya City, for example, is promoting green infrastructure development to prevent inland flooding by providing support for such efforts.

Minami-Machida Grandberry Park is a large integrated retail development with retail and park space that opened in 2019 in front of a suburban transit station. In addition to the smooth integration of the station, retail facilities, and park, it includes a range of planted vegetation of mainly native species, open spaces such as plazas, and rainwater infiltration including "rain gardens." The project is one of the most progressive examples of green infrastructure in development planning and has received various awards in addition to LEED ND Gold certification.

Many other initiatives can be observed in the region, such as park space, green walls, rooftop vegetable gardens, and de-culverting. The green infrastructure that has been created is generally small-scale and local, and it often is adapted along with land-use changes and building renovations or redevelopment projects. Thus, on the grand regional scale of land management and resilience these initiatives may not seem so significant, but they do contribute to the improvement of the quality of life for residents of the surrounding communities. New opportunities to expand green infrastructure may arise as cities shrink or become less dense due to declining birthrates, the aging of society, and declining population.

4 Rediscovering Land Resources for Green Infrastructure

Green infrastructure plays a crucial role in future visions for cities. Initiatives need to include not only short-term but also as medium- and long-term consideration. Thus, an important question is where cities can find the spatial resources needed for green infrastructure.

4.1 Small-Scale Changes

For small-scale changes, vacant and underused land can be converted to green infrastructure. Examples include vacant lots, parking lots, and rooftop spaces. They play roles mainly in mitigating heat island, networking ecological patches as "stepping stones" for living things, and healing our minds.

One of the potential land resource is parking lots. Tokyo has many flat paved and multi-story parking lots for vehicles, and many flat parking lots for bicycles. Flat parking lots that tend to be vacant for long hours of the day could potentially be greened. Recent years, they are gradually increasing in retail and public facilities; however, Tokyo has not yet seen many greening projects for parking lots. Figure 5 shows a large percentage of urban areas are occupied by parking lots, excluding

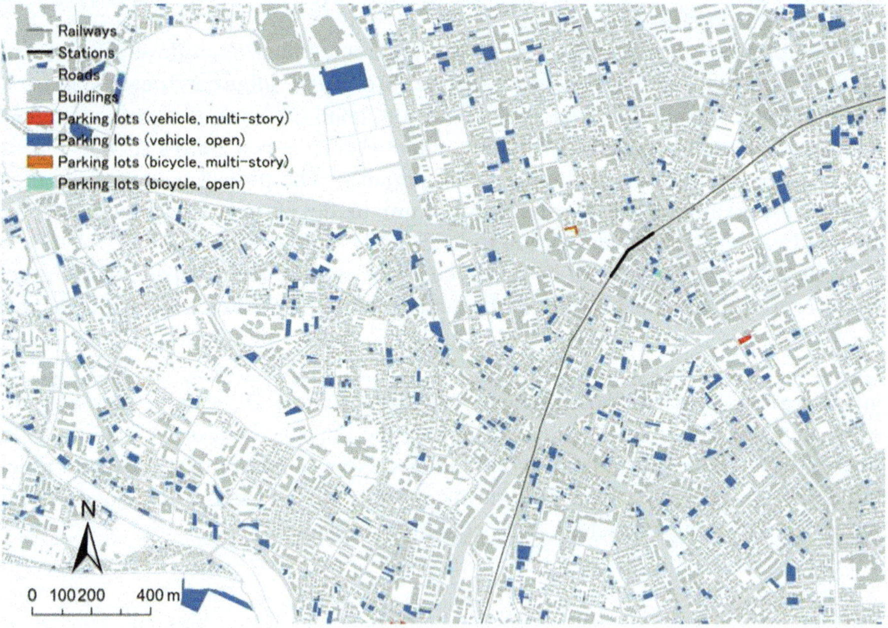

Fig. 5 Parking lots in Futako-Tamagawa. (Source: Author created using the land-use data by Setagaya City 2016)

parking lots for detached houses, condominiums, supermarkets, etc. It seems that there is still much room for improvement in the greening of parking lots in Tokyo.

4.2 *Large-Scale Change*

There are many challenges in building green infrastructure on a large scale in cities, but more and more possibilities are emerging to include it as a new strategy to cope with natural disasters. Japan has seen an increase in the severity of flood damage in recent years, especially in low-lying land and other areas exposed to high disaster risk.

Correspondingly, the Japanese government enacted legislation on watershed management in 2021, which designates areas with a high risk of flooding and regulates land-use changes and building use in these areas. It also expands schemes for induce of relocation of housing for disaster prevention, recommends frameworks to secure land for water retention and recreational functions, and promotes the preservation of urban green spaces. A supplementary resolution by the House of Representatives stated that "the idea of green infrastructure should be promoted to make use of the diverse functions of the natural environment and contribute to the formation of an ecological network by actively conserving or restoring the functions of ecosystems that contribute to the reduction of disaster risks" (House of Representatives 2021). These legislative references to the concept of watershed and water management strengthen the rationale for preserving existing urban green spaces and also provide opportunities to discuss efforts to establish water retention and recreational areas in high-risk areas and land use after relocation for disaster prevention. Although these efforts will not produce major changes in the short term, they can be a step forward in promoting green infrastructure initiatives from a long-term perspective.

5 Discussion and Conclusion

There is a huge gap between the global environmental crisis and the world's actual goals in response. In 2015, the Paris Agreement on climate change included the goal of limiting the global temperature rise to 2 °C above preindustrial levels. The same year, the Sendai Framework for Disaster Risk Reduction called for a significant reduction in disaster-related deaths, victims, and economic losses. By 2019, however, Japan had only reduced its CO_2 emissions by 14%, although it announced the goal of reducing emissions by 46% by 2030 (both relative to 2013) (Ministry of the Environment 2020). Meanwhile, Japan has experienced more natural disasters from extreme weather, including a major landslide in Hiroshima City in 2014, and heavy rains and powerful typhoons in East Japan in July 2018. Japan also experiences earthquakes, such as the 2018 Hokkaido Eastern Iburi Earthquake, the 2016

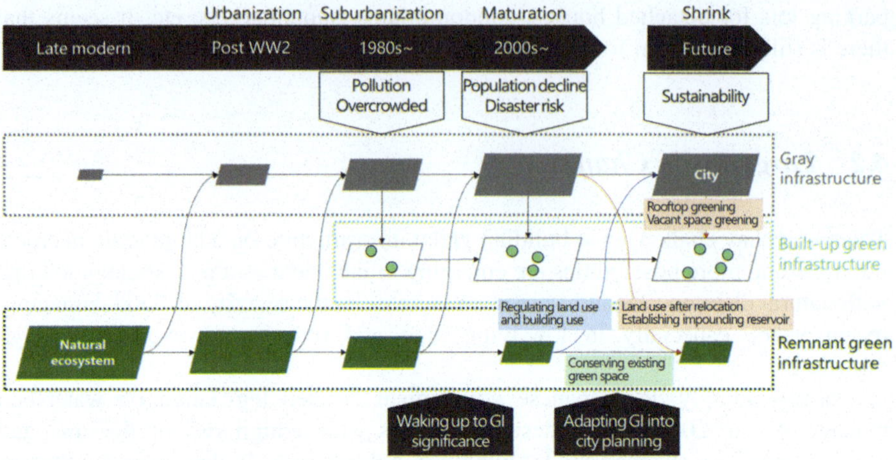

Fig. 6 The future of green infrastructure in shrinking cities. (Source: authors)

Kumamoto Earthquake, the 2011 off the Pacific coast of Tohoku Earthquake, etc. Green infrastructure can help mitigate damage from such disasters.

On the positive side, technological advances have made cars more fuel-efficient and eco-friendly, home appliances more energy-efficient, and power generation more renewable energy-based. Japan has also made social capital improvements such as drainage channels and higher embankments to redirect and hold back floods. Investment in technological developments and social infrastructure needs to continue, but there are limitations to what can be accomplished with these types of investments. Japan also faces aging infrastructure, so it must spend a large amount of money just to replace existing infrastructure, but the population is also declining while financial burdens are increasing.

Against that background, this chapter has described how Tokyo planners realized that, while natural ecosystems have suffered as a result of urbanization, if properly utilized, the various ecosystem services can help not only replace but also enhance conventional technological development (Fig. 6). Tokyo started working on green infrastructure later than some other countries, and its strategy focuses on Tokyo's own unique green infrastructure by utilizing traditional spaces such as shrine and temple forests and new built-up spaces. This attitude toward green infrastructure is an attempt to adapt to social change and risk, which is an important aspect of the title of this book.

In fact, this is not the first time that Tokyo has tried to use greening in the adaptive city; inspired by London's example, green belts were introduced after the Second World War, and while some remain, many have been lost to urbanization. As the problems being faced become increasingly complex and urgent, we cannot afford to repeat the mistakes of the past.

In a report on green infrastructure, the Science Council of Japan (2020) identified current issues as (1) overall planning for the construction of an "adaptive city" using green infrastructure, (2) development of the necessary technologies, and (3) research, development, and decision-making for multiple protections. These could be restated as (1) a strategy for the development of green infrastructure for the entire Tokyo metropolitan area; (2) remote-sensing methods for the development of green infrastructure, the downscaling of climate change prediction to the municipal level, and other technological developments; and (3) the creation of urban multi-protection green infrastructure that can mitigate disasters such as earthquakes and tsunamis and improve biodiversity (Science Council of Japan 2020). It is hoped that Tokyo, with its limited habitable area relative to population and its exposure to annual monsoon-related and geological-related disasters, can be a leader in the research and development of green infrastructure and promote it throughout Asia and the world.

Acknowledgments The authors would like to thank the Bureau of Urban Development of Tokyo Metropolitan Government and Setagaya City. This chapter was supported by the JSPS KAKENHI [grant number JP22J13885] and the Belmont Forum's Sustainable Urbanization Global Initiative (SUGI): Food-Water-Energy Nexus/M-NEX project [grant number 11314551, Japan Science and Technology Agency (JST): 1009781].

References

Bolund P, Hunhammar S (1999) Ecosystem services in urban areas. Ecol Econ 29:293–301. https://doi.org/10.1016/S0921-8009(99)00013-0

Cabinet Secretariat (2014) Fundamental plan for national resilience

Cabinet Secretariat (2015) National spatial planning

Fujita N (2007) Study on evaluation of "Shasoh" space as a green space in urban areas (2): the quantitative analysis on "Shasoh" in urban areas by setting different spatial scales. Bull Univ Tokyo For 117:21–64

House of Representatives (2021) Tokutei-toshi-kasen shinsui-higai-taisaku-hou-tou-no ichibu-wo kaiseisuru houritsuan-ni-taisuru futai-ketsugi (Supplementary resolution for partial revision of the Act on Countermeasures against Flood Damage of Specified Rivers Running Across Cities). https://www.shugiin.go.jp/Internet/itdb_rchome.nsf/html/rchome/Futai/kokudoB82B6401ACD9B759492586B0001DCB58.htm. Accessed 8 Dec 2022

Ibáñez I, Primack RB, Miller-Rushing AJ, Ellwood E, Higuchi H, Lee SD, Kobori H, Silander JA (2010) Forecasting phenology under global warming. Philos Trans R Soc Lond Ser B Biol Sci 365(1555):3247–3260. https://doi.org/10.1098/rstb.2010.0120

Ichikawa K, Okubo N, Okubo S, Takeuchi K (2006) Transition of the Satoyama landscape in the urban fringe of the Tokyo metropolitan area from 1880 to 2001. Landsc Urban Plan 78:398–410. https://doi.org/10.1016/j.landurbplan.2005.12.001

IPCC (2014) Climate change 2014: synthesis report. In: Core Writing Team, Pachauri RK, Meyer LA (eds) Contribution of working groups I, II and III to the fifth assessment report of the Intergovernmental Panel on Climate Change. IPCC, Geneva, 151 pp.

Iwachido Y, Uchida K, Ushimaru A, Yokota S, Sasaki T (2020) Nature-oriented park use of satoyama ecosystems can enhance biodiversity conservation in urbanized landscapes. Landsc Ecol Eng 166:163–172. https://doi.org/10.1007/s11355-020-00413-y

Iwasa Y, Nishida T (2017) Social potential of green infrastructure in the population reduction and mature society. Jpn J Ecol 67:239–245. https://doi.org/10.18960/seitai.67.2_239

Kuramochi T, Shinohara A, Ono H, Nomura S, Jinbo U, Saito H, Hasegawa K, Nishiumi I, Kawada S, Tomokuni M, Owada M, Kiyoshi T (2014) Flora and Fauna of the Imperial Palace, Tokyo. Fauna Mem Natl Mus Nat Sci 50:1–7

Ministry of Agriculture, Forestry and Fisheries and Ministry of Land, Infrastructure, Transport and Tourism (2016) Plan for promotion of urban agriculture

Ministry of Land, Infrastructure, Transport and Tourism (2015) Priority plan for social infrastructure development

Ministry of Land, Infrastructure, Transport and Tourism (2019) Green infrastructure *suishin-senryaku* (Strategy for promotion of green infrastructure)

Ministry of Land, Infrastructure, Transport and Tourism (2020) Naze ima green infrastructure nanoka (Why green infrastructure now?). https://www.mlit.go.jp/sogoseisaku/environment/sosei_environment_fr_000143.html. Accessed 8 Dec 2022

Ministry of the Environment (2006) Kokyo-ni-okeru cool island koka-no kansoku-kekka-ni-tsuite (Observations of the Cool Island Effect at the Imperial Palace). https://www.env.go.jp/press/7564.html. Accessed 8 Dec 2022

Ministry of the Environment (2012) The national biodiversity strategy of Japan 2012–2020

Ministry of the Environment (2016) Report of comprehensive assessment of biodiversity and ecosystem services in Japan

Ministry of the Environment (2018) The basic environment plan

Ministry of the Environment (2020) 2019-no onshitsu-koka gas haisyutsu-ryo-ni tsuite (Greenhouse gas emission in 2019). https://www.env.go.jp/press/108734.html. Accessed 8 Dec 2022

Motegi N, Yanai S (2004) A study on the characteristics of birds distribution in rooftop vegetation in Tokyo ward. J Jpn Inst Landsc Architect 68:597–600. https://doi.org/10.5632/jila.68.597

Narita K, Mikami T, Sugawara H, Honjo T, Kimura K, Kuwata N (2004) Cool-Island and cold air-seeping phenomena in an urban park, Shinjuku Gyoen, Tokyo. Geogr Rev Jpn 77:420. https://doi.org/10.4157/GRJ.77.403

Numata S (2014) Current status and perspectives of biodiversity in Tokyo. J Geogr (Chigaku Zasshi) 123:497–515. https://doi.org/10.5026/jgeography.123.497

Primack RB, Ibáñez I, Higuchi H, Lee SD, Miller-Rushing AJ, Wilson AM, Silander JA (2009) Spatial and interspecific variability in phenological responses to warming temperatures. Biol Conserv 142(11):2569–2577. https://doi.org/10.1016/j.biocon.2009.06.003

Rudd H, Vala J, Schaefer V (2002) Importance of backyard habitat in a comprehensive biodiversity conservation strategy: a connectivity analysis of urban green spaces. Restor Ecol 10:368–375. https://doi.org/10.1046/j.1526-100X.2002.02041.x

Science Council of Japan (2020) Kiko-hendo-ni-tomonai gekijinka-suru saigai-ni-taishi green infrastructure-wo katsuyo-shita kokudo-keisei-ni-yori “inochi-machi”-wo tsukuru (Creating “inochi-machi” through land management using green infrastructure to cope with severe disasters by climate change)

Setagaya City (2018) Green Master Plan: 2018–2027

Takeuchi K, Brown RD, Washitani I, Tsunekawa A, Yokohari M (2003) Satoyama: the traditional rural landscape of Japan. Springer Japan, Tokyo

The Economics of Ecosystems and Biodiversity (2011) TEEB manual for cities: ecosystem services in urban management

Tokyo Metropolitan Government (2017) Tokyo agriculture promotion plan

Tokyo Metropolitan Government (2020) The changing face of Tokyo: from Edo to today, and into the future. Shirakabashashinkogei, Chiba

Tokyo Metropolitan Government (2022) Okujo-ryokka jisseki (Achievements of rooftop greening). https://www.kankyo.metro.tokyo.lg.jp/nature/green/roof_plant/actual.html. Accessed 8 Dec 2022

Uchida K, Koyanagi TF, Matsumura T, Koyama A (2018) Patterns of plant diversity loss and species turnover resulting from land abandonment and intensification in semi-natural grasslands. J Environ Manag 218:622–629. https://doi.org/10.1016/j.jenvman.2018.04.059
Yokota S, Niwa Y (2020) Development and application of the evaluation tool for the citizen-participated monitoring on runoff control utilizing private gardens. AIJ J Technol Des 26:1270–1275. https://doi.org/10.3130/aijt.26.1270

Calculating the Demand for Food, Energy, and Water in the Spatial Perspective

Shun Nakayama, Wanglin Yan, and William Galloway

Abstract Diverse spatial patterns of cities, to a large extent, reflect the quality of life of an area's inhabitants and the correspondent environmental load their lifestyles impose. This study aims to develop a method to express a design baseline that enables the quantitative evaluation and comparison of the demand and supply of food, energy, and water (FEW), as well as their contribution to environmental load with regard to FEW at the city block level. The demand for and consumption of FEW associated with residences are strongly influenced by household type. The potential production and supply of FEW for a given household are mostly determined by the capacity to install solar panels and home gardens, which are themselves strongly determined by building forms and land use. Therefore, the baseline of a city block for FEW demand and supply can be assessed by identifying household types and building forms. This method was applied to the Tokyo-Yokohama metropolitan area as a case study. This knowledge is helpful to identify the starting point for urban planning and design aimed at zero-carbon emissions and other forms of sustainable urbanism.

Keywords Low carbon · FEW supply and demand · Urban typology

S. Nakayama (✉)
Graduate School of Media and Governance, Keio University, Fujisawa, Kanagawa, Japan
e-mail: shun-nakayama@keio.jp

W. Yan
Faculty of Environmental Information Studies, Keio University, Fujisawa, Kanagawa, Japan
e-mail: w_yan@keio.jp; yan@sfc.keio.ac.jp

W. Galloway
Department of Architectural Science, Toronto Metropolitan University, Toronto, ON, Canada
e-mail: william@frontofficetokyo.com; will.galloway@ryerson.ca

W. Yan et al. (eds.), *Resilient and Adaptive Tokyo*,
https://doi.org/10.1007/978-981-99-3834-6_7

1 Background

Urbanization has accelerated. It has been reported that in 2018 urban populations accounted for 55% of the world's population and will account for 68% by 2050 (United Nations 2018). While some argue that the importance of space and distance has diminished due to the development of Internet technology, especially during the recent pandemics, cities still provide citizens with goods, services, and employment opportunities, efficiently providing for a potentially abundant lifestyle. With such a high concentration of human activity, cities have been considered the spatial form that most efficiently promotes social and economic development.

Conversely, cities are seen as a major contributor to climate change (Rosenzweig et al. 2010). Cities account for 78% of the world's energy demand and are associated with more than 60% of greenhouse gas emissions (UN-HABITAT 2011), even though they occupy only about 2% of the Earth's surface. Thus, cities have a high demand for ecosystem services and environmental resources (Mori and Christodoulou 2012). With urban population overcrowding and rapid urbanization expected to continue, the growing demand for urban resources poses a major challenge for sustainable urban development (Zhang et al. 2019; Galli et al. 2020).

In response to this increased use of resources, the household sector is attracting attention and is expected to play a major role in decarbonization on the consumer side. As a factor, the household sector is expected to become increasingly important, as urbanization progresses and living standards are projected to rise accordingly (Kennedy et al. 2009). In addition, the household sector also has a significant environmental impact, as greenhouse gas emissions from the household sector have been reported to account for more than 60% of global greenhouse gas emissions (Nejat et al. 2015; Ivanova et al. 2016). By country, more than 80% of carbon dioxide (CO_2) emissions in the USA, more than 40% in China, and more than 50% in Japan are attributed to household consumption activities (Bin and Dowlatabadi 2005; Liu et al. 2011; Nakamichi et al. 2015). In particular, more than 80% of CO_2 emissions from the household sector in Japan are related to food, energy, and water (Nakamichi et al. 2015). Therefore, understanding demand and supply gaps for food, energy, and water of the household sector is a common starting point for carbon neutral and sustainable urban planning and urban design.

These gaps largely reflect population density, building form, and household form, as well as the quality of life of the inhabitants of the area. This is closely related to the high environmental burden caused by food, energy, and water demand and supply (Ramaswami et al. 2021). This aim of this study was to develop methodologies that can relate spatial patterns to food, energy, and water supply and demand at the city block scale, in order to obtain a detailed understanding of actual demand for resources in the household sector. It is anticipated that these methodologies will be transferable to higher spatial scales.

2 Typologizing Social and Physical Patterns for FEW

Cities form diverse spatial patterns as a result of their composition, including population density, architectural form, and household form. These spatial patterns are known to influence the movement of people, goods, and services and have a significant impact on energy demand and CO_2 emissions. Numerous studies have been conducted on how to evaluate these spatial patterns, based on "urban morphology," a discipline that deals with the typology of urban space.

Urban morphology analyzes forms and their transformation at each scale, from architectural to urban, and has achieved results in elucidating the mechanisms of transformation of urban space (Min and Satoh 2007). Researchers have conducted studies that typify the physical form of buildings in Japan. Saito et al. typified buildings by height and shape, while Min and Satoh typified buildings by roof shape in addition to height and shape, both of which were used to describe the urban form (Saito et al. 1995; Min and Satoh 2007). In addition, typologies of building uses in basic urban planning surveys and census surveys have been used in many studies as typologies of the social roles of buildings (Ai et al. 2010; Yokoi et al. 2010). The upper-scale typology of city blocks is often based on these physical and social building typologies. In such cases, the typology is mainly combined with road typology (Wakabayashi et al. 2017; Yoshihara and Tanaka 2020).

Conversely, it has been considered difficult to relate spatial patterns to energy demand and CO_2 emissions due to the diverse scale and resolution of environment-related data and the large amount of non-spatial data (Kellett et al. 2013). However, with recent advances in the sophistication of statistical information and modeling methods, findings from urban morphology have been transferred to the estimation of environmental impacts at multiple spatial scales, ranging from architectural to urban scales (Iino et al. 2002). In addition, research is accumulating on the impact of spatial patterns on the potential for resource use to mitigate environmental impacts (Natanian et al. 2019).

Urban morphology can also be utilized to estimate the demand for food, energy, and water. Specifically, in order to closely represent the impact of spatial patterns on the supply and demand for food, energy, and water, two typologies are used: physical and social. First, the demand and consumption for food, energy, and water associated with residential life are strongly influenced by the type of households. Second, the potential of production and supply of food, energy, and water for a household to alleviate the demand is mostly determined by the capability to install solar panels and home gardens, which is in turn strongly determined by building forms. Therefore, the baseline of a city block for FEW demand and supply can be assessed by identifying the type of households and the form of buildings. We categorized the "household type" in terms of the number of family members, the groupings of age and gender, and the "building form" in terms of detached houses and apartments. The basis for this classification depends on the available FEW consumption intensity data.

When this work is expanded to the district or city level, data size can become extremely large and unmanageable, but GIS can be used to lighten the workload. Information on household structures is available using census data and other sources. Useful GIS data is readily available for quantification of FEW demand on the metropolitan scale. For building typology, the number of buildings in each city block can be counted easily by using categories that match building attributes in GIS data. Two types of GIS data that includes building attributes are available in Japan: the Basic Urban Planning Survey (maintained by the government) and building point data (maintained by Zenrin Corporation). The former is available free of charge upon request for information disclosure, although classifications differ among local governments. The latter is available for a fee, but classifications are uniform and more detailed.

3 Calculating FEW Demand

3.1 Food Demand

The quantification of food demand is largely dependent on age and sex of a population. In other words, the demand for foodstuff (k) in the target area (FD_k) can be calculated by multiplying the food demand intensity by age and sex ($FD_{k,\ age,\ sex}$) by the population by age and sex ($P_{age,\ sex}$), as shown in Eq. (1).

$$FD_k = FD_{k,age,sex} \times P_{k,age,sex} \tag{1}$$

In urban areas, a declining labor force due to the aging of agricultural workers has led to the conversion of ordinary farmlands into allotment gardens. Home vegetable gardens are also gaining popularity because they can be easily managed. While it is important to understand the actual situation of such small-scale vegetable production, it is difficult to identify production in home and allotment gardens without careful fieldwork. When estimating for a large area, the work becomes quite unmanageable. However, by using the cultivated area of home vegetable gardens by type of houses (SKG_{houses}) and the rate of the families with such home vegetable garden in the type (RKG_{houses}), we can obtain the production area.

$$L_{FP,farm} = \sum_{houses} SKG_{houses} \cdot RKG_{houses} \tag{2}$$

3.2 Energy Demand

The quantification of energy demand depends largely on household size. Energy demand (ED) of the target area can be calculated by multiplying energy demand intensity ($ED_{household}$) by the number of households per household size ($H_{household}$), as shown in Eq. (3).

$$ED = ED_{household} \times H_{household} \tag{3}$$

It is also necessary to consider the existence of autonomous decentralized energy supply systems. Solar panels are a decentralized power generation system possible at the household level. The amount of electricity generated by solar panels can be estimated from the size and number of panels. The annual electricity production (EP) from solar panels in the target area can be expressed as Eq. (4) using the system capacity intensity of solar panels by building type (P_{bldg}) and the solar panel installation ratc by building typc (RPV_{bldg}). Parameters (H), (P), and (K) refer to average solar radiation, system capacity of solar panels, and loss factor, respectively.

$$EP = H \cdot K \cdot 365 \cdot \sum_{bldg} P_{bldg} \cdot RPV_{bldg} \tag{4}$$

Thus, net energy demand (ED) is expressed as shown in Eq. (5), taking into account the amount of electricity generated by solar panels (EP).

$$ED = ED_{household} \times H_{household} - EP \tag{5}$$

3.3 Water Demand

The quantification of water demand depends largely on the number of household members. Water demand (WD) in the target area can be calculated by multiplying water demand intensity ($WD_{household}$) by the number of households per household size ($H_{household}$), as shown in Eq. (6).

$$WD = WD_{household} \times H_{household} \tag{6}$$

Meanwhile, the installation of rainwater storage tanks is promoted in some areas to create a resilient water supply environment in residential areas. Rain that falls on the roof is channeled into a water tank to store the rainwater. If we can determine the roof area of houses that have a water tank installed, we can calculate annual rainwater supply based on annual rainfall. The amount of water available for self-sufficiency (WP) of the target area can be expressed as Eq. (7), using positive radiant area of the roof of each building (RA_x), water storage tank installation rate by building type (RWS_{bldg}), and annual precipitation (RW).

$$WP = RW \cdot \sum_{bldg} \sum_{x} RWS_{bldg} \cdot RA_x \tag{7}$$

Thus, net water demand (WD) can be expressed as in Eq. (8), taking into account the amount of water available for self-sufficiency (WP).

$$WD = WD_{household} \times H_{household} - WP \tag{8}$$

4 Case Studies in Tokyo

4.1 Features of the Study Area

The target areas for this study are Tokyo and Kanagawa Prefecture, the latter being located in the southwest part of the Tokyo Metropolitan Area, which is the world's most populous metropolitan area. Tokyo and Kanagawa together are home to more than 23 million people in an area of about 4600 sq. km. Many of the urbanized, built-up areas were developed during the period of high economic growth after the Second World War and fan out widely in suburban and extra-urban areas built up along radially distributed railways. Similar to all Japanese cities, this region is facing declining birth rates and an aging population. Governments are trying to remake the cities in a more compact form by attracting services and residents to the walkable areas near railway stations. Built-up areas are approaching a time when infrastructure and other upgrades will be needed, as seen in the vulnerability witnessed from devastating disasters like the Great East Japan Earthquake in 2011, a major heat wave in 2018, and super typhoon Hagibis in 2019. In consideration of its size and environmental impacts, the metropolis has a responsibility to act to reduce CO_2 emissions.

We identified three area types to represent the core social and spatial characteristics of the study area (Fig. 1).

High density housing is typical in central Tokyo like inside Yamanote Line, where city blocks were developed without systematic planning and oversight after the Second World War, and the size of all type of houses is extremely small. These areas are generally more popular among younger generations for reasons such as proximity to places of employment. In suburban areas a bit further out, the size of individual houses is somewhat larger, but these towns generally also developed without systematic planning and are interspersed with farmland. In suburbs further out and up to 30 km from the center of Tokyo, planned townscapes prevail. Here, the general pattern is large apartment complexes and relatively large detached houses

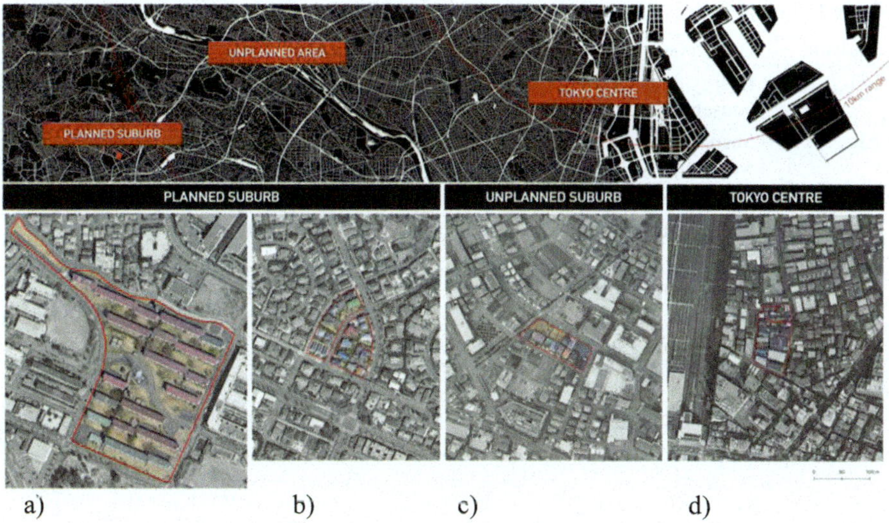

Fig. 1 Representative study areas selected in the Tokyo Metropolitan Area. (**a**) Medium-rise apartment, (**b**) Low-rise residential, (**c**) Mixed use area, (**d**) Dense urban area

built from near rail stations. Today, residential buildings can be found even in areas a couple of kilometers away from the station. There is very little boundary between residential areas based on different stations due to the close distance between stations. As a current issue, some housing complexes are approaching an age that will need replacement or major upgrades, and in some cases, residents are moving out. Detached homes are typically located further from stations and do not tend to match the housing preferences of young people, and population here are also aging.

4.2 Data Management

The respective data on age and number of households by household size required to determine social typology for this study was obtained from the 2015 national census. Other data required in the process of analysis was compiled as shown in Table 1.

4.3 Results

Using methods described in Chap. "Climate Change in Global Cities" and data collected in Sect. "Data Management" above, we presented FEW demand visually as shown in Fig. 2. Tokyo's urban center has an extremely high population density of

Table 1 List of sources of data needed for calculating FEW demand

Variable	Description	Data Source
$ED_{household}$	Energy demand intensity by household size	Inoue et al. (2006)
ER_m	Share of energy sources (*m*) in water supply	Agency for Natural Resources and Energy (2017)
ERR_l	Share of energy source (*l*)	
$FD_{k,\ age,\ sex}$	Food demand intensity by gender, by age	Ministry of Health, Labour and Welfare (2019)
$H_{household}$	Number of households by household size	Ministry of Internal Affairs and Communications (2015)
$P_{age,\ sex}$	Population by age and sex	
RW	Annual precipitation	Japan meteorological agency

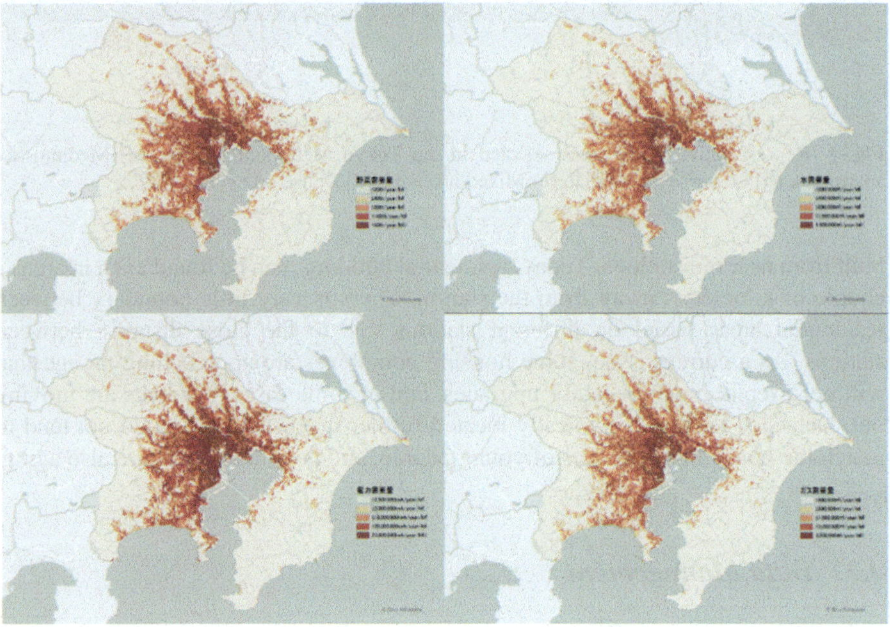

Fig. 2 Estimation of FEW demand in the Tokyo Metropolitan Area with the statistical unit: *Above left*, annual demand for vegetables; *right*, annual demand for water; *below left* annual demand for electricity; *right*, annual demand for gas

357 persons/ha, suggesting high demand for FEW. Unplanned areas have a density of 76 persons/ha, about 20% that of the city center, and demand for electricity, gas, and water is much lower. For example, the annual demand for electricity is 56,662 kWh/ha in the city center, but in unplanned areas only 11% of that amount, at 6163 kWh/ha. This trend is also true in planned suburban residential areas.

Table 2 Existing efforts to produce food, energy, and water

	(a) Medium-rise apartment	(b) Low-rise residential	(c) Mixed-use area	(d) Dense urban area
Population density (persons/ha)	80	102	76	357
Distance to nearest station (m)	600	1000	1350	150
Distance to central business district (km)	22	22	15	8
FEW demand				
Vegetables (t/ha/year)	7.20	9.76	7.08	32.74
Electricity (kWh/ha/year)	7480.7	11103.3	6162.9	56,662.2
Gas (m^3/ha/year)	871.2	1289.2	682.3	5823.0
Water (m^3/ha/year)	5179.8	7674.0	3956.1	32,805.4
FEW production				
Farmland or home garden (m^2)	0.0	0.0	1463.0	0.0
PV (solar) panels (m^2)	0	102.8	91.9	102.5
Roof with rainwater tank (m^2)	0	0	0	0

The differences in the results obtained in Table 2 are due to the variables considered this time. For example, for electricity demand, we evaluated the demand from two perspectives: whether the house is a single-family home or an apartment complex, and how many people are in the household. However, since we did not take into account production within a city block, actual demand may be different. For example, the unplanned area has small farmland parcels. Also, some houses that appear to have been recently rebuilt have solar panels installed. As shown in Table 2 such autonomous decentralized FEW production functions can be evaluated together to understand the actual demand.

5 Discussion and Conclusion

In this chapter, we were able to quickly evaluate the food–energy–water (FEW) baseline in any specific conditions by identifying a set of dietary tables, households, and building types. We found that evaluating FEW demand in conjunction with the production function of FEW improves the accuracy of FEW demand calculations, and that GIS can be used to calculate FEW demand with a certain degree of accuracy over a wide area.

The typology of space reveals differences in FEW demand within a region. This brings great value to the field of urban design for decarbonization. Urban design method is a process of integrating knowledge and information with different landscapes: architectural, geographic, policy, etc. Each landscape is expressed as a factor, each one including various objects, with social and physical features related to the supply and demand of FEW components. The demand for FEW can be assessed at the household or building unit level and then scaled to the neighborhood, city, or regional level. Identifying demand for food, energy, and water has been described as a foundation for designing sustainable cities (Yan and Roggema 2019).

There are always gaps between supply and demand. Shortfalls may be a matter of quality and/or quantity, and both may be affected by the uneven spatial distribution of urban services. Future studies will aim to identify these gaps and propose solutions through the mobilization of social-ecological resources and governmental policy. The outcomes would typically take the form of broad concepts in the political landscape such as the compact cities, smart cities, eco cities, resilient cities, and zero-carbon cities.

Acknowledgments This study was done as part of M-NEX (11314551), a Belmont Forum granted project under "Collaborative Research Area on Sustainable Urbanization Global Initiative (SUGI)/ Food-Water-Energy Nexus." We would like to thank the Tokyo Metropolitan Government and Kanagawa Prefectural Government for providing its basic survey on urban planning.

References

Agency for Natural Resources and Energy (2017) Energy consumption statistics by prefecture [Data set]

Ai H, Sadahiro Y, Asami Y (2010) Categorization of shopping malls in mega cities by their change pattern of floor area and building allocations. J City Plan Inst Jpn 45(2):23–29

Bin S, Dowlatabadi H (2005) Consumer lifestyle approach to US energy use and the related CO2 emissions. Energy Policy 33(2):197–208

Galli A, Iha K, Pires SM, Mancini MS, Alves A, Zokai G, Lin D, Murthy A, Wackernagel M (2020) Assessing the ecological footprint and biocapacity of Portuguese cities: critical results for environmental awareness and local management. Cities 96:102442

Iino A, Hoyano A, Yoon S, Tsukamoto K, Marui M (2002) Classification of urban blocks for evaluation of urban thermal environment with regard to land use and land cover: evaluation of urban thermal environment in the tropical regions part 1. J Archit Plan 67(554):55–62

Inoue T, Mizutani S, Tanaka T (2006) Energy consumption in housing on the basis of National Scale Questionnaire: analysis of influence of various factors on annual energy consumption part 2. J Environ Engin 71(606):75–80

Ivanova D, Stadler K, Steen-Olsen K, Wood R, Vita G, Tukker A, Hertwich EG (2016) Environmental impact assessment of household consumption. J Ind Ecol 20(3):526–536

Kellett R, Christen A, Coops NC, van der Laan M, Crawford B, Tooke TR, Olchovski I (2013) A systems approach to carbon cycling and emissions modeling at an urban neighborhood scale. Landsc Urban Plan 110:48–58

Kennedy C, Steinberger J, Gasson B, Hansen Y, Hillman T, Havránek M, Pataki D, Phdungsilp A, Ramaswami A, Mendez GV (2009) Greenhouse gas emissions from global cities. Environ Sci Technol 43(19):7297–7302

Liu L-C, Gang W, Wang J-N, Wei Y-M (2011) China's carbon emissions from urban and rural households during 1992–2007. J Clean Prod 19(15):1754–1762

Min S-H, Satoh S (2007) A study on the analysis of block with typology of urban architecture: a study on the theory of urban architectural block design with methodology of urban morphology part 1. J Archit Plan 72(621):53–60

Ministry of Health, Labour and Welfare (2019) National health and nutrition survey [Data set]
Ministry of Internal Affairs and Communications (2015) National census [Data set]
Mori K, Christodoulou A (2012) Review of sustainability indices and indicators: towards a new city sustainability index (CSI). Environ Impact Assess Rev 32(1):94–106
Nakamichi K, Yamagata Y, Hanaoka S, Wang X (2015) Estimation of indirect missions in each municipality and comparison to direct emissions. J Japan Soc Civil Eng 71(5):I_191–I_200; Ser. D3 (Infrastructure Planning and Management)
Natanian J, Aleksandrowicz O, Auer T (2019) A parametric approach to optimizing urban form, energy balance and environmental quality: the case of Mediterranean districts. Appl Energy 254:113637
Nejat P, Jomehzadeh F, Taheri MM, Gohari M, Abd MZ, Majid. (2015) A global review of energy consumption, CO2 emissions and policy in the residential sector (with an overview of the top ten CO2 emitting countries). Renew Sust Energ Rev 43:843–862
Ramaswami A, Tong K, Canadell JG, Jackson RB (2021) Carbon analytics for net-zero emissions sustainable cities. Nature Sustainability 4(6):460; https://www.nature.com/articles/s41893-021-00715-5
Rosenzweig C, Solecki W, Hammer SA, Mehrotra S (2010) Cities Lead the way in climate-change action. Nature 467(7318):909–911
Saito C, Hachiya K, Sakamoto K (1995) The town typology based on spatial composition in Tokyo: a study on arrangement of buildings in urban blocks. J Archit Plan 60(474):123–131
UN-HABITAT (2011) "Hot cities: battle-ground for climate change"
United Nations (2018) "The World's cities in 2018." Department of Economic and Social Affairs, Population Division, World Urbanization Prospects, p 1–34
Wakabayashi K, Eom S, Suzuki T (2017) Urban classification by combining building and road characteristics in Tokyo Ward area. J City Plan Inst Jpn 52(3):711–716
Yan W, Roggema R (2019) Developing a design-led approach for the food-energy-water nexus in cities. Urban Plan 4(1):123–138
Yokoi T, Yamamoto Y, Tokai A, Morioka T (2010) Development of decision support system to integrate block renewal and energy planning towards low-carbon city. Doboku Gakkai Ronbunshuu G/JSCE J Environ Syst 66(1):17–34
Yoshihara T, Tanaka T (2020) Locational and physical categorizing of densely built-up areas. J City Plan Inst Jpn 55(3):274–281
Zhang Y, Shen L, Shuai C, Bian J, Zhu M, Tan Y, Ye G (2019) How is the environmental efficiency in the process of dramatic economic development in the Chinese cities? Ecol Indic 98:349–362

Identifying Gaps Between Food Supply and Demand Under Compact City Policies

Shun Nakayama, Wanglin Yan, and Rajib Shaw

Abstract Metropolitan areas typically experience urban sprawl, but some are actually beginning to face population declines, at the same time as planners have also been promoting the transition to compact cities with the goal of creating healthy cities. With all these changes, gaps in food supply and demand can emerge, as evidenced in Japan by the withdrawal of supermarkets from their catchment areas. This chapter defines the gap between food supply and demand as a spatial gap between *accessibility* on the demand side and food environment on the supply side. This study uses empirical analysis based on time-series data to examine where and how such gaps can arise. With Yokohama suburbs as the study area, two driving factors are examined: the decline of resident mobility in an aging society and changes in supermarket location that affect the availability of products and services. The study discovered a decrease in *accessibility* throughout the suburbs, especially in areas that were further away from train stations. Meanwhile, compact city policies promoted the development of more supermarkets in the vicinity of stations, but residential areas further away saw a rapid decrease in the number of supermarkets. An obvious conclusion is that the transition to compact cities may not necessarily contribute to the goal of creating healthy cities.

Keywords Compact city · Healthy city · Gap analysis

S. Nakayama (✉)
Graduate School of Media and Governance, Keio University, Fujisawa, Kanagawa, Japan
e-mail: shun-nakayama@keio.jp

W. Yan
Faculty of Environmental Information Studies, Keio University, Fujisawa, Kanagawa, Japan
e-mail: w_yan@keio.jp; yan@sfc.keio.ac.jp

R. Shaw
Graduate School of Media and Governance, Keio University, Fujisawa, Kanagawa, Japan
e-mail: shaw@sfc.keio.ac.jp

W. Yan et al. (eds.), *Resilient and Adaptive Tokyo*,
https://doi.org/10.1007/978-981-99-3834-6_8

1 Introduction

Begun in 1986 by the European office of the World Health Organization (WHO), the Healthy Cities project has expanded rapidly and become a key concept in urban planning worldwide (Barton and Grant 2013). Cities gather resources to develop and implement cross-sectoral approaches to health and are often identified as the lowest administrative level with a political mandate and authority (Ashton et al. 1986). In Japan, a program called "Health Japan 21" was launched in 2000 by the Ministry of Health, Labour and Welfare (MHLW), which defined access to food as an important foundation for health and quality of life and pointed out the importance of urban planning as shown in Fig. 1. The Ministry of Land, Infrastructure, Transport and Tourism (MLIT), which is responsible for infrastructure and urban policies, formulated "Guidelines for the Formation of Healthy Cities," which specifically states the need to shift to compact cities in order to form healthy cities.

This concept of transition to compact cities has been attracting attention even more attention than healthy cities, as it contributes not only to health formation but also to more efficient administrative services and lower carbon emissions. Once the usefulness of compact cities in the Japanese urban context was recognized around 2000, the transition to compact cities was first presented by the government in 2007 by the Social Infrastructure Development Council of MLIT. Initially, the transition was mainly targeted at regional cities that have rapidly declining populations, but as the prospect of population decline in the suburbs of large cities became more apparent, the transition to compact cities became a topic in metropolitan areas as well.

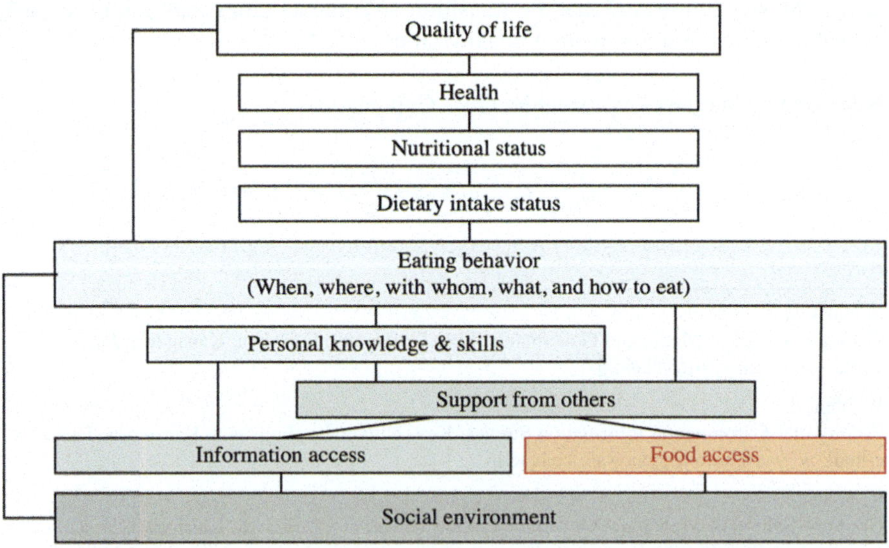

Fig. 1 Concept of "Health Japan 21" by Ministry of Health, Labour and Welfare, Japan

At first glance, there is some consistency between the concept of access as a basis for health and efforts to achieve health through urban policies of transitioning to compact cities. It has been pointed out, however, that unanticipated store closures led to a waning of early efforts to transition to compactness in smaller regional cities (Taniguchi et al. 2011). It has also been pointed out that such cross-cutting approaches run the risk of being counterproductive, especially if they do not take into account the effects of market-oriented regulatory processes and the uneven dynamics of policy formation (Cole et al. 2017). Therefore, it is necessary to clarify whether healthy cities can truly be formed through the transition to compact cities. Thus, this study aims to examine whether or not the transition really creates healthy cities and to clarify the relationship between efforts to transition to compact cities and changes in the food supply-and-demand environment. The findings of this study will provide insight into the complex impacts of efforts to transition established urban areas to compact cities.

2 Identifying Gaps Between Demand and Supply for Food

There is a growing body of research on issues connected to the period of transition toward compact cities, especially with regard to access to grocery stores, a key component for maintaining a high quality of life. For instance, a number of simulations and scenario analyses on the risk of supermarkets withdrawing from regional cities as they become more compact. One study found that the risk of withdrawal of lifestyle services is particularly high during the transition period (Adachi et al. 2012). However, this research topic is still in its infancy and there are still several issues to be addressed. First, the Store Location Model used by Adachi et al. is inadequate because it only considers store closures and does not take into account new store openings. Second, these studies covered only regional cities, so it is still not clear what happens in large cities as they try to become more compact. A healthy food environment is formed from the interrelationships among (1) food environment exposure, (2) accessibility, and (3) interventions (Hu et al. 2020). Previous studies have focused on the relationship between food environment exposure and interventions and lacked the mutually interactive perspective of food environment exposure (food supply side) and accessibility (both food supply *and* demand).

Therefore, this study was designed to empirically examine the impact on the supply and demand of food during the transition to a compact city typology in a large city, making use of time-series data. The empirical analysis will aim to (1) capture the actual transition to compact cities and identify where and how urban compactness is occurring, separately evaluate (2) food environment exposure and (3) accessibility, and (4) reveal where and how the food demand–supply gap is linked to urban downsizing. Compact cities in Japan's metropolitan suburbs, which are the focus of this study, have urban functions concentrated mainly in the vicinity of train stations. Most studies define this *vicinity* as the area within a 500-m radius from a station.

That 500-m radius is used as the station "vicinity" for this study as well. A station's larger "sphere of influence" is identified using commuter pass data, which provides information on where people live and the stations they most frequently use. In other words, whether or not a point is within a station's "sphere of influence" is determined by the station that is most frequently used by the people who live there. Following previous studies, we define the food demand side as "accessibility" and the food supply side as "food environment exposure." Accessibility is assessed in terms of "walkable" distances, based on the transportation mix used during shopping activities. Food environment exposure is evaluated from three easily available data points, namely location, product assortment, and product supply capacity. For each train station, we define the station area and area of influence as above and aggregate data on accessibility and food environment exposure in time series based on whether or not compact city policies are in effect. Finally, the Mann-Whitney U test is used to examine the statistical significance of the impact of compact city policies on accessibility and food environment exposure.

2.1 Evaluation of Accessibility

The "mobility" of residents from the food demand side perspective differs greatly depending on the age of users, whether or not they own a private car, and the level of development of the public transportation network. Therefore, in defining the mobility of residents, it is necessary to look at the transportation mix that is available for residents to access grocery stores.

The transportation mix for accessing grocery stores can be identified using what is called a Person Trip (PT) survey. Figure 2 shows the percentage of respondents who use walking as a means of transportation to access grocery store, by region, using publicly available data from sixth Tokyo PT Survey conducted in 2018. Figure 2 shows that walking is the most common mode of transportation for accessing grocery stores in the suburbs of Tokyo. Therefore, this study focuses on walking and defines the walking mobility of a region as "the average value of the ease of walking of local residents based on the age structure of the residents." The ease of walking is defined in terms of the fatigue associated with walking. To evaluate this, we use the concept of "generalized cost" as a qualitative evaluation of fatigue (Goodwin 1976). This is a relative comparison of different modes of transportation and indicates the degree to which fatigue is felt.

The equivalent time coefficient of Mori et al. was used to evaluate the fatigue associated with walking. Tables 1 and 2 summarize the symbols and parameter values used in the calculation of walking mobility, respectively.

The walking mobility M of an area with a total population of n is calculated using individual k's walking mobility M_k as follows:

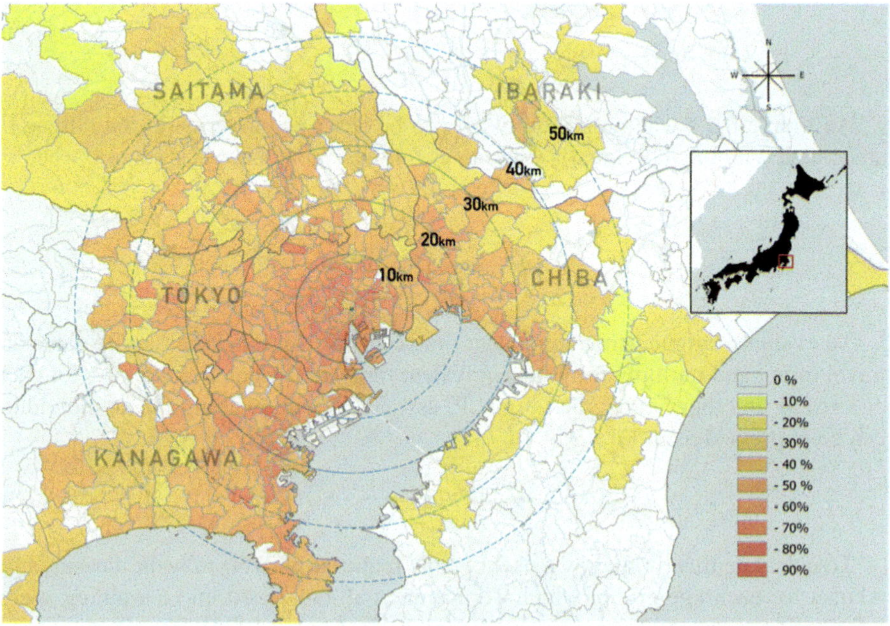

Fig. 2 Percentage of respondents who use walking as a means of transportation to access grocery stores in the Tokyo metropolitan region, based on sixth Tokyo Person Trip survey conducted in 2018

Table 1 Symbol legend

Symbol	Description	Unit
k	An individual	–
F_k	Fatigue associated with person k's walking mobility	–
M	Local walking mobility	–
M_k	Walking mobility capacity of person k	–
n	Total population in the region	Persons
t	Time required to walk 1 m	Min/m
v	Walking speed	Min/m
x	Equivalent time coefficient	–

Table 2 Parameters for calculations

Age	x	v
Twenties	2.215	80.24
Thirties	2.385	80.76
Forties	2.305	79.12
Fifties	2.205	70.19
Sixties	2.205	63.18
Seventies	2.205	56.22

$$M = \frac{1}{n}\sum_{k=1}^{n} M_k \tag{1}$$

An individual's walking mobility M_k can be described in terms of the fatigue F_k associated with the individual's walking as follows. Note that M_k is set to 0 when F_k is 0.

$$M_k = \frac{1}{F_k}\,(F_k \neq 0) \tag{2}$$

To evaluate fatigue using equivalent time coefficients, it is necessary to calculate travel time t and multiply it by the equivalent time coefficient x of walking for each age group. In other words, the fatigue F_k associated with walking for an individual can be expressed as in Eq. (3).

$$F_k = x \cdot t \tag{3}$$

To calculate travel time t, we use the time required for a 50-m walk measured by Akutsu for each age group, which Terayama et al. converted into a walking speed v for each age group (Terayama and Odani 2015; Akutsu 1975). For simplicity, we assume that the path distance of the road is 1 m and divide this by the walking speed v above to obtain the travel time t. The fatigue F_k associated with individual walking can be transformed from Eq. (3) as shown in Eq. (4).

$$F_k = x \cdot \frac{1}{v} \tag{4}$$

2.2 *Evaluation of "Food Environment Exposure"*

From the viewpoint of food suppliers, the following four factors have conventionally been used to define food environment exposure: (1) store location, (2) economic cost of food products offered, (3) types of food products offered, and (4) nutritional value of food products offered (Walker et al. 2010). Of these, (1) store location is indispensable because it captures store location consolidation and store withdrawal. However, since the factors (2), (3), (4) are also influenced by regional characteristics, we will examine them closely to determine whether they should be included in this evaluation.

First, we examine factor (2), economic cost. Nakamura et al. point out that the impact on consumers based on economic factors is a phenomenon unique to large urban centers (Nakamura and Asami 2019). They found that the presence of many high-end supermarkets creates consumers who do not have access to general supermarkets (those that do not target high-end consumers). The target area in the present study is limited to the suburban area of Yokohama City and does not include the central area, so economic costs were not considered in this study.

Next, we consider food types in factor (3). For example, it is clear that there is a difference between small and large supermarkets in terms of product assortment. This was also revealed in a preliminary survey conducted by the authors. However, no detailed data was available on the types of foods offered in stores. Therefore, in this study, we evaluate product assortment based on sales floor area.

Finally, we examine nutritional value in factor (4). Previous studies have indicated that nutritional status tends to deteriorate for older people living in areas with poor access to stores (Iwama et al. 2016). Conversely, no bias in nutritional value among stores has been noted.

Grocery stores included in this study were limited to supermarkets. The reason for this is that most food access in Japan is concentrated in supermarkets. Many previous studies have also focused on supermarkets in evaluating food accessibility (Adachi et al. 2012).

3 Case Studies in the Tokyo Metropolitan Area

3.1 Study Area

After World War II, the demand for workers in Tokyo rose rapidly due to the need for labor for postwar reconstruction and special procurement associated with the Korean War. As a result, a serious housing shortage persisted from the end of the war to the period of rapid economic growth that continued into the 1960s, making the need to expand housing construction an important issue. The government adopted supply-oriented policies to help the middle class acquire owner-occupied housing. Housing sites were developed by railway companies along private rail lines that radiated out from city centers, and the metropolitan area expanded in concentric circles (Bagan and Yamagata 2012). Privately developed residential areas along rail lines have played an important role in the realization of postwar policies that promoted owner-occupied housing. Private railway companies also needed to recoup their investments in rail line construction, so it was necessary to make the new lines economically viable by establishing a certain population at an early stage. To do so, private railway companies attracted medium- and high-rise residential complexes in the form of apartments in the vicinity of stations. Housing complexes were built on land owned by railway companies and also by other landowners. As a result, suburban areas in the Tokyo metropolitan area feature an urban structure with an order of commercial areas, housing complexes, and detached houses built from stations.

For this study, the suburban area of Yokohama City, located in the Tokyo metropolitan area, was chosen as the target area as shown in Fig. 3, because it has

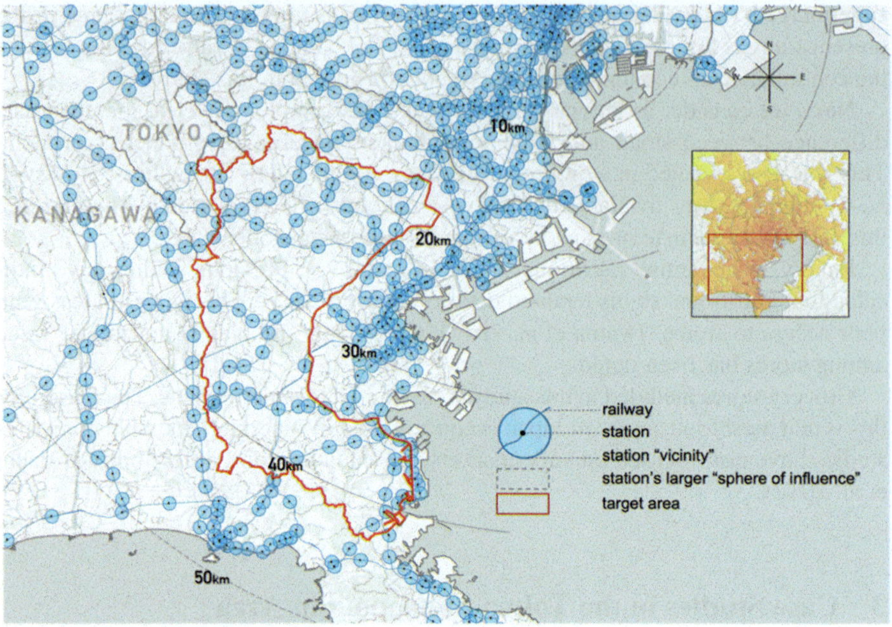

Fig. 3 Location and people flow in the target area

a station-centered urban structure that is typical of large metropolitan areas in Japan, and because its urban master plan calls for a shift to a compact city.

In recent years, redevelopment of train stations and their surrounding areas has been progressing due to the aging of stations and the shift to a station-centered intensive urban structure. The City of Yokohama has used ordinances to mandate "land readjustment projects" and "urban redevelopment projects" as "statutory" projects and has created a designation to encourage "excellent building development projects" as "non-statutory" projects. In addition, the city is accelerating the pace of development by establishing new "urban development consultation districts" where consultation among stakeholders is encouraged for the purpose of consolidating commercial and business functions. As of 2020, of the 68 stations located in the suburbs of Yokohama, 15 were in the process of transitioning toward compact cities through such initiatives as shown in Fig. 3.

3.2 Data Management

The supermarket store data used in this study is the Points of Interest (POI) on supermarkets provided by marketing agency NTT Town page, and the Japan Supermarket Directory by Shogyokai Publishing. The POI data includes only addresses, but the Japan Supermarket Directory contains detailed information such as sales floor space, number of parking spaces, date of store opening, and product categories handled, in addition to store names and addresses. For this study, we were able to obtain the POI datasets as of September 1998, September 2008, and September 2018, but were only able to obtain the Japan Supermarket Directory for fiscal year 2017. For this reason, we compiled Japan Supermarket Directory information into the same three time periods (September 1998, September 2008, and September 2018) based on date of store opening and used this in our analysis. However, since only data on stores as of 2017 was available, we used the directory only for reference. In order to conduct the GIS-based analysis in this study, we converted address data to latitude and longitude using the CSV address matching service of the Center for Spatial Information Science (CSIS) at the University of Tokyo.

For the evaluation of walking mobility based on population composition, we used population data from 1995, 2005, and 2015 in order to match the data in the POI datasets. Since regional mesh data were not available from 2000, we used subregional data that has been available since 1995. The e-Stat of the Statistics Bureau of the Ministry of Internal Affairs and Communications (MIC), which releases census data, includes boundary data for these regional mesh data and subregional data, but not for the 1995 subregional boundary data. Therefore, we used the 2000 subregional boundary data in this study.

3.3 Result

Data for (a) accessibility, (b) number of stores, (c) sales floor area, (d) number of parking spaces were collected for 1998, 2008, and 2018 and aggregated for each area and each compact city policy. Based on this aggregation, we graphed changes as of 2008 and 2018, relative to 1998, which was set as 1. The results are shown in Fig. 4, with the horizontal axis as time, and the vertical axis representing each aggregate item. The results of statistical tests of the impact of policies on these changes are summarized in Table 3 (1998–2008) and Table 4 (2008–2018).

In Fig. 4, we can see that the number of grocery stores is increasing around stations where there is a compact city policy to create that outcome, while the number of stores is decreasing rapidly in the larger *sphere of influence* of the stations where such policy is being implemented. Tables 3 and 4 present the results of statistical validation performed using the Mann-Whitney U test. The p-value is used as an indicator of the test results; values of $p < 0.05$ and $p < 0.01$ indicate a significant difference at 95% and 99% confidence intervals, respectively. In Tables 3 and 4, regarding the number of stores, we can also see that the changes from 1998 to 2008 in *vicinity* of stations and from 2008 to 2018 in stations' *spheres of influence* are statistically significant.

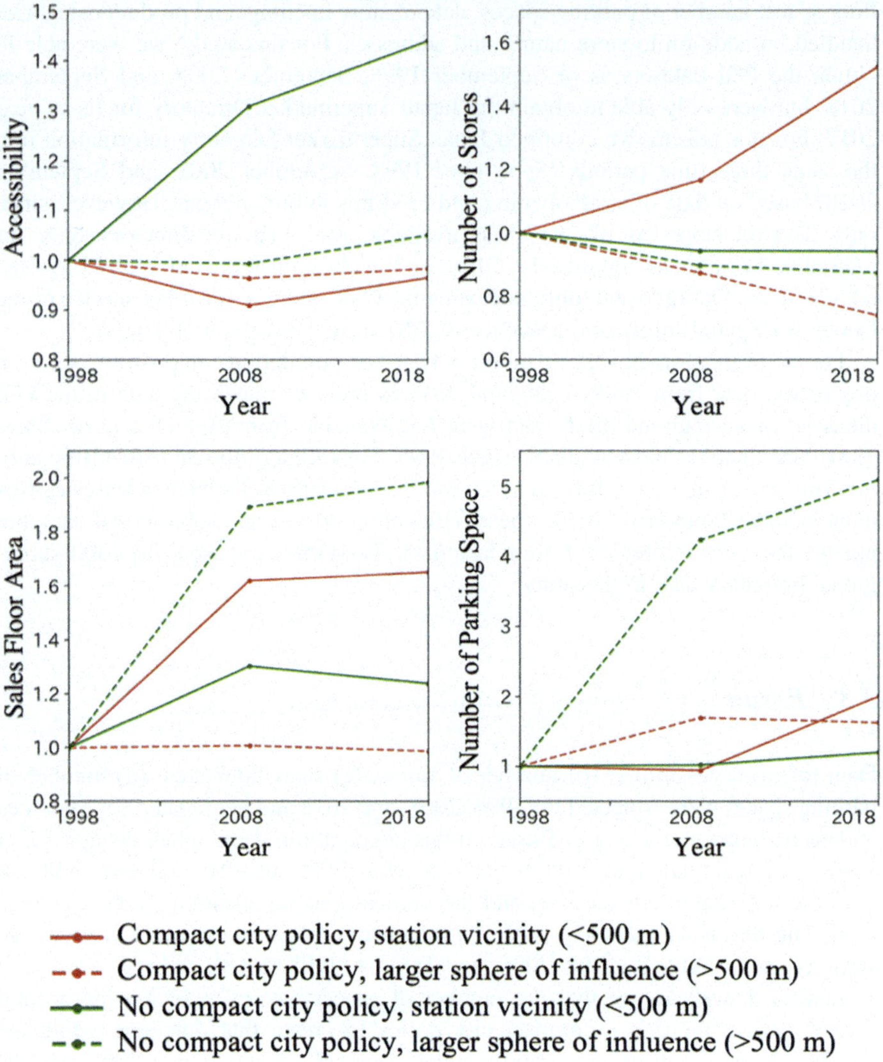

Fig. 4 Time series change of each evaluation item

Table 3 P-value of Mann-Whitney U test (from 1998 to 2008)

Aggregate items	(a)	(b)	(c)	(d)
Area in *vicinity* of station (<500 m)	0.315	0.008	0.276	0.605
Station's *sphere of influence* (>500 m)	0.277	0.621	0.861	0.415

Key: (a) accessibility, (b) number of stores, (c) sales floor area, (d) number of parking spaces

Table 4 P-value of Mann-Whitney U test (from 2008 to 2018)

Aggregate items	(a)	(b)	(c)	(d)
Area in *vicinity* of station (<500 m)	0.544	0.715	0.436	0.637
Station's *sphere of influence (>500 m)*	0.902	0.024	0.841	0.780

Key: (a) accessibility, (b) number of stores, (c) sales floor area, (d) number of parking spaces

4 Discussion and Conclusion

The study described in this chapter sought to determine whether or not compact city policies contribute to the formation of healthy cities. The specific focus was on whether or not compact city policy-induced changes affect the food supply–demand gap in the study area of the suburbs of Yokohama, within the Tokyo metropolitan area. Here we discuss demographic and regulatory factors related to compact cities.

Regarding accessibility, previous studies in Japan have shown that the intention to move is especially low among owner-occupied households (Nakamichi et al. 2019). As mentioned in 3.1, many of these households are located somewhat far from train stations in Tokyo's suburban areas. Therefore, it can be assumed that the ratio of older people in the community is increasing in the station's *sphere of influence* (i.e., beyond the 500 m vicinity of the station). This is seen as a contributing factor to the decline in food accessibility in the station's *sphere of influence* as shown in Fig. 4 (red lines). In response, the city government and developers have made some attempts to encourage older adults to move to buildings in the *vicinity* of stations and to attract young people outward to single-family residential areas away from stations. However, authors' interviews with developers revealed that older adults are not moving, and that buildings in the *vicinity* of stations are occupied mainly by young people.

As for food environment exposure, the transition toward a compact city design in the study area has upset the food market balance between in the *vicinity* of station (<500 m) and station's *sphere of influence* (>500 m). Figure 4 (number of stores, sales floor area) suggests that factors causing this market imbalance include differences in store size. In the *vicinity* of stations where compact city policies were implemented, it is observed that there is not only an increase in the number of stores, but also an increase in the sales floor area and number of parking space. It is pointed out that this increases shopping behavior at commercial facilities in the *vicinity* of stations that offer a variety of transportation options and a full range of products and services and encourages the withdrawal of neighborhood stores around residential areas (Chikaraishi et al. 2016). In fact, within the larger *sphere of influence* of stations where compact city policies were implemented, the number of stores has decreased while the sales floor space has remained unchanged. The time lag between the periods of statistical significance also suggests that compact city policies first had impacts in the *vicinity* of stations, and this change may have led later to adverse impacts stations' larger *spheres of influence*. This is seen as another factor to the decline in food accessibility in the station's *sphere of influence* as shown in Fig. 4 (red lines).

This chapter specifically examines the gaps between food supply and demand that are associated with efforts to create compact cities. On the food demand side, we found that accessibility has slightly deteriorated. One possible reason for this is the increasing ratio of older people in the community. On the food supply side, the number of supermarkets in the vicinity of stations where compact city policies are implemented is statistically lower than that in the vicinity of stations where no such policies are in place. It was also suggested that there is a relationship between the vicinity of stations where compact city policies are implemented and the station's sphere of influence where these policies are implemented. Given the above findings, the transition to compact cities may not necessarily contribute to the goal of creating healthy cities.

Acknowledgments This study was done as part of the M-NEX project (No.11314551), a Belmont Forum grant project of the Collaborative Research Area on the Sustainable Urbanization Global Initiative (SUGI)/Food-Water-Energy Nexus. We would like to thank the Jisedai Kogai Machidukuri (Next Generation Suburban Planning) WISE Living Lab for their cooperation as a research base.

References

Adachi M, Suzuki T, Taniguchi M (2012) Perspective of new risk arises at the process of urban compaction. J Jpn Soc Civ Eng 68(2):70–83; *Ser. D3 (Infrastructure Planning and Management)* (in Japanese), 68, 70

Akutsu, K. (1975). The Science of Walking: Overcoming Physical Inactivity. Fumaido. (In Japanese)

Ashton J, Grey P, Barnard K (1986) Healthy cities—WHO's new public health initiative. Health Promot Int 1(3):319–324

Bagan H, Yamagata Y (2012) Landsat analysis of urban growth: how Tokyo became the World's largest megacity during the last 40 years. Remote Sens Environ 127:210; https://www.sciencedirect.com/science/article/pii/S0034425712003653

Barton H, Grant M (2013) Urban planning for healthy cities. J Urban Health 90(1):129–141

Chikaraishi M, Nishikawa F, Seya H, Fujiwara A, Zhang J (2016) Modeling shopping destination choice behavior incorporating non-market interactions among neighborhood inhabitants. J Jpn Soc Civ Eng 72(5):I_595–I_605; *Ser. D3 (Infrastructure Planning and Management)* (in Japanese), 72, I_595

Cole H, Shokry G, Connolly JJT, Pérez-Del-Pulgar C, Alonso J, Anguelovski I (2017) Can healthy cities be made really healthy? Lancet Public Health 2(9):e394–e395

Goodwin PB (1976) Human Effort and the Value of Travel Time. Journal of Transport Economics and Policy 10(1):3–15

Hu L, Zhao C, Wang M, Shiliang S, Weng M, Wang W (2020) Dynamic healthy food accessibility in a rapidly urbanizing metropolitan area: socioeconomic inequality and relative contribution of local factors. Cities 105:102819

Iwama N, Tanaka K, Komaki N, Ikeda M, Asakawa T (2016) Mapping residential areas of elderly people at high risk of undernutrition. J Geogr 125(4):583–606; (in Japanese)

Nakamichi K, Kiriyama K, Hanaoka S (2019) A study on the target household type of residential relocation to promote compact cities focusing on life stage changes. J City Plan Inst Jpn 54(3):680–687; (in Japanese)

Nakamura E, Asami Y (2019) The Food Desert caused by the economic difficulty of access. J Archit Plan 84(756):437–445; (in Japanese)

Taniguchi M, Hashimoto S, Fujii K, Kanai T, Ochiai J (2011) Study on new risks lurking in future urban layout: the problem of retreating Urban Services. J Jpn Soc Civ Eng 67(5): 67_I_263–67_I_269; *Ser. D3 (Infrastructure Planning and Management)* (in Japanese), 67, 67_I_263

Terayama K, Odani M (2015) Study on a Method for Measuring Individual Mobility Using Indies of Availability of Travel Modes. Journal of Japan Society of Civil Engineers, Ser. D3 (Infrastructure Planning and Management), 71(5), I_481– I_491. (In Japanese)

Walker RE, Keane CR, Burke JG (2010) Disparities and access to healthy food in the United States: a review of food deserts literature. Health Place 16(5):876–884

Urban Agriculture as a Tool for Adapting Future Cities

Keidai Kishimoto and Wanglin Yan

Abstract Urbanization, industrialization, and improvements in transportation have meant that most produce (fruits and vegetables) no longer needs to be produced within the cities where they are consumed. It can be more efficient to grow food intensively and on a large scale in rural areas. The separation of agriculture and cities raises issues about energy consumption for transport and massive use of pesticides and fertilizers in rural areas while the loss of farmland reduced food security in overcrowding cities. The benefits of urban agriculture have been attracting more attention in recent years, especially since the COVID-19 pandemic in 2020. While the capacity of urban agriculture is limited yet, it can indeed provide beneficial ecosystem services such as supplying healthy and fresh vegetables to urban dwellers, mitigating heat island and climate change, offering recreational opportunities, and alleviating disaster risks, etc. Japan's Basic Act on the Promotion of Urban Agriculture, enacted in 2015, recognized the importance of urban agriculture. This chapter will first review the trends and related policies of urban agriculture in Japan, and then introduces the services and functions of urban agriculture in Tokyo from the urban planning perspective.

Keywords Urban agriculture · Land use change · Ecosystem services

K. Kishimoto (✉)
Graduate School of Media and Governance, Keio University, Fujisawa, Kanagawa, Japan
e-mail: keidai@keio.jp

W. Yan
Faculty of Environmental Information Studies, Keio University, Fujisawa, Kanagawa, Japan
e-mail: w_yan@keio.jp; yan@sfc.keio.ac.jp

W. Yan et al. (eds.), *Resilient and Adaptive Tokyo*,
https://doi.org/10.1007/978-981-99-3834-6_9

1 Policies and Trends of Urban Agriculture in Tokyo

1.1 Basic Act on the Promotion of Urban Agriculture

Under Japan's Basic Act on the Promotion of Urban Agriculture, enacted in 2015, central and local governments are mandated to take concrete measures for conserving and promoting urban agriculture. In the past six decades, city planning in Japan was based on the assumption that agricultural land in cities should be converted to residential land uses. However, the Basic Act shifted the paradigm (Ishihara 2019), to the view that urban agriculture is a valid land use, and that agriculture does belong in cities (Ministry of Agriculture, Forestry and Fisheries and Ministry of Land, Infrastructure, Transport and Tourism 2016). The Basic Act conveys the messages that (1) urban agriculture should be conserved in order to deliver various ecosystem functions, (2) city planning should aim to make a better environment through coordination of built-up land uses and agricultural land uses via the multifaceted functions of urban agriculture, and (3) cities should promote urban agriculture in ways that are compatible with local conditions (Ministry of Agriculture, Forestry and Fisheries and Ministry of Land, Infrastructure, Transport and Tourism 2016).

Accompanying the Basic Act, the Basic Plan on the Promotion of Urban Agriculture listed six functions of urban agriculture: (1) supplying agricultural production, (2) preparing for disasters, (3) forming favorable landscapes, (4) maintaining the national land and environment, (5) providing opportunities for farming experience, learning, and communication, and (6) enhancing understanding of agriculture (Table 1). Some functions are unique to Japan (Sect. "Services of Urban Agriculture in Tokyo"), but most functions have been demonstrated by researchers around the world who have studied urban agriculture (e.g., Lovell 2010; Peng et al. 2015; Shinji 1995; Azuma 1995).

Accordingly, local governments have reacted with ten measures as listed below.

Table 1 Expected benefits of urban agriculture in Japan. (Source: Ministry of Agriculture, Forestry and Fisheries and Ministry of Land, Infrastructure, Transport and Tourism (2016))

Benefits	Examples
Agriproducts	Fresh local produce
Disaster	Fire prevention, shelters, temporary housing sites
Scenery	Relaxation
Environment	Rainwater infiltration, groundwater recharge, biodiversity conservation
Society	Experiencing and learning about agriculture, exchanges between producers and residents
Understanding	Fostering understanding of agriculture and agricultural policy

1.1.1 Improving the Supply of Agricultural Products and Attracting/Fostering Farmers

Governments support the matching of farmland lessors and lessees and the entry into agriculture by social services and education. Some farmers employ persons with disabilities in an effort to support agricultural management and encourage their participation in society, while they also gain a sense of purpose and confidence by engaging in agriculture. Governments also support the introduction of screens, greenhouses, hedges, and appropriate waste disposal to prevent pesticide spraying, odors, noise, and dust from affecting surrounding nearby residential areas. For example, Tokyo Metropolitan Governments supported 63, 56, and 39 farm households in 2013, 2014, and 2015, respectively, to introduce facilities (Tokyo Metropolitan Government (a) (n.d.), (b) (n.d.), (c) (n.d.)). Some agricultural products have brand names unique to the region, as branding can help spotlight and preserve the local culture.

1.1.2 Promoting Disaster Prevention, Landscape Enhancement, Land/Environmental Conservation

Governments conclude agreements with related organizations and including farmland as part of regional disaster prevention plans. Farmland can play a role in fire prevention and as temporary evacuation sites.

Governments also promote agriculture-related landscapes (e.g., traditional groves around farmhouses, irrigation watercourses, etc.) in landscape plans, special green space preservation areas, and green master plans.

1.1.3 Developing Accurate Land Use Plans

Governments incorporate farmland (large areas, long-term use) into Urbanization Control Areas. They also include farmland into the site optimization plans and master plans to control future land use. In 2018, "Countryside Residential Districts" is introduced in Japanese zoning system, which previously consists of 12 categories to regulate the types of land use in urban areas. In Countryside Residential Districts, developments are prohibited without governmental permission to control the balance between agriculture and living environment. This is expected to support managing urban agriculture (Akashi 2018).

1.1.4 Tax Measures

Farmers let their farmland registered inside Urbanization Promotion Area as Productive Green Space for reaping favorable tax, if they intend to keep farming for

30 years. When farmland was not so designated, farmers would pay taxes equivalent taxes on residential land. The Act on Productive Green Space was enacted in 1991 and revised for the first time in 2018. The government has relaxed the conditions for registration and allowed the construction of farmer restaurants and farmer's market inside Productive Green Space.

1.1.5 Promoting Local Consumption

Governments support the expansion of sales channels, including fresh produce stands and restaurants. They also promote the use of local products in school lunches and social services facilities. For example, in Tachikawa City, vegetables grown in the local account for 16.8% of the total amount of vegetables purchased for school lunches in 2020 (Tachikawa City (n.d.)).

1.1.6 Developing Programs/Conditions for Agricultural Experience

Governments dispatch experts to create programs, and provide supports for maintenance of allotment gardens and experience farms. From the 1960s to the 1980s, today's allotment gardens began to develop in Japan by utilizing vacant farmland, and in 1990 various acts were enacted to manage allotment gardens for improving urban environment (Kudo 2009). The 432 allotment gardens were provided in the end of March 2018 by local governments (84.8%), farmers (10.7%), etc. (Tokyo Metropolitan Government (e) (n.d.)). Efforts to expose urban residents to urban agriculture provide opportunities for them to experience and learn about farming.

1.1.7 Enhancing Opportunities for Agricultural Experience in School Education

Some local governments incorporate year-round agricultural experience into curricula and extracurricular activities and develop teaching materials and manuals. (In Japan, agricultural experience in school education is limited to planting and harvesting; there is room to improve the effectiveness of learning opportunities).

1.1.8 Promoting Public Understanding and Interest

Governments advertise urban agriculture through the media and events and provide opportunities for urban residents to experience agriculture and agricultural products through agricultural festivals and farmers' markets.

1.1.9 Promoting Agricultural Learning and Skills Development for Urban Residents

Governments promote initiatives to allow urban residents to participate in urban agriculture as volunteer farmers, attracting urban residents who would like to be involved in agriculture due to a growing interest in farming and health promotion.

1.1.10 Promoting Research

Governments promote the development of basic data and statistics on urban agriculture, as well as empirical evaluation of the effects of measures and model cases.

1.2 A Look at Tokyo and Municipalities

Plans for urban agriculture promotion are formulated in each prefecture and municipality. In Tokyo, because urban agriculture includes not only allotment gardens and educational farmlands, but also many professional farming operations, professional farmers play an important role in determining whether or not urban agriculture is conserved and promoted.

The Tokyo Metropolitan Government encouraged urban agriculture and promoted the concept to the national government even before the adoption of the Basic Act on the Promotion of Urban Agriculture. The Tokyo Metropolitan Government aims to have urban agriculture coexist with cities and contribute to the lives of a large urban population (Tokyo Metropolitan Government 2017). It has basic guidelines that promote (1) fostering new farmers and developing strong management styles, (2) preserving farmland and fulfilling its multifunctional roles, (3) promoting sustainable agricultural production and local production for local consumption, and (4) promoting agriculture that takes advantage of regional characteristics (Tokyo Metropolitan Government 2017).

Each municipality positions its agricultural promotion plan in response to national and Tokyo Metropolitan Government plans, higher-level plans such as urban master plans, and related plans such as green master plans and basic plans for industrial promotion. In Tokyo, 27 out of 49 cities had agricultural promotion plans or similar plans as of January 2020. Most of these plans are to be reviewed every 5 years and revised every 10 years. Some municipalities that do not have a plan specifically for agriculture do have an equivalent plan (such as a detailed policy for the agricultural sector as part of an industrial development plan).

Comparing the plans of some municipalities in Tokyo on the promotion of urban agriculture, we found that the major cities emphasized the relationship between agriculture and urban/community development, while the suburban municipalities emphasized support for agriculture as an industry, such as local consumption and agricultural management support (Arakawa and Akita 2021). Urban areas with large populations and fragmented farmlands of small area tended to focus on the significance of agriculture for their citizens, while suburban areas tended to focus on developing the agricultural industry. In basic policies as well, urban areas tended to focus on the general public, while suburban areas tended to focus on agriculture as an industry. In addition, while value-adding through branding is more prevalent in large cities, value-adding in suburban areas is promoted through the integrated promotion of the growing, processing, and consumption of produce (referred to as "sixth industry").

1.3 Evolution of Urban Agriculture Policy in Japan

In Sects. "Basic Act on the Promotion of Urban Agriculture" and "A Look at Tokyo and Municipalities", we confirmed that the Basic Act on the Promotion of Urban Agriculture established a policy for promoting urban agriculture and related policies. There was a paradigm shift in the policy approach toward urban agriculture in Japan, replacing the perspective that farmland in urban areas should be converted to residential uses with the current perspective that farmland should belongs in the urban environment. Historically, urban agriculture was actively excluded and converted to residential areas, but the past attitudes were gradually relaxed, and today, urban agriculture is being conserved and promoted. These processes resulted in the current landscape of urban agriculture in Tokyo. It is necessary to understand this as the background for future developments. This section outlines the historical evolution of urban agriculture policy in Japan.

During the period of high economic growth after World War II, Tokyo Metropolitan Area had concentrated industries and populations extensively, leading to a proliferation of sprawling development. Due to a serious shortage of housing in urban areas, farmland was being converted to residential land. Satoyama (traditional Japanese rural landscape with a mosaic type of land uses including semi-natural ecosystems) (Ichikawa et al. 2006; Takeuchi et al. 2003) were also being modified to accommodate new towns and housing complexes all over Tokyo. In 1968, the City Planning Act introduced a zoning system to limit urbanization. This zoning system divided urban areas into Urbanization Promotion Areas and Urbanization Control Areas, encouraged the conversion of Urbanization Promotion Areas to residential land within 10 years, and encouraged the conversion of agricultural land.

However, with rapid urbanization, the cities were getting overcrowded, and green space sharply declined in the 1970s and 1980s, and living environment worsened. Furthermore, the so-called bubble economy (asset inflation) encouraged the development of suburban areas and dramatically raised land prices. Based on the increasing demand for land, the attention of land use policy has shifted to the interests for better living environment. The Act on Production Green Space enacted in 1991 set the scene for a new urban agriculture scheme by dividing urban agriculture into (1) what should be converted to residential land, and (2) what should be conserved as

Productive Green Space. In order to take advantage of the new system, many farmlands were designated as Productive Green Space (PGS) in 1992. Those contractors of PGS are now facing either to renew or release to residential market because of land succession.

After the collapse of the bubble economy, the Tokyo Metropolitan Area faced a new phase. Urbanization process has being shifted to re-urbanization (Kanda et al. 2020; Ushijima 2012) as described by Klaassen's model (Klaassen et al. 1981). In these shrinking urbanization stages, society came to expect urban agriculture multiple functions rather than paved built-up areas. As a precursor, the Food, Agriculture, and Rural Areas Basic Act in 1999 indicated that the government should implement the necessary measures to promote agriculture to meet urban demand by taking advantage of proximity between farmland and the areas where farm output would be consumed.

2 Agricultural Land Use Patterns in Tokyo

As described above, along with social maturity, the approach toward urban agriculture in Japan has gradually changed from viewing farmland as something that should be converted into residential land, to viewing farmland as belonging in cities. However, the evolution of attitudes toward land in cities has produced a fragmented distribution of small-scale farmland, and weak urban agriculture. The distribution pattern of farmland has changed dramatically in accordance with individual circumstances, social conditions, and government policies toward urban agriculture. To get fruitful services from urban agriculture, it is important to protect existing farmland. In this section, we will examine the situation of existing farmland, the significant decrease of farmland in the past, and current distribution patterns.

2.1 Farmland Loss and Conversion

The area of agricultural land in Tokyo has decreased by about half over the past 40 years (Fig. 1). In particular, the area of rice paddy has decreased significantly, from 7360 ha in 1975 to just 249 ha in 2015, and the area of field has decreased from 13,200 ha in 1975 to 6470 ha in 2015. In the meantime, despite several policies changed as mentioned above, the loss has continued steadily. Figure 2 shows that the area of field designated as Production Green Space decreased by about 20% over the 20 years from 1995 to 2015, while two-thirds of the field not designated as Productive Green Space was converted to residential land. In other words, farmland loss continues even after the amendment of the Act on Productive Green Space, however, the more significant decrease is attributed to non-registered farmland as Productive Green Space (Ishihara 2014).

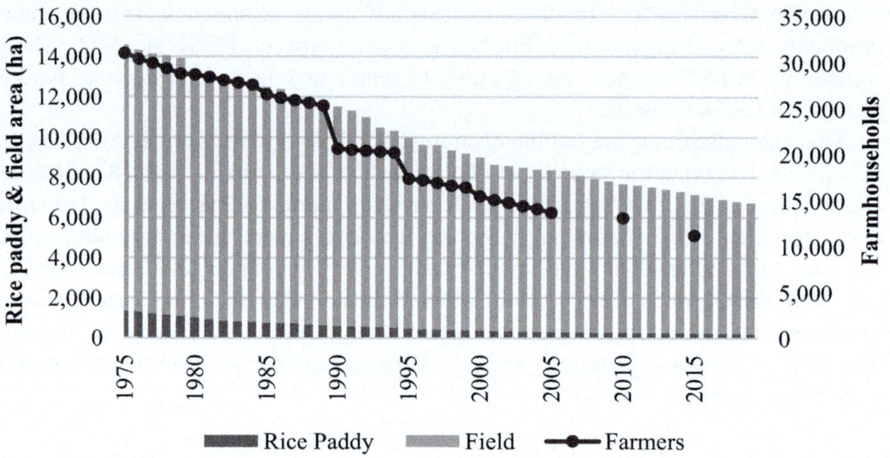

Fig. 1 Changes in rice paddy area, field area, and farm households in Area in Tokyo Metropolitan Government (1975–2019) (data source: Kanto Regional Agricultural Administration Office (n.d.))

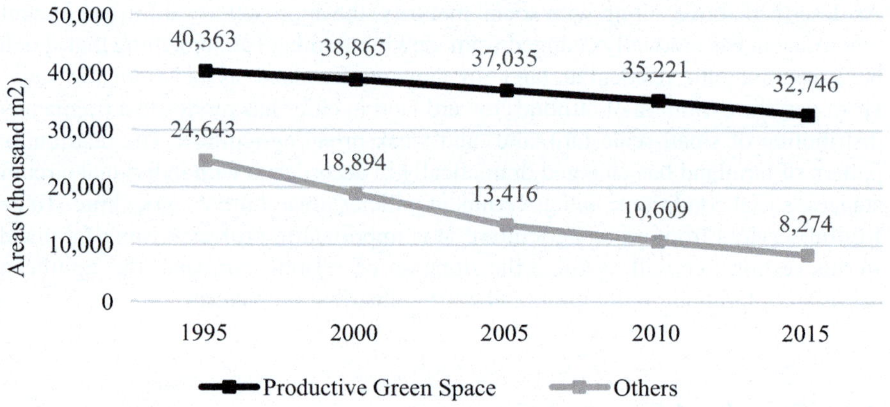

Fig. 2 Changes in Productive Green Space and other farmland inside urbanization promotion area in Tokyo (1975–2015) (data source: Tokyo Metropolitan Government (f) (n.d.), (g) (n.d.))

The factors behind farmland loss are probably related to competition with other land uses based on economic rationality, and the decreasing number of farmers due to aging and a lack of successors. In terms of competition with other land uses, there is high pressure to develop land into residential and commercial facilities, especially around railway stations. As a result, farmlands with high land prices and small parcel area are sold and converted mainly to housing (Nakahara and Hoshino 2006) as shown in Fig. 3. In Nerima City, a Tokyo suburb, agricultural land was being converted for infrastructure development in the 1960s, condominium development in the 1980s, and transitional land uses such as parking lots and storage lots, as well as supermarkets and convenience stores since the 1990s (Takatori 2000). In recent

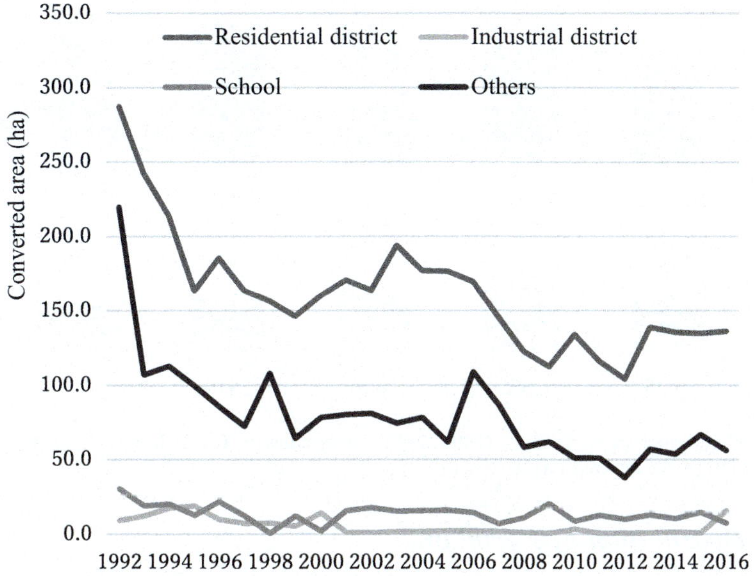

Fig. 3 Land use converted from farmland (1992–2016) (data source: Tokyo Metropolitan Government (h) (n.d.))

years, some land has been converted to parks and welfare facilities, reflecting the demand for green space and facilities to meet the societal needs of an aging population (Satake and Saio 2018).

As shown in Fig. 1, the number of farming households decreased by about one-third over 40 years to 2015. Looking at the age composition of the farming population, the percentage of farmers aged 65 and above increased from 48% to 54% between 1995 and 2015, while the percentage of farmers under 30 years old was steady at only 2% (Fig. 4). Furthermore, there are many reports of farmers giving up their farmland due to this aging of the farming population. In some cases, individual farmers leave their farmlands due to physical problems or injuries, and in other cases, entire families give up their farmland when farmers die. As for inheritance, "mini-development" (subdividing land to liquidate the property or settle inheritances) often occurred around cities, leading to fragmentation into smaller parcels, and loss of farmland (Nakahara and Hoshino 2006).

In 2022, many farmland parcels that were registered in 1992 as PGS will be exit from the 30-year obligation to cultivate and the benefit of tax incentives. There are concerns that many farmers will not want to re-designate their farmlands as PGS and will seek to convert them to residential uses. Various measures have been implemented, such as easing the conditions for the designation of PGS and approving of farmer-run restaurants, but it is likely that the effectiveness of these measures will not be immediately evident.

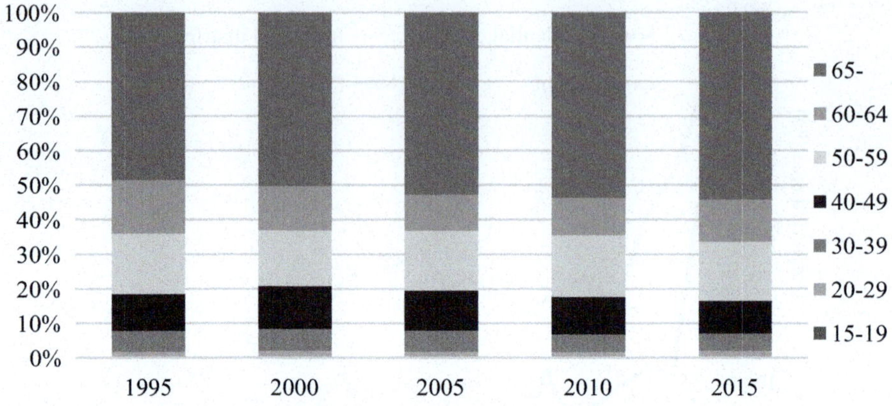

Fig. 4 Demographics of farmers (1995–2015) (data source: Tokyo Metropolitan Government (i) (n.d.), (j) (n.d.), (k) (n.d.), (l) (n.d.), (m) (n.d.))

2.2 *Distribution of Farmland*

In Tokyo, the Japan Railway (JR) Yamanote Line is an elongated circular line that connects the central areas and sub-centers of Otemachi, Shinjuku, Ikebukuro, and Shibuya, a number of railways radiate out from the JR Yamanote Line toward the suburbs. Commuters walk or bus to suburban stations, and take trains to central areas. Therefore, by looking at the distribution of agricultural land in Tokyo in terms of distance from the JR Yamanote Line and distance from the nearest railroad station, one can begin to grasp the relationships between land uses in the urban structure.

Within 10 km outside the JR Yamanote Line, urbanization is significant except for the coastal areas and rivers, while farmland only exists far from stations. In this zone, agricultural land has been fragmented into small farmland parcels due to high urbanization pressures due to proximity to the city center. Meanwhile urban agriculture provides cities with leafy vegetables, flowers, and many varieties of plants, benefiting from proximity to the city center. Figure 5a shows a typical farmland landscape in this zone. Mini-developments have often resulted in the conversion of agricultural land to housing, parking lots and supermarkets (Nagae 2007; Takatori 2000), but many farm parcels are still located in this zone.

Within about 20 km of the JR Yamanote Line, there are densely populated areas around the railroad lines and in some surrounding areas. As shown in Fig. 5b, urban agriculture produces not only leafy vegetables but also trees and plants that might be more typical in suburban areas.

Within a 30 km radius of the JR Yamanote Line, urbanization is progressing in the vicinity of railroad stations, and factories and distribution centers are located in the suburbs. Although this area is in competition with urbanization and development, the mix of agricultural products from here is somewhat different from areas within the 10 km or 20 km radius because of the relatively large parcels and the

Fig. 5 The faces of urban agriculture in Tokyo. Changes in landscapes from city center to the suburban areas are depicted from photos (**a**–**f**). Aerial photos are from Geospatial Information Authority of Japan and cropped by the authors

distance from city center. Depending on the region, root crops, fruit trees, and flowers are cultivated. However, as shown in the example of Kiyose City in Fig. 5c, agricultural land is being converted to other uses due to inheritance issues and urbanization encroaching from the city center.

Within a 40 km radius of the JR Yamanote Line, urbanization is advancing around railroad stations, while large parks, military bases, and factories are located in the surrounding areas. The surrounding flatland and hilly areas are densely populated. In the northern part of the region, farmlands are located on the flatlands to keep separate from factories and bases. In the western region of Tokyo, in the southern part of the Tama Hills, rice production is done in small valleys while vegetable cultivation, mainly tomatoes and eggplants, is done mainly in the northern part of Machida City (Fig. 5d). Urban agriculture in this region varies significantly based on regional factors and topographical features such as hills and river valleys, and large-scale developments such as a military base and Tama New Town.

The area within 50 km includes foothills and the western edge of the Musashino Plateau. Urbanization has taken place along the JR Ome Line and around Hachioji City, and farmland is distributed on flat and extensive hills and terraces that skirt cities and mountains. In Akiruno City, shown in Fig. 5e, farmland is distributed on terraces and hills along the Akigawa River, and tomatoes, sweet corn, and chestnuts are the major crops. In these areas, leafy vegetables account for less of the farm produce by weight than in other distance zones, while the cultivation of fruit vegetables and potatoes is large.

Outside the 50 km radius, the area is predominantly mountainous and has low population density. Farming is conducted on mountain slopes and terraces, but most of it is outside of built-up areas, so it is not considered to be urban agriculture (Fig. 5f).

As described above, an examination of urban agriculture in Tokyo based on distance from the city center reveals differences in the farming environment depending on the degree of urbanization. Within a 20 km radius, farmland appears fragmented into small parcels due to high demand for land, while within a 30 km and 40 km radius, relatively large parcels remain, although those areas are facing challenges in finding farming successors and gradual urbanization. Within a 50 km radius, large parcels of farmland remain, with landscape features of suburban farming.

3 Services of Urban Agriculture in Tokyo

We have confirmed that urban agriculture, which has been regarded as belonging in cities, is shrinking due to internal and external pressures such as competition for land use and a decrease in the number of farmers. Therefore, to ensure that urban agriculture survives in cities, it is important to evaluate the various services that urban agriculture provides, articulate its significance to society, and take concrete actions such as urban planning and individual projects.

3.1 Production and Resource Use by Urban Agriculture in Tokyo

In Tokyo, 78,895 tons of vegetables were produced in 2019, yielding an agricultural output value of 18.9 billion yen. The population of Tokyo is about 14 million, so vegetable output per capita works out to 5.6 kg. Since the average daily consumption of vegetables per Japanese person is 269.8 g (Ministry of Health, Labour and Welfare, 2020), vegetables produced in Tokyo are equivalent to 20.9 days of vegetable consumption. The national health promotion policy Health Japan 21 set a target daily vegetable consumption of 350 g; in that context, Tokyo produced enough vegetables for the equivalent of 16 days.

Next, we compare the production of vegetables by category (Fig. 6). The following 14 vegetables are commonly consumed in Japan: barley, carrot, cabbage, Chinese cabbage, cucumber, eggplant, leek, lettuce, onion, radish, spinach, sweet pepper, taro, and tomato. Tokyo produces many of the common vegetables such as cabbage, eggplant, radish, and tomato, but does not widely produce bell pepper, lettuce, onion, and taro.

Meanwhile, leafy vegetables such as komatsuna (sometimes referred to as Japanese mustard spinach), udo (sometimes referred to as Japanese spikenard or mountain asparagus), and wasabi (sometimes referred as Japanese horseradish) are specialty products in Tokyo. Areas surrounding the urban area of the Edo period (1603 to 1867) consumed large amounts of vegetables, and to this day, the areas surrounding the JR Yamanote Line still cultivate komatsuna, spinach, and garnishes, for which freshness is important (Adachi City 2011). In fact, komatsuna originated from Edo and spread throughout Japan, and although present-day Tokyo is gradually ceding production to other prefectures (Ishihara 2015), komatsuna is still one of Tokyo's most important agricultural products. Udo is a vegetable native to Japan, mostly cultivated in holes in the ground where sunlight does not reach (Koito et al. 2018). In Tokyo, udo is mainly cultivated in Tachikawa City and is branded as Tokyo udo (Koito et al. 2018). Although udo is not a common vegetable, it is known as an ingredient in Japanese cuisine. In Tokyo, Okutama Town is famous for wasabi production. Wasabi is a crop used as a condiment for soba noodles and sushi. The cultivation of wasabi requires a very specific environment including clear water, so the number of suitable growing areas is limited even in Japan. Many of these Tokyo specialty vegetables have been produced since the Edo period and are branded as Edo-Tokyo vegetables.

It is important to note, however, that these vegetables consume a large amount of water and energy in their life cycle. The total amount of energy and water used for fertilizers, pesticides, and cultivation facilities amounts to 835.1 terajoule (TJ) and 18.9 megaliter (ML), respectively, or in terms of farmland, 19.1 MJ/m^2 and 0.434 L/m^2, respectively (Kishimoto and Yan 2021). Water- and energy-intensive agriculture is often located close to urban centers, putting a strain on the environment (Kishimoto and Yan 2021). This research is still in its infancy, so it will be important to further investigate and quantify these aspects of urban agriculture in the future.

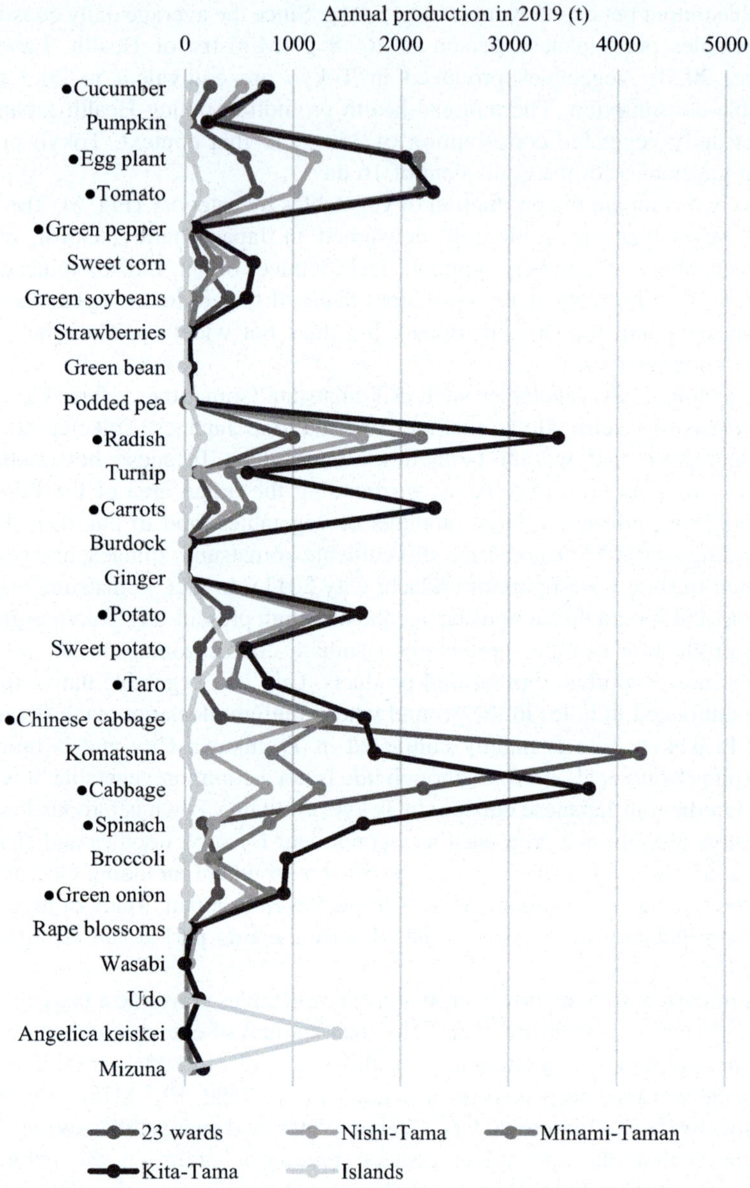

Fig. 6 Vegetable production in each selected areas in 2019. Commonly consumed 14 vegetables are with •. (data source: Tokyo Metropolitan Government (n) (n.d.))

3.2 Fresh Produce Stands for Supplying Local Vegetables and Enhancing Understanding of Farming

Agricultural products produced in Tokyo are consumed via direct sales, in school lunches at elementary and junior high schools, through distribution at produce markets, and through contract shipments to mass merchandisers. Tokyo has a large number of fresh produce stands that sell locally-produced vegetables. Fresh produce stands are operated by a variety of entities, and farmers themselves account for 26% of total flows of vegetables, and Japan Agricultural Cooperatives (JA) and municipalities 20% (Tokyo Metropolitan Government 2017). About 70% of citizens reportedly want to buy local vegetables (Tokyo Metropolitan Government (d) (n. d.)). Fresh produce stands operated directly by farmers often use vending machines or are unmanned stands with payments done on an honor system (Fig. 7). Farmers and residents interact either directly or indirectly in the processes of packaging, pricing, and selling vegetables. Meanwhile, fresh produce stands operated by JA and

Fig. 7 Fresh produce sales in Tokyo (unmanned stand, vending machine)

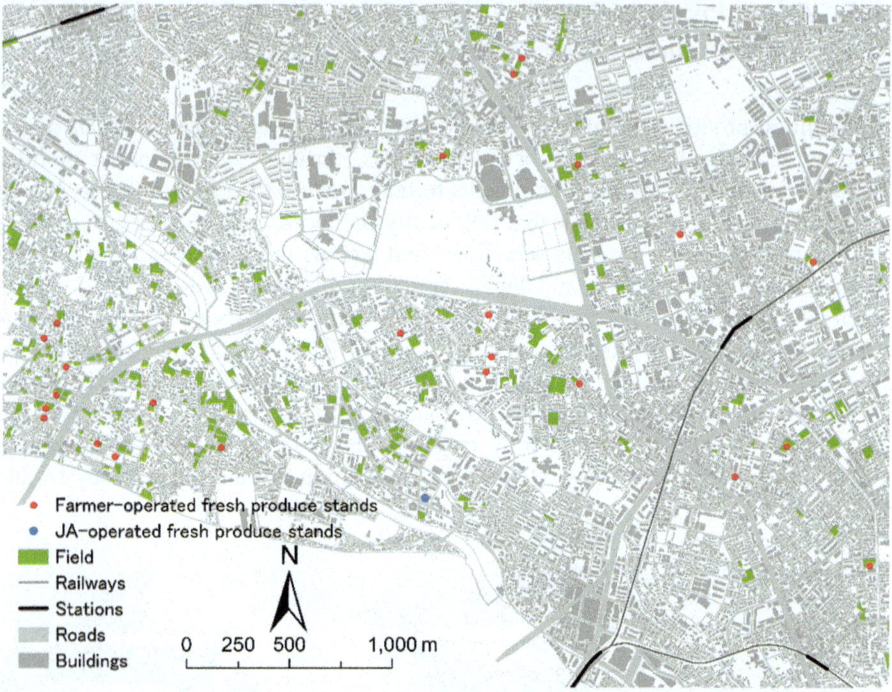

Fig. 8 Distribution of fresh produce stands in southern Setagaya area (source: Authors created using land-use data from Setagaya City)

municipalities have large shop areas where people can enjoy a variety of local vegetables.

Fresh produce stands are clearly an important nexus connecting citizens and urban agriculture. They enhance awareness and understanding of urban agriculture and local vegetables. In many cases in Japan, the location of fresh produce stands is published on local government websites or in official booklets. Figure 8 shows the distribution of fresh produce stands in the southwestern part of Setagaya City. This area is located about 15 km from Tokyo's major sub-center of Shibuya, or about 15 min by train, which is very convenient for access. Even though the area is highly urbanized, we identified about ten fresh produce stands within only a few kilometers of stations. Each farmer sets up a simple store or vending machine on farmland or attached to the house, to sell harvested vegetables. Individual farmers typically have different production schedules, harvest different vegetables, and use their own techniques to make vegetables tasty. JA operates two fresh produce stands in Setagaya City, which sell locally-produced vegetables, sandwiches and jams, thereby promoting primary, secondary, and tertiary industries (so-called sixth industry).

3.3 Farmland Based Disaster Risk Reduction

Urban agriculture can also play a role in coping with disaster risks, since the farmland itself is valuable open space in the midst of cities and agricultural facilities can be utilized in emergencies. Farmland can be utilized as sites for temporary evacuation or temporary housing. Urban farmland can provide quickly accessible areas where people can escape to safety in the event of an earthquake or a fire. As of the end of March 2015, 56 municipalities in Japan's three major metropolitan areas (Nagoya, Osaka, Tokyo), including 29 in Tokyo, had signed agreements with farmers on cooperation in the event of a disaster (Ministry of Agriculture, Forestry and Fisheries 2015). A study in Sakai City in Osaka reported that the city has the capacity to temporarily accommodate all residents who live within a 400-m radius of farmland (Hara et al. 2016).

Groundwater wells on farmland can also provide a valuable source of water in times of disaster. The so-called disaster cooperation wells are part of an initiative to make wells accessible to local residents free of charge in the event of a disaster. For example, Yokohama City has 2529 and Chiba City 151 disaster cooperation wells (Endo et al. 2020), some of which are probably originally for agricultural use.

Because Japan has experienced so many natural disasters throughout its history, the disaster response function has been integrated into urban agriculture.

4 Discussion and Conclusion

This chapter introduced Tokyo's approach of urban agriculture to resilient and adaptive cities, which is based on the idea that agriculture does indeed belong in cities. We described the evolution of policies and society in terms of urban agriculture, the actual situation of farmland distribution and status of farmers, and services provided by urban agriculture. Here we discuss three key aspects of how urban agriculture will need to adapt to social, economic, and ecological change in the future.

The first issue is the need to adapt to social change in terms of population decline and aging. For centuries, the Tokyo Metropolitan Area has attracted a growing population, but is projected to enter a phase of population decline in 2035 (National Institute of Population and Social Security Research 2018). Meanwhile, the ratio of seniors (aged 65 and older) is projected to increase from 22.7% in 2015 to 30.7% in 2045 (National Institute of Population and Social Security Research 2018). These changes will lead to a shortage of successors for farmers. Previous urban planning to address demographic change included optimization of land uses and compact cities as a part of land use plans, with a focus on housing and city centers. Previous schemes were not able to conserve farmland effectively, so city planning from now on requires more discussion about issues such as land use and successors of farming.

Second, urban agricultural land is increasingly being recognized as multiple-use public space for local communities rather than mainly being for economic activities by farmers (Miyachi 2006). Currently, farmers own the land and do the farming, but urban residents also benefit from urban agriculture and wish to preserve it. There may be insufficient linkages between the supply and demand sides of these services. As it stands today, the conservation of farmland depends on the will of farmers, but if farmland is to be used more broadly as a "public space" for the improvement of the urban environment, systems of cooperation and support will be necessary to encourage farmers to continue farming, combined with policies and initiatives by local governments to conserve farmland. For example, Multifunctional Payment Systems is an initiative to support activities to maintain farmland providing multiple functions and improve the quality of local resources (Ministry of Agriculture, Forestry, and Fisheries, 2022). It includes mowing the slopes of farmland, mucking up waterways, repairing cracks in waterways, and planting trees. There may also be ways to combine the activities of the private sector and non-profit and community organizations rather than relying solely on local governments. In response to public demand in recent years from people who want to enjoy farming accessibly, an increasing number of private allotment gardens has appeared, where prices may be high but extensive guidance and support is also provided.

Third, the situation surrounding urban agriculture has been changing since the beginning of the COVID-19 pandemic. Some newspapers have reported that more and more people have been visiting produce stands to obtain fresh and inexpensive vegetables. Some people have moved out of the crowded Tokyo central area to suburban and rural areas where they can work remotely from home while also enjoying more comfortable living environments. As a result of these changes of perspective and behavior, many people have been benefitting from urban agriculture. Further research is needed to clarify what role urban agriculture has played in responses to the COVID-19 pandemic, but it is likely that urban agriculture will attract more attention as an adaptive strategy for the "new normal" in Tokyo.

Acknowledgments This chapter was supported by the JSPS KAKENHI [grant number JP22J13885]; and the Belmont Forum's Sustainable Urbanization Global Initiative (SUGI): Food-Water-Energy Nexus/M-NEX project [grant number 11314551, Japan Science and Technology Agency (JST): 1009781].

References

Adachi city (2011) Adachi nogyo shinko plan (plan for promotion of nogyo in Adachi) [translated from Japanese]

Akashi T (2018) Current situation of property tax on mingled farmland in urban area and availability of new established "garden residential zone". Urban Hous Sci 101:22–27; (in Japanese with English abstract)

Arakawa A, Akita I (2021) Characteristics of policies based on basic law for urban agricultural promotion formulated by local governments located in urbanization promotion area of Tokyo. J Instit Landsc Archit 84(5):575–580; (in Japanese with English abstract)

Azuma M (1995) Nouchi-Wa ryokuchi-ka (is farmland green space?) [translated from Japanese]. Toshi Mondai 86(12):41–52

Endo T, Mori Y, Oki T (2020) Current situation and future challenges for disasater emergency wells in ordinance-designated cities in Japan. J Jpn Soc Hydrol Water Resour 33(4):144–155; (in Japanese with English abstract)

Hara Y, Yoshii T, Tsujimura K, Sampei Y (2016) Land use and land ownership characteristics of designated disasater evacuation farmland in Sakai City, Japan. Memories of the Center for Research and Education of Disaster Reduction, Wakayama University 2:31–37

Ichikawa K, Okubo N, Okubo S, Takeuchi K (2006) Transition of the satoyama landscape in the urban fringe of the Tokyo metropolitan area from 1880 to 2001. Landsc Urban Plan 78(4): 398–410. https://doi.org/10.1016/j.landurbplan.2005.12.001

Ishihara H (2014) Change of the agriculture in the city of Tokyo after 1990. Bull Geo-Environ Sci 16:21–36

Ishihara H (2015) Continuing strategy of market shipment type Komatsuna (Brassica campestris L. rapifera group) production center in Edogawa-ku, Tokyo. Bull Geo-Environ Sci 17:83–100; (in Japanese with English abstract)

Ishihara H (2019) Toshinougyou-Wa minnade sasaeru jidai-e (toward an era of supporting urban agriculture together) [freely translated from Japanese]. Kokin Shoin

Kanda H, Isoda Y, Nakaya T (2020) The transition of urban structure in Japan in the phase of decreasing population. Quart J Geogr 72:91–106. https://doi.org/10.5190/tga.72.2_91; (in Japanese with English abstract)

Kanto Regional Agricultural Administration Office (n.d.) Shuyo nougyo shihyo-no ruinen toukei (Cumulative Statistics of Major Agricultural Indicators) [freely translated from Japanese]

Kishimoto K, Yan W (2021) Performance of urban agriculture in Tokyo: a geospatial perspective of the food-water-energy nexus. *REAL CORP 2021 Proceedings*, p 577–586. https://doi.org/10.48494/REALCORP2021.1087

Klaassen LH, Bourdrez JA, Volmuller J (1981) Transport and reurbanisation. Gower Press, Aldershot

Koito Y, Miyashita C, Numajiri K, Yano T, Oura Y (2018) Blanched udo produced in Tokyo: field research for the promotion of production and consumption. Bull Tokyo Metropol Agric Forest Res Center 13:81–112; (in Japanese with English abstract)

Kudo Y (2009) Historical change of the Japanese community garden and its publicness. J Archit Plan 74(643):2043–2047; (in Japanese with English abstract)

Lovell ST (2010) Multifunctional urban agriculture for sustainable land use planning in the United States. Sustainability 2:2499–2522

Ministry of Agriculture, Forestry and Fisheries (2015) Sandai-toshiken tokuteishi-ni-okeru bousai-kyoryoku-nochi-tou-no torikumi-jokyo (Status of initiatives for designated disasater evacuation farmland in ordinance-designated cities in three metropolitan cities) [freely translated from Japanese]

Ministry of Agriculture, Forestry and Fisheries (2022) Nihon-gata chokusetsu shiharai seido (Japanese Direct Payment System) [freely translated from Japanese]. Accessed 5 Sep 2022. https://www.maff.go.jp/j/nousin/tamen5/tameniinkai/attach/pdf/haihushiryou-15.pdf

Ministry of Agriculture, Forestry and Fisheries and Ministry of Land, Infrastructure, Transport and Tourism (2016) Plan for promotion of urban agriculture

Ministry of Health, Labour and Welfare (2020) The National Health and Nutrition Survey in Japan, 2019

Miyachi T (2006) Kaisei-seisanryokuchihou-ka-no toshi-nogyo-no tenkai (dynamics of urban agriculture under the rivised production green space law) [freely translated from Japanese]. Geogr Rep 103:1–16; (in Japanese)

Nagae M (2007) Development of farmland for housing lot in Setagaya ward and agricultural committee. 1960-1975. J Soc Sci 58(3–4):101–120

Nakahara Y, Hoshino S (2006) The present condition and problems on urban farmland: case study of the west Ward in Kobe City. J Rural Plan Assoc 25:437–442

National Institute of Population and Social Security Research (2018) Regional population projects for Japan: 2015-2045. Population Research Series, 340 (in Japanese)

Peng J, Liu Z, Liu Y, Hu X, Wang A (2015) Multifunctionality assessment of urban agriculture in Beijing City, China. Sci Total Environ 537:343–351

Satake H, Saio N (2018) Study on the conservation of farmland in urban areas based on transformation of productive green spaces and diversification of agricultural management: a case study of Nerima Ward, Tokyo. J City Plan Inst Jpn 53(3):522–528. https://doi.org/10.11361/JOURNALCPIJ.53.522

Shinji I (1995) Toshi-ni-okeru nougyo nouchi-no yakuwari (role of the agriculture and farmland in city) [freely translated from Japanese]. Toshi Mondai 86(12):3–15

Tachikawa City (n.d.) Jimoto yasai-no gakkou-kyusyoku-eno katsuyo-no suii (Trends of local vegetables use in school lunches) [freely translated from Japanese]. Accessed 2 Sep 2022. https://www.city.tachikawa.lg.jp/gakkokyushoku/kosodate/kyoiku/iinkai/gakko/kyushoku/documents/jimotoyasai28-r2.pdf

Takatori Y (2000) The multi-functional system of urban agriculture in the suburbs of Tokyo: a case study in the Nishi-Ooizumi area of Nerima Ward, Tokyo. J Geogr 109(3):401–417

Takeuchi K, Brown RD, Washitani I, Tsunekawa A, Yokohari M (2003) Satoyama: the traditional rural landscape of Japan. Springer, Tokyo

Tokyo Metropolitan Government (2017) Tokyo agriculture promotion plan

Tokyo Metropolitan Government (a) (n.d.) Heisei-25-nendo toshi-nogyo keiei power-up jigyo (Power-up program for managing urban agriculture in 2013) [freely translated from Japanese]. Accessed 2 Sep 2022. https://www.sangyo-rodo.metro.tokyo.lg.jp/nourin/pdf/powerup/powerupfiles/25jisseki.pdf

Tokyo Metropolitan Government (b) (n.d.) Heisei-26-nendo toshi-nogyo keiei power-up jigyo (Power-up program for managing urban agriculture in 2014) [freely translated from Japanese]. Accessed 2 Sep 2022. https://www.sangyo-rodo.metro.tokyo.lg.jp/nourin/26jisseki.pdf

Tokyo Metropolitan Government (c) (n.d.) Heisei-27-nendo toshi-nogyo keiei power-up jigyo (Power-up program for managing urban agriculture in 2015) [freely translated from Japanese]. Accessed 2 Sep 2022. https://www.sangyo-rodo.metro.tokyo.lg.jp/nourin/27jisseki.pdf

Tokyo Metropolitan Government (d) (n.d.) Tosei monitor questionnaire kekka (Questionnaire survey results about agriculture in Tokyo in 2015) [freely translated from Japanese]. Accessed 12 Aug 2021. https://www.metro.tokyo.lg.jp/tosei/hodohappyo/press/2015/08/60p8a100.html

Tokyo Metropolitan Government (e) (n.d.) Heisei-30-nendo shimin-nouen-tou chousa kekka (Survey results of allotment gardens and community gardens at the end of March, 2018) [freely translated from Japanese]. Accessed 2 Sep 2022. https://www.agri.metro.tokyo.lg.jp/files/shimin/3120E5B882E6B091E8BEB2E59C92.pdf

Tokyo Metropolitan Government (f) (n.d.) Takuchika nouchi-no suii (Trends in field area) [freely translated from Japanese]

Tokyo Metropolitan Government (g) (n.d.) Seisan ryokuchi chiku-no suii (Trends in Productive Green Space) [freely translated from Japanese]

Tokyo Metropolitan Government (h) (n.d.) Nouchihou dai-4-jou oyobi dai-5-jou-niyoru youtobetsu nouchi tennyou menseki-no suii (Trends in converted field area by land use based on the Agricultural Land Act no.4 and no.5) [freely translated from Japanese]

Tokyo Metropolitan Government (i) (n.d.) Chiiki, nenrei, oyobi danjobetsu kikanteki nougyou juujisyasuu in 1995 (The number of farming populations by area, age, and gender in 1995) [freely translated from Japanese]

Tokyo Metropolitan Government (j) (n.d.) Chiiki, nenrei, oyobi danjobetsu kikanteki nougyou juujisyasuu in 2000 (The number of farming populations by area, age, and gender in 2000) [freely translated from Japanese]

Tokyo Metropolitan Government (k) (n.d.) Chiiki, nenrei, oyobi danjobetsu kikanteki nougyou juujisyasuu in 2005 (The number of farming populations by area, age, and gender in 2005) [freely translated from Japanese]
Tokyo Metropolitan Government (l) (n.d.) Chiiki, nenrei, oyobi danjobetsu kikanteki nougyou juujisyasuu in 2010 (The number of farming populations by area, age, and gender in 2010) [freely translated from Japanese]
Tokyo Metropolitan Government (m) (n.d.) Chiiki, nenrei, oyobi danjobetsu kikanteki nougyou juujisyasuu in 2015 (The number of farming populations by area, age, and gender in 2015) [freely translated from Japanese]
Tokyo Metropolitan Government (n) (n.d.) Tokyo-to nousakumotsu seisan joukyou chousa in 2019 (Survey of crops in Tokyo in 2019) [freely translated from Japanese]
Ushijima C (2012) The urban life cycle in Tokyo 60 km zone and urban expansion and contraction. J Facult Lett Komazawa Univ 70:117–135; (in Japanese)

Assessing Urban Resource Consumption and Carbon Emissions from a Food–Energy–Water Nexus Perspective

Xujie Hu and Wanglin Yan

Abstract This chapter aims to analyze the carbon emissions in the food–energy–water (FEW) nexus from an urban supply-and-demand perspective by quantifying resource consumption flows in FEW supply chains so that carbon dioxide (CO_2) emissions from supply and consumption, as well as the contributions of FEW to urban systems, can be calculated quantitatively. By using monetary input–output tables of the Tokyo Metropolitan Government's Statistics Division, we tracked the flows of water, energy, and CO_2; visualized energy and water consumption and carbon emissions in each sector; and assessed resource efficiency in different sectors. We then defined the elements in the food nexus system and established the relationships among the elements of supply and demand, in which the supply-side includes agriculture, animal husbandry, fisheries, and food manufacturing, while the demand-side includes food wholesale and retail, catering, and households. We also calculated water, energy, and carbon flows in the food system. Finally, we allocated reduction targets to specific sectors according to the results of emissions at different scales through embodied flow analysis. The results show that in Tokyo, services have the greatest food-energy-water consumption and carbon emissions. There is considerable potential to conserve resources and reduce emissions in these sectors. This research shows that FEW makes a significant contribution to carbon emissions, and more effort is required to achieve the "Zero Emission Tokyo Strategy" for 2050.

Keywords Food-energy-water nexus · CO_2 emissions · Environmental input-output analysis · Tokyo

X. Hu (✉)
Graduate School of Media and Governance, Keio University, Fujiswa, Kanagawa, Japan
e-mail: huxujie@keio.jp

W. Yan
Faculty of Environmental Information Studies, Keio University, Fujisawa, Kanagawa, Japan
e-mail: yan@sfc.keio.ac.jp

W. Yan et al. (eds.), *Resilient and Adaptive Tokyo*,
https://doi.org/10.1007/978-981-99-3834-6_10

1 Introduction

Food, energy, and water are essential for human social development, and the food–energy–water (FEW) nexus has huge impacts on resource consumption, CO_2 emissions, and the security of FEW supplies (Ramaswami et al. 2017). Currently, about 70% of freshwater consumption (Gleick 2003) and 30% of carbon emissions (Vermeulen et al. 2012) are associated with the supply of food. According to UN World Urbanization Prospects projections (https://population.un.org/wup/), more than half the world's population is living in cities now, and by 2050, the urban population will reach 68%. This means that cities will face significant challenges from the rapidly growing demand for food, energy, and water, and a range of urban environmental issues, such as resource depletion, environmental pollution, garbage, and ecosystem damage (Xu et al. 2021). Tokyo, the largest metropolis in the world, with high population density and a concentration of industry, also faces urban environmental problems such as insufficient local food supplies due to an aging population and labor shortages, therefore being unable to meet all the food needs of the local population. Global warming has had a negative impact on Tokyo's water quality and quantity, while dramatic changes have occurred in the structure of its energy supply because of the Great East Japan Earthquake in 2011. Final energy consumption dropped by about 12% from 2010 to 2015, but Tokyo's carbon emissions returned to 2011 levels in 2015.

In mega-cities like Tokyo, sectoral linkages inevitably affect the city's inputs and outputs (Wang et al. 2019). Food, energy, and water are the main areas of physical and energy flow in the inputs and outputs through cities. Thus, the functioning of these sectors is key to resource use efficiency and reducing carbon emissions. It remains critical to find approaches to quantify material and energy flows; that is to assess the resource consumption and carbon emissions in terms of sectoral linkages, in order to identify key sectors and productive ways to achieve resource savings and emissions reductions. In recent years, input–output analysis (IOA) has become a primary method used to analyze these problems, as it is useful for illustrating the supply and consumption processes of FEW resources in various economic systems (Zhang et al. 2014; Chen et al. 2017). In general, IOA can translate economic and environmental data into physical flows using value flows (Dong et al. 2014; Tan et al. 2018), and can reflect the resource consumption and emissions embodied in the trade of goods (Zheng et al. 2020).

An increasing amount of research focuses on input–output analysis at the national, urban agglomeration, and city scales. At the national scale, Xu et al. (2021) assessed water use, energy consumption, and carbon emissions in China based on environmental input–output analysis and ecological network analysis for the years 2007, 2012 and 2017. They found that advanced manufacturing and agriculture played important roles in resource conservation and emissions reduction. Advanced manufacturing was the largest energy consumer and carbon emitter, and agriculture was the bigger water user. Owen et al. (2018) explored the food-energy-water nexus in the UK by input–output analysis and structural-path analysis. They found that demand-side policies can help governments to reduce food, energy, and water consumption. At the urban agglomeration scale, Zhang et al. (2016) identified regional energy consumption features and their ecological roles in the energy metabolic system to advance coordinated development by combining multi-region input–output (MRIO) and ecological network analysis (ENA) in Beijing-Tianjin-Hebei (Jing-Jin-Ji). Liu et al. (2020) compiled an urban multi-regional input–output table for analyzing the energy–water nexus in the Pearl River Delta urban agglomeration. At the interprovincial level, many suggestions and measures for energy conservation and emissions reduction have been put forward. At the city scale, Chen and Chen (2016) proposed a system-based framework for assessing water and energy use in Beijing. They found that services consumed the most energy and that manufacturing was the largest water consumer. Yang et al. (2018) assessed the energy–water–carbon nexus in Shanghai and Beijing in 2012 using an environmental input–output model, and found that the services sector consumes the most resources and emits the most carbon in Shanghai and Beijing. Compared to Beijing, Shanghai faced greater environmental challenges, they found.

Previous studies, whether at the national or city scale, have tended to focus on the energy–water nexus at the sectoral level, including agriculture, commerce, industry, so they had limited value for developing resource conservation and emissions reduction policies specific to local conditions, such as what products or services in the agricultural sector need to have energy efficiency and emissions reduction targets. Detailed analysis (e.g., agriculture sector includes grain, meat, and vegetables) could support policymaking. In food research, previous work tended to concentrate on the food supply (agriculture and food manufacturing), ignoring the contribution of food demand, such as catering, because current statistics in cities are generally not detailed enough to characterize specific products. In addition, some studies were only snapshots of one statistical year, whereas dynamic assessments of multiple years over time could be more useful for macro-level policymaking.

Thus, the objective of this research was to assess the food–energy–water nexus and CO_2 emissions through an environmental input–output model based on Tokyo's input–output tables (2011 and 2015). The study area for this research consists of the 23 wards covered by the Tokyo Metropolitan Government. First, we calculated direct and indirect consumption and emissions at the sector level to identify the relationship between economic sectors and energy consumption, water use and CO_2 emissions. Then, we visualized resource utilization and carbon emissions in the food system, from food supply to food demand, and identified the vital food products which have high consumption and emission levels based on an input–output table at the product level. Finally, we used embodied flow analysis to identify key elements for resource conservation and emissions reduction, and to facilitate policy recommendations to effectively reach the goals of the "Zero Emission Tokyo Strategy" for 2050.

2 Methods

2.1 *Environmental Input–Output Analysis*

Environmental input–output models can be used to calculate the resource consumption of different industries based on economic input–output tables. They can be used for assessing the ecological footprint (Bicknell et al. 1998; Wiedmann et al. 2006; Galli et al. 2013), the water footprint (Hoekstra and Chapagain 2006; Hoekstra and Mekonnen 2012), energy consumption (Chen and Chen 2016), and CO_2 emissions (Xu et al. 2021). The calculation is shown in Eq. (1):

$$C + \varepsilon A = \varepsilon O \tag{1}$$

where C represents a $1 \times n$ vector that indicates the FEW consumption and carbon emissions for each sector, respectively. The units are petajoules (PJ), cubic meters (m^3) and tonnes (t), respectively. A is the economic input–output matrix and O is the economic output in each sector, in unit of one million yen. Meanwhile, ε is a $1 \times n$ embodied resources intensity vector used to calculate the embodied resource flows from sector i to sector j.

2.2 *Direct Consumption and Emissions*

Different types of energy consumption can be calculated directly in each sector. The types of energy in the Final Energy Consumption and Greenhouse Gas Emissions report (https://www.kankyo.metro.tokyo.lg.jp/en/climate/index.html) mainly include electricity, city (natural) gas, liquified petroleum gas (LPG), and fuel oil. In this chapter, carbon emissions were calculated based on these four types of energy resources, as shown in Eqs. (2) and (3):

$$e_i = E_{ele,i} \times \frac{T_{ele,i}}{\sum_{i=1}^{m} T_{ele,i}} + E_{gas,i} \times \frac{T_{gas,i}}{\sum_{i=1}^{m} T_{gas,i}} + E_{oil,i} \times \frac{T_{oil,i}}{\sum_{i=1}^{m} T_{oil,i}} \tag{2}$$

$$c_i = C_{ele,i} \times \frac{T_{ele,i}}{\sum_{i=1}^{m} T_{ele,i}} + C_{gas,i} \times \frac{T_{gas,i}}{\sum_{i=1}^{m} T_{gas,i}} + C_{oil,i} \times \frac{T_{oil,i}}{\sum_{i=1}^{m} T_{oil,i}} \tag{3}$$

where e_i and c_i are direct energy consumption (PJ) and CO_2 emissions (t) of each sector. $E_{ele,i}$, $E_{gas,i}$, and $E_{oil,i}$ represent total industrial consumption of electricity, city gas, and oil, respectively (units are PJ). $T_{ele,\ gas,\ oil,\ water,\ i}$ are the intermediate use (million yen) of sector m (production and distribution of electricity, gas, oil and water) in sector i. $C_{ele,i}$, $C_{gas,i}$, and $C_{oil,i}$ represent the total CO_2 emissions from industrial electricity, city gas use, and oil use.

According to the Tokyo Metropolitan Government' Bureau of Waterworks, domestic water consumption consists of industrial consumption, urban activities consumption, and daily life consumption. Sub-sectoral direct water use can be obtained from the total amount based on the economic output of the production and distribution of water (Xu et al. 2021). The direct consumption and emissions are formulated as in Eq. (4):

$$w_i = \mathrm{W}_{\mathrm{water},i} \times \frac{\mathrm{T}_{\mathrm{water},i}}{\sum_{i=1}^{m} \mathrm{T}_{\mathrm{water},i}} \tag{4}$$

where w_i is direct water use (m^3) in each sector and $\mathrm{W}_{\mathrm{water},\ i}$ is the total water consumption of industry *i* in Tokyo.

2.3 Indirect Consumption and Emissions

In addition to direct usage of energy and CO_2 emissions, production sectors will impact consumption and emissions through sector interlinkages called indirect or embodied consumption and emissions. This represents the nexus impact of urban systems from a consumption perspective. The embodied consumption and emissions are triggered by urban final demand (Chen and Chen 2016). First, direct intensities are calculated as shown in the Eqs. (5) and (6),

$$\mathrm{ewc}_j^d = \mathrm{ewc}_j / x_j \tag{5}$$

$$a_{ij} = z_{ij} / x_j \tag{6}$$

where ewc_j^d is the direct intensity of energy, water, and CO_2 in sector j. Finally, a_{ij} is the coefficient of direct consumption and x_j is the economic output of sector j, as shown in Eq. (7),

$$\mathrm{EWC}_{em} = \mathrm{EWC}_d (\mathrm{I} - \mathrm{A})^{-1} \mathrm{Y} \tag{7}$$

where EWC_{em} is the row vectors of direct intensities, A is the coefficient matrix of direct use, I is the identity matrix, and Y is the final demand.

2.4 *Embodied Flow Analysis*

By measuring the embodied flows between urban sectors, the efficiencies and mechanisms of the urban system can be quantified and analyzed (Zhang 2013). We can determine the vital sectors and pathways for resource conservation and emissions reduction by analyzing embodied flows of energy, water, and carbon in the urban system (Li et al. 2017; Cai et al. 2019). Therefore, sectoral embodied flows of consumption and emissions must be considered (Fang and Chen 2017). The embodied flows reflect the resource consumption and carbon emissions embodied in the trade of goods. Embodied flows can be quantified based on environmental input–output analysis (Tang et al. 2019; Xu et al. 2019; Wang et al. 2019), as shown in Eq. (8),

$$F_{ij}^{ewc} = \left[f_{ij}^{ewc}\right] = \varepsilon^{ewc} Z \tag{8}$$

where f_{ij} is the embodied flows from sector i to sector j and ε is the diagonal matrix of embodied intensity.

3 Data Processing

To assess resource consumption and CO_2 emissions, we used the Tokyo Metropolitan Government Statistics Division's economic input–output table (IO table) with different structures (38 sectors and 191-sector structures) for 2011 and 2015. The IO tables include intermediate demand, intermediate supply, final demand, imports, gross value-added, and total production value (Fig. 1). Intermediate sectors include many industries, such as agriculture, mining, and manufacturing. Final demand includes households, government, and export. Energy use data, including four types of fuel, were gathered from the Tokyo Metropolitan Government's Bureau

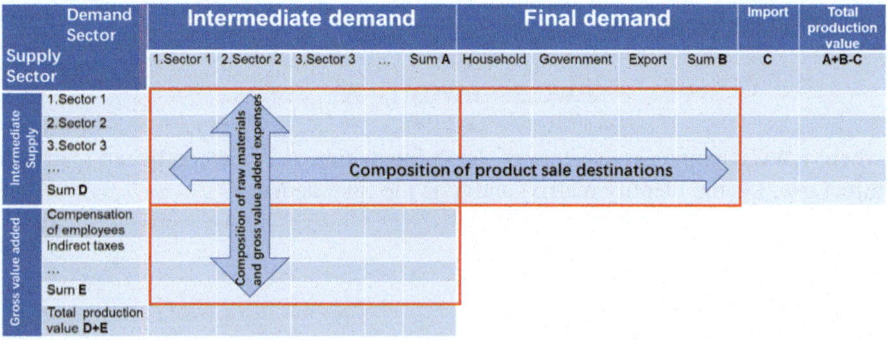

Fig. 1 Tokyo economic input–output table

of Environment. Information on water use of each sector was acquired and analyzed from the Tokyo Metropolitan Government's Waterworks Bureau.

Tokyo's monetary input–output tables (38-sectors) were aggregated to an 11-sector table for assessing urban energy–water–CO_2 emissions, as follows: Agriculture (Agr), Mining (Min), Manufacturing (Man), Construction (Con), Electricity, Gas and Water (EGW), Wholesale and Retail Trade (WR), Finance and Insurance (FI), Real Estate (RE), Transport and Post office (TP), Telecommunications (Tel), and Services (Ser).

4 Results

4.1 *Energy and Water Consumption and CO_2 Emissions in Tokyo*

Figure 2 shows sectoral energy consumption, water utilization, and CO_2 emissions in 2011 and 2015. Figure 2a presents the direct and indirect energy consumption of each sector. It shows that Service (Ser), Transport and Post (TP), and Manufacture (Man) were the top three energy consumers. However, only the energy demand of Service (Ser) has increased. The biggest proportion of energy is directly consumed by Transport and Post (TP) (more than 30% of total energy consumption). In addition, Manufacture (Man) has extremely high indirect consumption, because products from this sector are typically the raw materials for other industries, and there is big demand in Tokyo. The percentage of indirect energy and water consumption in Agriculture (Agr) and Manufacture (Man) remained at more than 80%. In addition, the proportion of embodied energy use in Mining (Min) and Electricity, Gas and Water (EGW) dropped by 35% and 28%, respectively, while at the same time it rose10% in Construction (Con). The reason for this situation was that the 2011 earthquake destroyed the energy supply system, and many buildings had to be repaired and rebuilt.

Figure 2b describes the direct/indirect water utilization of each sector. In 2011, about 2 billion m^3 of the water is consumed by Services (Ser) due to high demand in the food and beverage services and public bath services, accounting for more than 55% of total water use. Electricity, Gas, and Water (EGW) was the second largest direct water consumer, using 31.98 million m^3 and 59.32 million m^3 in 2011 and 2015, respectively. The total embodied water utilization triggered by final demand was more than 4 billion m^3, in which the consumption in Services (Ser), Wholesale and Retail Trade (WR), and Manufacture (Ma) were 57.5%, 11.95% and 11.8%, respectively, in 2011.

Figure 2c shows the direct/indirect carbon emissions of each sector. Transport and Post (TP) was the biggest direct carbon emitter, at more than 11 million t. Apart from Agriculture (Agr) and Manufacture (Man), carbon emissions have increased in the rest of the industries, especially Ser, which increased by more than six million

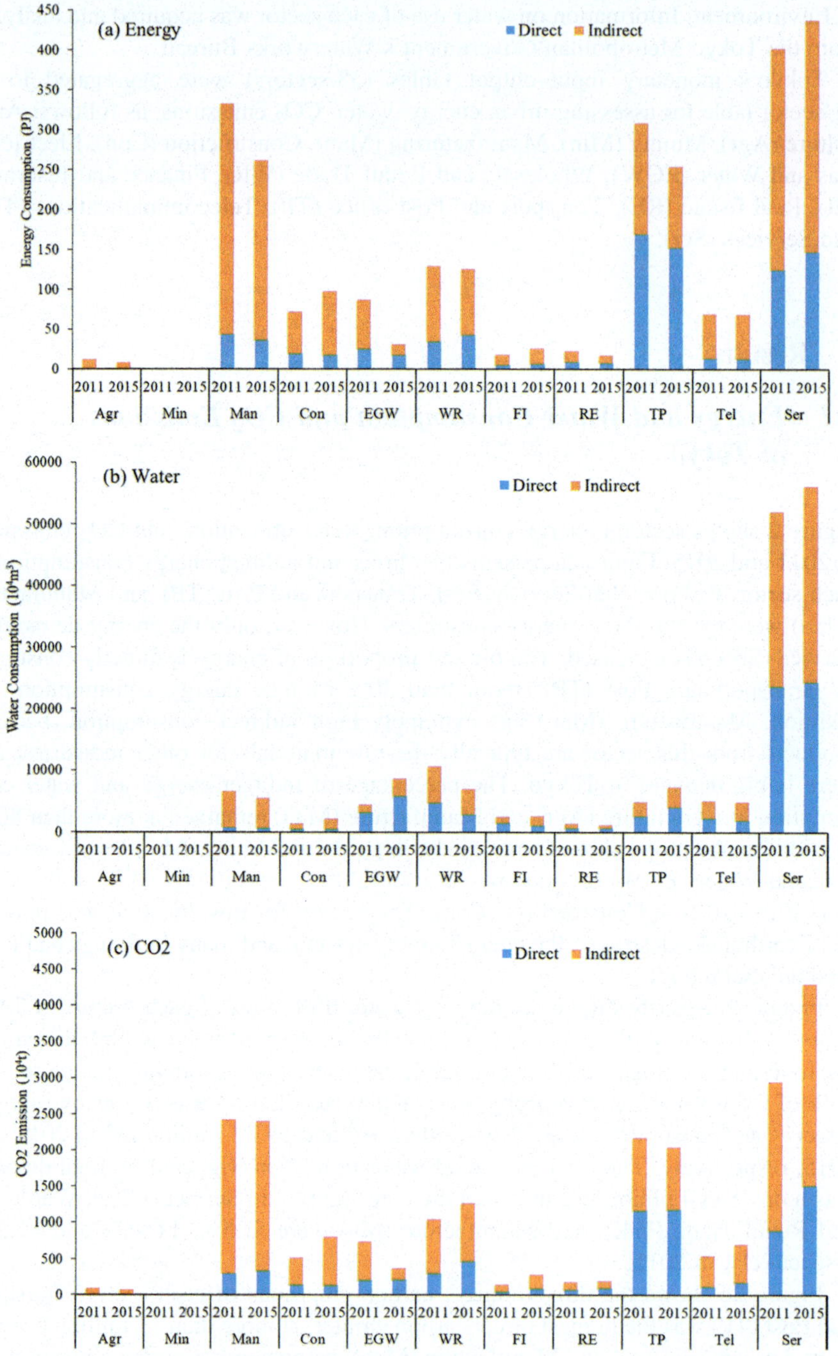

Fig. 2 Sectoral energy consumption (**a**), water resources utilization (**b**), and CO_2 emissions (**c**) in Tokyo

t. Moreover, the indirect emissions in Man remained at the highest level, because other sectors had the largest demand for manufacturing supply.

4.2 Energy and Water Consumption and CO_2 Emissions in the Food System

Figure 3 shows resource use and carbon emissions in the food system in 2011 and 2015. In 2011, the food supply consumed 15.3% of energy and 3.9% of water and emitted 15.2% of total carbon emissions, and it was a similar situation in 2015. In addition, we find that more than 80% of energy use and carbon emissions, and more than 90% of water use, came from food demand.

Figure 4 shows direct consumption and emissions in Agriculture and Food Manufacturing. Marine fisheries are the industries with high-energy consumption and high emissions, as they consume more than 50% of the energy and emit more than 40% of CO_2. Non-edible foods consumed 40% of the water in 2011, because seeds, flowers, and tobacco require much water to grow.

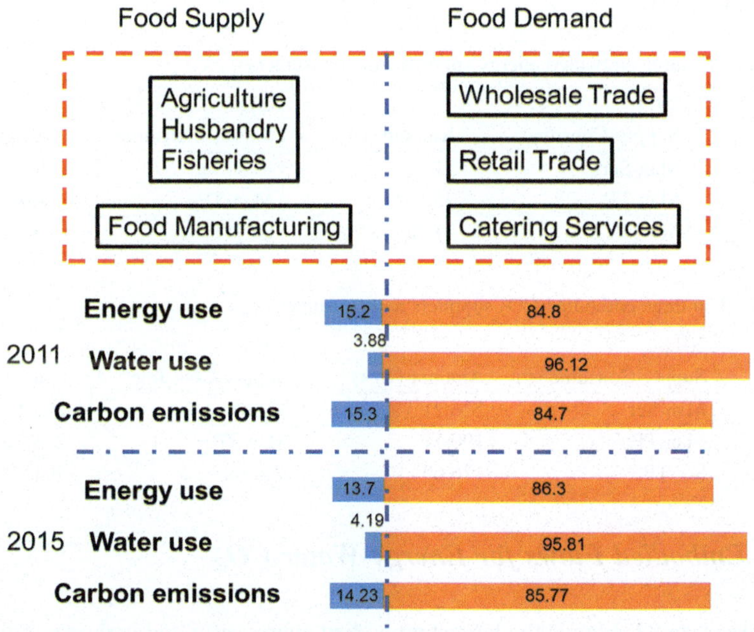

Fig. 3 Direct and embodied consumption and emissions in the food system

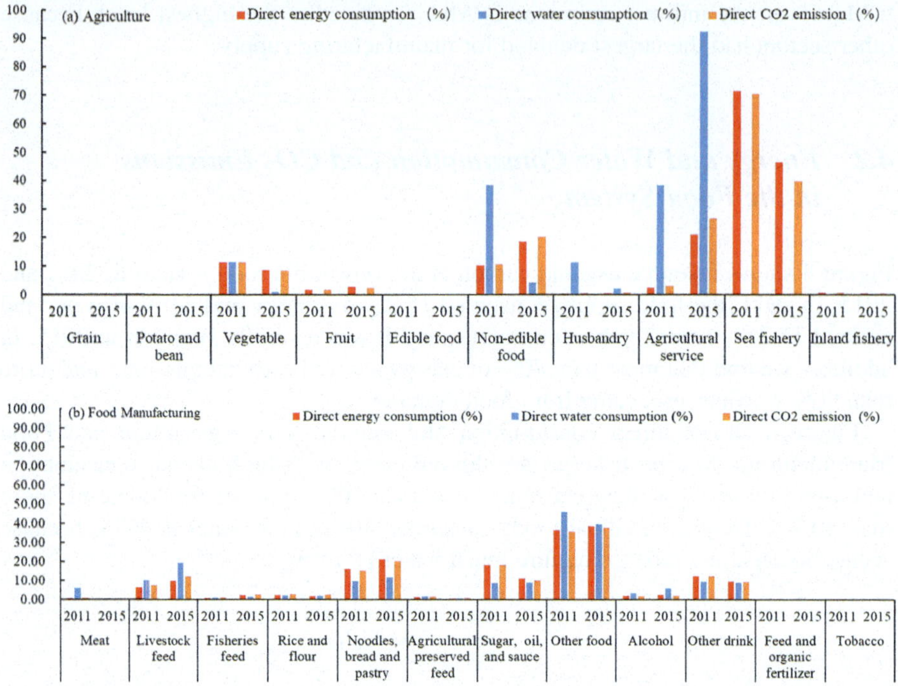

Fig. 4 Direct energy use, water consumption, and CO2 emissions in agriculture (**a**) and food manufacturing (**b**)

Table 1 Top three embodied energy suppliers and consumers

Energy	2011		2015	
No.	Supply-Consume	Amount (PJ)	Supply-Consume	Amount (PJ)
1	Man-Ser	31.55	Man-Ser	30.18
2	Man-TP	25.55	Man-TP	30.04
3	Ser-TP	21.28	Ser-TP	24.67

Table 2 Top three embodied CO_2 suppliers and consumers

CO_2	2011		2015	
No.	Supply-Consume	Amount (10^4t)	Supply-Consume	Amount (10^4t)
1	Man-Ser	226.21	Man-Ser	293.16
2	Man-TP	190.11	Man-TP	238.24
3	Ser-TP	176.65	Tel-Ser	201.98

4.3 *Embodied Flows for Energy-Water-CO_2*

In Tables 1 and 2, Man is the largest embodied energy user and carbon emitter from 2011 to 2015. Manufacture (Man), Services (Ser), and Transport and Post (TP) have

Table 3 Top three embodied water suppliers and consumers

Water	2011		2015	
No.	Supply-Consume	Amount (10^4m^3)	Supply-Consume	Amount (10^4m^3)
1	Man-Ser	3423.28	Man-Ser	3518.29
2	Tel-Ser	2308.38	Man-TP	2424.02
3	Ser-Tel	1178.2	Min-EGW	1661.21

Table 4 Resource use efficiency

	Energy use efficiency (GJ/million yen)			Water use efficiency (m^3/million yen)		
Sectors	2011	2015		2011	2015	2015–2011
Agr	13.59	7.23	−6.35	0.30	1.03	0.73
Min	14.80	19.02	4.22	2.53	2.14	−0.39
Man	5.01	4.03	−0.98	0.96	0.80	−0.16
Con	2.90	2.23	−0.67	0.77	0.70	−0.07
EGW	11.18	9.03	−2.15	13.81	28.99	15.18
WR	1.81	1.80	−0.01	2.47	1.21	−1.26
FI	0.55	0.66	0.11	1.42	1.01	−0.41
RE	0.68	0.58	−0.11	0.50	0.48	−0.02
TP	27.65	20.33	−7.33	4.31	5.41	1.11
Tel	0.76	0.62	−0.13	1.18	0.85	−0.33
Ser	2.61	2.54	−0.07	4.94	4.18	−0.76

always been the top three consumers of embodied energy and carbon emissions. The largest embodied energy user and carbon emitter in 2011 was Services (Ser), followed by Transport and Post (TP) and Manufacture (Man), because of huge demand for the goods trade. In Table 3, Service (Ser) and Electricity, Gas and Water (EGW) were the biggest water users because of the functions of providing public bath services and cooling water for power generation facilities, respectively. The top three embodied energy and carbon supply-consumption flows pairs all existed in Man-Ser, Man-TP, and Ser-TP. Ser and TP were the biggest source of demand for intermediate products from Man, in line with their sectoral features and raw materials demand. As a major embodied carbon emitter, Man has traditionally been the largest sector. The biggest supply-consumption flows pair is Man-Ser in embodied water.

4.4 Resource Use Efficiency and Its Change

Resource use efficiency is calculated by taking resource consumption divided by total production value. This is calculated for 2011 and 2015, respectively. In Table 4, we find that the energy use efficiency of Transport and Post (TP), Agriculture (Agr) and Electricity, Gas, and Water (EGW) decreased quickly by 7.33, 6.35, and 2.15, respectively. However, the intensity of Mining (Min) and Finance and Insurance

Table 5 Resource use efficiency per thousand employees

	Energy use efficiency per thousand capita (GJ/million yen/thousand employees)			Water use efficiency per thousand capita (m^3/million yen/thousand employees)		
Sectors	2011	2015	2015–2011	2011	2015	2015–2011
Agr	3.3940	1.7524	−1.6416	0.0756	0.2491	0.17342
Min	6.4796	9.1026	2.62294	1.1069	1.0253	−0.08160
Man	0.0054	0.0056	0.00021	0.0010	0.0011	0.00008
Con	0.0058	0.0048	−0.00100	0.0015	0.0015	−0.00004
EGW	0.3083	0.3322	0.02398	0.3808	1.0665	0.68562
WR	0.0009	0.0009	−0.00004	0.0013	0.0006	−0.00068
FI	0.0013	0.0016	0.00028	0.0035	0.0025	−0.00097
RE	0.0020	0.0017	−0.00029	0.0014	0.0014	−0.00005
TP	0.0543	0.0426	−0.01171	0.0085	0.0113	0.00289
Tel	0.0009	0.0007	−0.00014	0.0014	0.0010	−0.00036
Ser	0.0006	0.0006	−0.00007	0.0012	0.0010	−0.00028

(FI) increased, especially Mining (Min), by 4.22. Water use intensity increased in Transport and Post (TP), Agriculture (Agr), and Electricity, Gas, and Water (EGW). Water use efficiency, particularly in Electricity, Gas, and Water (EGW) was also greatly increased, because the electricity supply of thermal power generation and the demand for cooling water increased rapidly.

Resource use efficiency also means how many resources are consumed per thousand employees for one million yen of production. It not only reflects resource efficiency but also automation level. In Table 5, energy use efficiency per thousand capita of Agriculture (Agr) and Transport and Post (TP) dropped quickly, especially Agriculture (Agr). The reason may be the decline in the urban agricultural population. And energy use efficiency per thousand capita in Mining (Min) and Electricity, Gas, and Water (EGW) increased. The water use efficiency per thousand capita and water use efficiency shows the same trend, except for Manufacture (Man). Compared with resource use intensity, resource use efficiency per thousand capita decreased in Manufacture (Man), because more than 50% of manufacturing companies with more than 100 employees have moved out of Tokyo to other areas (Bureau of Industrial and Labor Affairs 2006).

5 Discussion

5.1 Resource Consumption and Emissions in Tokyo

According to the Fig. 5, we find that from 2011 to 2015, resource use had a negative correlation with Tokyo's economic development, while carbon emissions had a positive correlation Tokyo's total sector production value increased by 16.1% from 2011 to 2015, but total direct energy consumption decreased by 3.6%, water

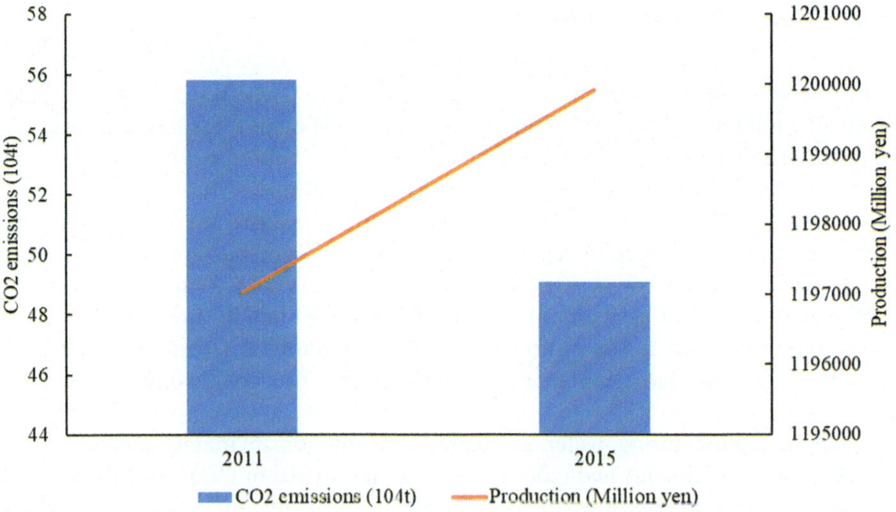

Fig. 5 Carbon emissions and production values in Tokyo from 2011 to 2015

use remained unchanged, and carbon emissions increased by 24.8%. These findings suggest that while Tokyo maintained economic growth, the energy structure changed because of the Great East Japan Earthquake, leading to increased carbon emissions. In addition, total indirect consumption and emissions are higher than direct consumption and emissions, and embodied energy consumption and carbon emissions are nearly two times the city's direct consumption and emissions. Factors in this phenomenon may be that traditional production sectors declined relatively as the population growth, tertiary industries developed rapidly, and increased citizen demand for goods results in increased indirect consumption, while Tokyo implemented its "Zero Emission Tokyo Strategy" policies.

More and more education and scientific research, medical services, and various service industries (accommodation, food and beverage, entertainment, etc.) have concentrated in Tokyo, resulting in increased electricity needs. It is a typical example of a railway-oriented city, where 76% of total fuel consumption comes from transport. In addition, there is a large environmental burden in agriculture and manufacturing; because the local food supply cannot meet local demand, so food must be imported from other areas. As for water consumption, thermal power generation increased due to the shutdown of nuclear power stations after the Great East Japan Earthquake. To maintain a steady supply of electricity, the thermal power generation needed a lot of water for cooling, resulting in a doubling of water consumption. And a large number of high-energy consumption products are consumed for manufacturing production in the typical process industries such as steel, brick and tile, and cement (Xu et al. 2021).

5.2 *Food System in Tokyo*

The average age of farmers has increased from 62 to 64 between 2000 and 2015, while the numbers of farmers and food manufacturing employees has decreased. As a result of this aging population and subsequent labor shortage, local food production decreased, while the population of Tokyo increased.

Although there is little cultivated land in Tokyo, there are still professional farmers, mainly producing vegetables and fruits. Vegetable growing consumed 11% of water and energy in 2011, as shown in Fig. 4a. And we can find that energy and water demand of agricultural service has increased rapidly from 2011 to 2015. This may be related to the promotion of smart agricultural technology in Tokyo, which, despite the large initial investment required, can greatly improve farming efficiency.

In food manufacturing, water and energy consumption are mainly concentrated in staple foods, condiments and other grocery products (frozen foods, fast foods, etc.), as shown in Fig. 4b. Because the food supply in Tokyo mainly depends on imports, in order to feed 14 million people, it is necessary to refrigerate a large number of vegetables, fruits, and meat. As a result, frozen foods and fast foods consume more than 36% of energy and water.

In the food system, especially in a mega-city, we should not only focus on the food supply, because local production is only a small part and the rest mainly comes from outside Tokyo. Food consumption is a vital research topic, because more than 80% of energy use, water use and carbon emissions comes from food demand, especially eating in the restaurant (Fig. 3). How to reduce resource consumption in the restaurant industry is crucial for carbon reduction in the Tokyo food system.

5.3 *Policy Recommendations*

According to the research results, the energy consumption of Service (Ser) and Manufacture (Man) showed a great difference: indirect consumption was more than two times direct consumption. A similar result was seen in the water utilization and carbon emissions of the two sectors. Therefore, in Service (Ser) and Manufacture (Man), there should be greater consideration of energy and water conservation and carbon emissions reduction methods. Although the energy consumption of Construction (Con) is not high, indirect consumption is 2.6 times that of direct consumption. There are many wooden houses in Tokyo, and research has showed that the average wooden house life expectancy is only 40 years, lower than buildings made of reinforced concrete (51 years). Because of the short average life expectancy of building structures, much energy and water are consumed, leading to high carbon emissions. A key objective to minimize consumption and emissions of the Construction (Con) sector should thus be to improve the technology and technical standards of housing construction.

In Electricity, Gas, and Water (EGW) supply, energy use and carbon emissions dropped, because the Tokyo Waterworks Bureau has been working to reduce energy consumption by replacing existing pumping equipment and lighting with high-efficiency equipment. At the same time, they have been reducing CO_2 emissions by installing power generation equipment that uses renewable energy sources such as solar power and small-scale hydroelectric power. For water consumption, Electricity, Gas, and Water (EGW) and Transport and Post (TP) display a tendency that is contrary to other sectors. For Electricity, Gas, and Water (EGW), the Great East Japan Earthquake is the main reason for the increase in water demand. Therefore, how to improve efficiency of electricity and water supply is a key point to reduce water use pressures.

The largest direct carbon emissions come from Transport and Post (TP). Relieving environmental stress from this sector can be effectively accomplished by promoting innovation in the automotive through various subsidy policies. According to the results of studying energy consumption, Transport and Post (TP), the largest the direct energy consumer, dropped 10% from 2011 to 2015. In 2010, the Japanese Ministry of Economy, Trade and Industry (METI) announced "Next-Generation Vehicle Strategy 2010," it promoted hybrid electric vehicles and electric vehicles. It was projected that through private-sector efforts alone, next-generation vehicles would make up less than 20% of new passenger car sales in 2020, and 30% to 40% in 2030. Meanwhile, government targets based on implementation of effective incentives aimed for 20% to 50% by 2020 and 50% to 70% by 2030. The national government reported that the number of electric vehicles, fuel cell vehicles and plug-in hybrid vehicles increased by 370% from 2011 to 2015. In addition, the Tokyo government proposed its "Automotive Environmental Management Plan" in 2011, in which companies in Tokyo that have a fleet of more than 30 vehicles should replace their traditional cars with low-emission and fuel-efficient vehicles. During this period (2011–2015), the carbon emissions of these companies decreased 8%.

From the standpoint of resource trading between sectors, suppliers and consumers have respective roles to save energy and water, and to reduce carbon emissions. For instance, the biggest embodied energy and carbon consumers in Tokyo have been Service (Ser), Transport and Post (TP), and Manufacture (Man). These three sectors therefore have the most opportunity to undertake steps to save energy and reduce carbon emissions. Meanwhile, Service (Ser), Electricity, Gas, and Water (EGW), Telecommunications (Tel), and Manufacture (Man) are the biggest embodied water consumers and have larger opportunities and responsibilities than other sectors for water saving. For embodied energy and CO_2, the main suppliers/consumers were Man-Ser, Man-TP, and Ser-TP, while Man-Ser, Tel-Ser, and Ser-Tel were the major supplier/consumer pairs for embodied water. The significant correlation impact between these sectors reveals the main paths to formulate energy conservation, water saving, and carbon-reduction methods.

6 Conclusions

This study evaluated energy, water, and CO_2 flows through environmental input–output analysis using 2011 and 2015 data at the sector and product level in Tokyo. Direct and indirect process analyses of the urban system showed that services have the greatest energy consumption and carbon emissions. Indirect water use by services is higher than by other sectors, so there is great potential to conserve resources and reduce CO_2 emissions from this sector, especially in the catering and public bath industries. Efforts targeting these services could help meet emissions reduction targets, guided by strategic policies established in the Zero Emissions Tokyo Strategy for 2050. In addition, direct and indirect energy consumption and carbon emissions from manufacturing showed a great variance. Most energy consumption and CO_2 emissions are indirect, as most manufactured products are needed by households, governments and for exports. Transportation dominated the list of direct energy consumption sectors, suggesting that a focus on promoting zero emission vehicles (ZEVs) could help reach emissions reduction goals. Manufacturing and services were the main embodied energy users and CO_2 emitters and embodied water consumers. These sectors are critical points for achieving the objective of conserving resources and reducing emissions.

In the food system, more than 80% of direct consumption and emissions come from food demand, especially in the catering industry. In the food supply sector, non-edible foods dominated direct water consumption. This finding suggests that a focus on increasing the efficiency of irrigation water use will help conserve water resources. Our study also found that marine fisheries had the highest percentages of energy consumption and CO_2 emissions in the food system. The top three items for consumption and emissions in food manufacturing were staple foods (noodles, bread, and rice), condiments, and other grocery products (frozen foods, fast foods, etc.). One important conclusion is that people should strive to purchase locally grown, seasonal produce, and eat a healthy balance of plant-based foods to relieve environmental stresses caused by food manufacturing.

These findings prove that the contribution of carbon emissions from FEW is very large. If Tokyo wants to achieve the "Zero Emission Tokyo Strategy" for 2050, the government needs to consider how to reduce carbon emissions in the use of FEW resources.

Acknowledgments The content of this chapter comes from part of the M-NEX (m-nex.net) project, which was under a grant of the Sustainable Urbanization Global Initiative (SUGI)—Food-Water-Energy Nexus by Belmont Forum and JST (Japan Science and Technology Agency). The authors are also grateful for financial support from the Taikichiro Mori Memorial Research Fund and Doctorate Student Grant-in-Aid Program of Keio University.

References

Bicknell KB, Ball RJ, Cullen R, Bigsby HR (1998) New methodology for the ecological footprint with an application to the New Zealand economy. Ecol Econ 27(2):149–160
Cai B, Zhang W, Hubacek K, Feng K, Li Z, Liu Y, Liu Y (2019) Drivers of virtual water flows on regional water scarcity in China. J Clean Prod 207:1112–1122
Chen S, Chen B (2016) Urban energy-water nexus: a network perspective. Appl Energy 184:905–914
Chen W, Wu S, Lei Y, Li S (2017) China's water footprint by province, and inter-provincial transfer of virtual water. Ecol Indic 74:321–333
Dong H, Geng Y, Fujita T, Fujii M, Hao D, Yu X (2014) Uncovering regional disparity of China's water footprint and inter-provincial virtual water flows. Sci Total Environ 500:120–130
Fang D, Chen B (2017) Linkage analysis for the water–energy nexus of city. Applied energy 189: 770–779
Galli A, Weinzettel J, Cranston G, Ercin E (2013) A footprint family extended MRIO model to support Europe's transition to a one planet economy. Sci Total Environ 461:813–818
Gleick PH (2003) Water use. Annu Rev Environ Resour 28(1):275–314
Hoekstra AY, Chapagain AK (2006) Water footprints of nations: water use by people as a function of their consumption pattern. In: Integrated assessment of water resources and global change. Springer, Dordrecht, pp 35–48
Hoekstra AY, Mekonnen MM (2012) The water footprint of humanity. Proc Natl Acad Sci 109(9): 3232–3237
Li H, Yang Z, Liu G, Casazza M, Yin X (2017) Analyzing virtual water pollution transfer embodied in economic activities based on gray water footprint: a case study. J Clean Prod 161:1064–1073
Liu Y, Hu Y, Su M, Meng F, Dang Z, Lu G (2020) Multiregional input–output analysis for energy-water nexus: a case study of Pearl River Delta urban agglomeration. J Clean Prod 262:121255
Owen A, Scott K, Barrett J (2018) Identifying critical supply chains and final products: an input–output approach to exploring the energy-water-food nexus. Appl Energy 210:632–642
Ramaswami A, Boyer D, Nagpure AS, Fang A, Bogra S, Bakshi B, Rao-Ghorpade A (2017) An urban systems framework to assess the trans-boundary food-energy-water nexus: implementation in Delhi, India. Environ Res Lett 12(2):025008
Tan LM, Arbabi H, Li Q, Sheng Y, Tingley DD, Mayfield M, Coca D (2018) Ecological network analysis on intra-city metabolism of functional urban areas in England and Wales. Resour Conserv Recycl 138:172–182
Tang M, Hong J, Liu G, Shen GQ (2019) Exploring energy flows embodied in China's economy from the regional and sectoral perspectives via combination of multi-regional input–output analysis and a complex network approach. Energy 170:1191–1201
Vermeulen SJ, Campbell BM, Ingram JS (2012) Climate change and food systems. Annu Rev Environ Resour 37:195–222
Wang X, Zhang Y, Yu X (2019) Characteristics of Tianjin's material metabolism from the perspective of ecological network analysis. J Clean Prod 239:118115
Wiedmann T, Minx J, Barrett J, Wackernagel M (2006) Allocating ecological footprints to final consumption categories with input–output analysis. Ecol Econ 56(1):28–48
Xu Z, Chau SN, Ruzzenenti F, Connor T, Li Y, Tang Y, Liu J (2019) Evolution of multiple global virtual material flows. Sci Total Environ 658:659–668
Xu W, Xie Y, Cai Y, Ji L, Wang B, Yang Z (2021) Environmentally-extended input–output and ecological network analysis for energy-water-CO2 metabolic system in China. Sci Total Environ 758:143931
Yang X, Wang Y, Sun M, Wang R, Zheng P (2018) Exploring the environmental pressures in urban sectors: an energy-water-carbon nexus perspective. Appl Energy 228:2298–2307
Zhang Y (2013) Urban metabolism: a review of research methodologies. Environ Pollut 178:463–473

Zhang Y, Zheng H, Fath BD, Liu H, Yang Z, Liu G, Su M (2014) Ecological network analysis of an urban metabolic system based on input–output tables: model development and case study for Beijing. Sci Total Environ 468:642–653

Zhang Y, Zheng H, Yang Z, Li Y, Liu G, Su M, Yin X (2016) Urban energy flow processes in the Beijing-Tianjin-Hebei (Jing-Jin-Ji) urban agglomeration: combining multi-regional input–output tables with ecological network analysis. J Clean Prod 114:243–256

Zheng X, Huang G, Liu L, Zheng B, Zhang X (2020) A multi-source virtual water metabolism model for urban systems. J Clean Prod 275:124107

Impacts of the 2011 Disaster on Food–Energy–Water Material Flows and Resource Use Efficiency in Yokohama City

Liu Hanyu and Wanglin Yan

Abstract With Yokohama City as the study area, this chapter uses input–output analysis to study impacts of the 2011 Tohoku earthquake and tsunami disaster on urban resource flows and inter-sectoral resource use efficiency. Quantifying resource flows among urban sectors, we assess resource use intensity for production. Input–output analysis of 2005 and 2011 data is used to consider self-sufficiency rates, distinguish resource sources geographically, and clarify impacts of production and consumption on local natural resource systems. Supply chain analysis is used to assess embodied energy use in the urban production layer to explore indirect impacts of sectoral resource consumption. The study found that in the production sector, the 2011 disaster had huge impacts on Yokohama's energy and water systems, with a significant decrease in resource use efficiency in the energy sector and a substantial increase in energy consumption in water and energy sectors. Supply path analysis showed strong production linkages between energy and water sectors, as well as energy and petrochemical basic product sectors. Demand in the water sector leads to more energy consumption in the energy sector, and further increases consumption of petrochemical basic products. Conversely, consumption in the household sector tends to be economical, resulting in a decline in total water and energy consumption.

Keywords Food–energy–water nexus · Great East Japan Earthquake · Input–output analysis · Resource use efficiency · Yokohama

---

L. Hanyu (✉)
Graduate School of Media and Governance, Keio University, Fujisawa, Kanagawa, Japan

W. Yan
Faculty of Environmental Information Studies, Keio University, Fujisawa, Kanagawa, Japan
e-mail: w_yan@keio.jp; yan@sfc.keio.ac.jp

W. Yan et al. (eds.), *Resilient and Adaptive Tokyo*,
https://doi.org/10.1007/978-981-99-3834-6_11

1 Introduction

The interdependencies inherent to food, water, and energy infrastructure systems make them highly susceptible to disasters, exposing infrastructure to crucial risk of failure (Dargin et al. 2020; Hameed et al. 2019), which in turn can lead to instability of physical provisioning systems. On March 11, 2011, a massive earthquake in the Pacific Ocean offshore of northeastern Japan damaged food, water, energy infrastructure systems and had a profound impact on society.

In terms of energy, infrastructure was severely damaged in the earthquake, nuclear and thermal power plant stoppages caused a significant drop in power supply capacity, resulting in power outages for nine million households in areas served by Tohoku Electric Power Company and Tokyo Electric Power Company (TEPCO). It took 3 months to fully restore power. Meanwhile, 16 companies in eight prefectures were forced to stop the supply of natural gas, and other parts of supply networks, such as oil refineries and oil supply stations were also damaged, resulting in a decline in energy supply capacity. This earthquake had a profound long-term impact on Japan's energy system.

As shown in Fig. 1, Japan's first commercial nuclear power plant began operating in 1966, and by 2010, national nuclear power generation capacity reached 288.2 billion kWh per year. However, due to the combined impacts of the 2011 Tohoku earthquake/tsunami and the subsequent accident at the TEPCO Fukushima Daiichi Nuclear Power Plant, the number of nuclear power plants out of service gradually increased, with the result that, and nuclear power generation plummeted to zero in

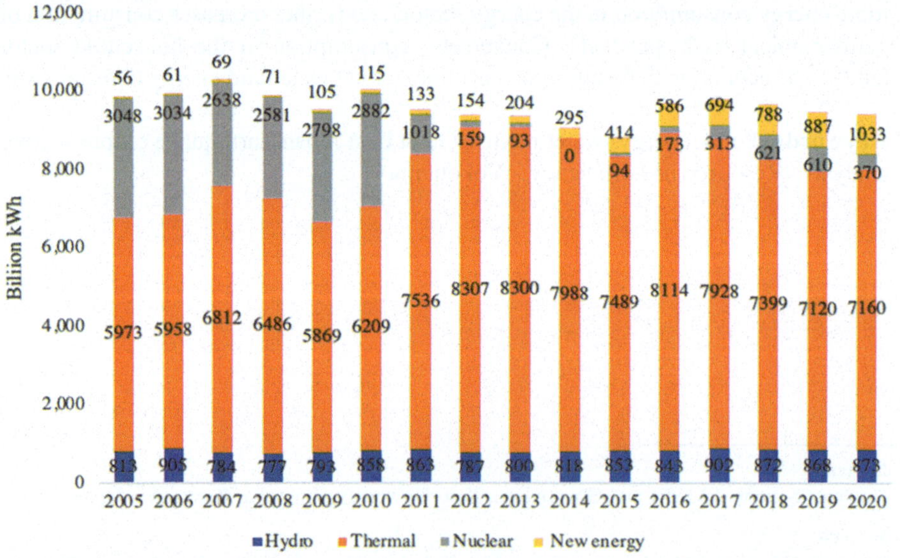

Fig. 1 Changes in electricity generated in Japan, by power source (1970 to 2019). (Data source: Bureau of Resources and Energy/Energy White Paper 2021)

2011. As a result of the shutdown of nuclear power plants, imports of oil and natural gas expanded, Japan's energy self-sufficiency ratio fell from around 20% during the 2000s to around 7% right after 2011 (IEA 2021). Coal generated electricity in 2014 was flat from 2013 at 282.4 billion kWh, about 16 times higher than in 1973 (BEIS 2020). Compared with nuclear power generation, thermal power generation produces more carbon emissions. Japan's total carbon emissions in 2011 increased by 4.4% compared with the previous year, while the energy conversion sector increased by 7.7% (Ministry of the Environment 2021).

In terms of water, sudden disasters have affected water supplies in disaster-stricken areas and increased pressure on water supply in the Kanto region. During Tohoku earthquake, at least 2.3 million households experienced a water supply cut off. Meanwhile, having faced a spate of extreme weather and weather-related disasters, Japan faces concerns about safety of the food supply. Besides impacts on agricultural production in disaster-affected areas, in the case of Fukushima, radioactive materials in the atmosphere adhered to vegetables through rainfall, and food production in the area was also affected by radioactive materials.

In general, damage to infrastructure from earthquakes threatens the security of resource supplies, especially the food, energy, and water that sustain urban social and economic systems, which inevitably affects production activities of relevant sectors. On the other hand, because of the complex links between FEW infrastructure systems and other sectors in a city, attention should be paid to interactions between FEW sectors (Arthur et al. 2019; Liu et al. 2018) and track resource consumption in each sector. Several studies have analyzed the impact of earthquakes, including disruptions of supply chains, damage to infrastructure, impacts on regional economic development, etc.(Ghobarah et al. 2006; Koshimura et al. 2014; Rose et al. 1997; Son et al. 2020). However, there have been few studies on the impact of earthquakes on resource flows in urban sectors, and most of the analyses on resource flows in urban sectors only stop at quantitative assessments without further evaluating changes in resource flows and their impacts.

Therefore, in order to understand the impacts of the earthquake on resource flows, this chapter aims to assess the change of energy and water flows between urban sectors, as well as resource use efficiency, through input–output analysis. Also, resource flows between sectors closely related to the urban energy and water sectors are further explored through supply chain analysis. The results may provide hints on how to achieve urban sustainable development by improving sectoral resource consumption and resource use efficiency.

2 Method to Assess Disaster Impacts by Input–Output Analysis and Supply Chain Analysis

Food, energy, and water are basic resources that residents rely on for survival. Also, the network characteristics of food, water, and energy make it easier for changes in one sector to affect other sectors, and this can include spreading risks in the event of a disaster. Research shows that the power industry is a key sector that spreads risks to other sectors (Haraguchi and Kim 2016). In a disaster, long-term damage to the infrastructure system may lead to changes in urban physical provisioning systems due to the inherently complex links between components of urban infrastructure (Dargin et al. 2020) and changes in resource supplies will impose the impacts on the amounts of resources used for production. At the same time, in the context of rapid economic development and expanding resource demand, improving resource use efficiency and ensuring the security of resource supply have become important tasks for the development of sustainable urban systems (Feng et al. 2019). Therefore, it is particularly important for industrial sectors in cities to improve their resource consumption patterns and use efficiency.

Existing research has applied various methods to analyze the FEW nexus from different perspectives. While analyzing the flow of a single element, nexus analysis pays more attention to the flow and trade-off relationships between elements (Hussien et al. 2017). Material flow analysis and input–output analysis based on the perspective of production and consumption are used to track and quantify the flows of resources in urban systems. Different from other methods, combined with the inventory analysis (Wang et al. 2019), input–output analysis can track the resource flow relationships between various sectors in a city (Albrecht et al. 2018; Feng et al. 2019; Li et al. 2019), so as to clearly point out the impacts of each sector's production and consumption on the FEW Nexus. The application of supply chain analysis can also clarify material flows between sectors involved in processes from production to consumption, so as to determine the key sectors that affect resource flows and focus on achieving energy conservation and emission reductions in key sectors (Owen et al. 2018).

Input–output analysis is an economic quantitative analysis method that analyzes "inputs" and "outputs" together. "Inputs" refer to the consumption of various production factors in production processes, such as the consumption of material products, the use of labor, and the acceptance of different types of productive services. "Outputs" refer to the distribution and use of outputs of production, that is, the direction of the physical movement of material products or the objects receiving the service (Leontief 1951). Input–output analysis can be carried out by combining input–output tables and input–output mathematical models.

Figure 2 depicts an input–output table, which shows the sources of inputs and the destinations of outputs of sectors in an economic system, reflecting in detail the interrelationships among the elements of the economic system. The column direction of the table indicates the consumption of inputs (elements) by sector, that is, the sources of the inputs. The row direction reflects the distribution and use of the

Output / Input		Intermediate output							Final output			Total output
		Sector 1	Sector 2	…	…	…	…	Sector n	Consumption	Investment	Export	
Intermediate input	Sector 1	q_{11}	q_{12}					q_{1n}		y_1		Q_1
	Sector 2	q_{21}	q_{22}					q_{2n}		y_2		Q_2
		…	…		A			…	B	…		…
	Sector n	q_{n1}	q_{n2}	…	…	…	…	q_{nn}		y_n		Q_n
Initial input	Depreciation											
	Payment				C							
	Income											
Total input												

Fig. 2 The general design of an input–output table

product, that is, the direction of the outputs. Intermediate inputs and intermediate output constitute area A, which is the core of the input–output table, reflecting the interconnections between intermediate sectors. Intermediate input refers to the input (purchase) of raw materials, etc. (goods and services) required for the production of goods and services from other sectors, while intermediate demand refers to the production (sale) of goods and services as raw materials, etc. for the production of other goods and services. The downward extension of intermediate inputs expands the input elements of material form to other forms, such as labor and fixed assets, constituting area C. The expansion of the intermediate sector to the right reflects that in addition to the expansion of intermediate products used in production, the demand of the final sector, such as household and business consumption, investment, and export, are separated from the production process of the year, constituting area B. The demands of the final sector refers to the consumption of produced goods and services as the final stage of transactions by households, governments, exports, etc. The entire input–output table is centered on area A, and combines column-wise inputs and row-wise outputs to describe the quantitative dependencies among sectors in the economic system.

Based on the input–output table, an input–output mathematical model can be established. The intermediate input and output sectors in Fig. 2 are divided into n sectors, whereby q12 in area A represents the quantity of products of sector 1 used by sector 2, y1 in area B represents the quantity of products provided by sector 1 for the final sector, and Q1 represents the total output of sector 1. From the horizontal line, area A represents the quantity of products consumed by each intermediate sector in a certain period, and area B represents the quantity of products consumed by the final sector. The sum of these two parts is equal to the total production Q of various products in a certain period of time. Therefore, we can establish the quantitative relation model Eq. (1).

$$\sum_{j=1}^{n} q_{ij} + y_i = Q_i \ (i = 1, 2, \ldots, \mathrm{n}) \tag{1}$$

The direct consumption coefficient εij is introduced, defined as the quantity of products provided by sector i to be consumed per unit of product production by sector j, which reflects the production technology structure and material use efficiency of the sector, as shown in Eq. (2).

$$\varepsilon_{ij} = \frac{q_{ij}}{Q_i} (i = 1, 2, \ldots, \mathrm{n}) \tag{2}$$

Combining Eqs. (1–3) can be obtained, where I is the unit matrix, (I-ε) reflects the relationship between the input and output of the unit physical product, each column in the matrix shows what sector j produces per unit of product and needs to invest in other product quantities.

$$\mathrm{Y} = (\mathrm{I} - \varepsilon)\mathrm{Q} \tag{3}$$

Here we introduce the resource consumption coefficient θi, defined as the quantity of a resource that needs to be used per unit of output, as shown in Eq. (4), where Vi represents the energy and water use of sector i.

$$\theta_i = \frac{V_i}{Q_i} \tag{4}$$

By combining Eqs. (3) and (4), we can track the flow of energy and water resources across urban sectors, as shown in Eq. (5), where P_i^f is the vector of energy or water consumption triggered by the final sector.

$$P_i^f = \theta(\mathrm{I} - \varepsilon)^{-1}\mathrm{Y}_i \tag{5}$$

In an open economic environment, the import of products and services also meets a part of the consumer demand within the city, which can be regarded as a leakage of the demand within the city. The resources consumed to meet this part of the demand come from outside the city, so by considering the self-sufficiency rate (1-M) of the sector, the consumption of local resources by local demand can be clarified, as seen in Eq. (6).

$$\mathrm{P}_i^f = \theta(\mathrm{I} - (\mathrm{I} - \mathrm{M})\varepsilon)^{(-1)}[(\mathrm{I} - \mathrm{M})\mathrm{Y}_i] \tag{6}$$

In Eq. (6), $\theta(\mathrm{I} - (\mathrm{I} - \mathrm{M})\varepsilon)^{-1}$ represents the resource consumption of sector i resulting from the production of one unit of final product by sector j. Therefore, the change of resource use efficiency of each sector can be observed through the change of $\theta(\mathrm{I}-(\mathrm{I}-\mathrm{M})\varepsilon)^{-1}$ over time, which implies the resource consumption required for the unit output of the sector.

Based on the basic input–output analysis, $(\mathrm{I}-\varepsilon)^{-1}$. is denoted as L, which can be expanded by Taylor's series, as in Eq. (7).

$$\mathrm{L}=\mathrm{I}+\varepsilon+\varepsilon^2+\varepsilon^3+\ldots+\varepsilon^n \tag{7}$$

Combining Eq. (6) with Eq. (7) gives us the following equation:

$$p=\theta I\boldsymbol{F}+\theta(I-M)\varepsilon\boldsymbol{F}+\theta[(I-M)\varepsilon]^2\boldsymbol{F}+\theta[(I-M)\varepsilon]^3\boldsymbol{F}+\ldots \\ +\theta[(I-M)\varepsilon]^n\boldsymbol{F} \tag{8}$$

Equation (8) is the Taylor's expansion, where $\theta[(I-M)\varepsilon]^n F$ is a calculation of the resource use of each production layer along the supply chain.

3 Study Area

Yokohama City is an ordinance-designated city with a population of 3.7 million in Kanagawa Prefecture. This prefecture has great industrial potential since it has been designated as a national strategic special zone to "enhance the international competitiveness of industry" and "promote the formation of international economic activity." In the region, the government can promote industrial development through targeted deregulation, enabling Yokohama to promote effective regional revitalization. Also, the city is popular for tourists thanks to its diverse tourism resources and facilities. Since Yokohama Station and Tokyo Station are only 30 min apart by rail, the city plays a role as a bedroom suburb of Tokyo as well.

Yokohama has a land area of 437.56 sq. km, accounting for 0.12% of the country's total area, although it supports 3% of the total Japanese population. As of 2013, Yokohama's total population was still increasing, although some districts were declining, and this could inevitably increase pressure on the supply of regional resources. Also, the population is getting older, and the number of households over 65 years of age is increasing, making them more vulnerable to earthquakes (Zhu and Sun 2017). Regarding changes in land use structure, the proportion of residential land, commercial land, and land used for public facilities in Yokohama as a whole has increased, while the proportion of industrial land, agricultural land, and wooded land has decreased due to an increase in the population. In particular, industries that require large-scale land use have been relocated in many cases in recent years, and the number of former sites becoming collective housing and large-scale shops is also increasing, which may affect local public infrastructure (Yokohama in 2015). Changes in Yokohama's land use hint at changes in the city's intermediate and the final sector. In the following, we take a detailed look at Yokohama's sectoral economy and resource consumption, and how they were affected by the 2011 Great East Japan Earthquake.

3.1 Yokohama's Industrial and Economic Development

Changes in city production value and share of Yokohama are given in Table 1. In 2015, Yokohama's production value accounted for 40.6% of the prefecture's share and 2.6% of the national share. With the negative growth of Kanagawa and the national production value in 2011, Yokohama's production value increased by 2.7% over 2005. In 2015, Yokohama's production value increased by 14.6% compared to 2011, which was much higher than the growth rate of Kanagawa and national production value.

In terms of sectoral development in Yokohama, as shown in Fig. 3, the sector that continues to grow is services and information communications, with production value increasing from 4.5 trillion yen in 1990 to 9.9 trillion yen in 2015, an increase of 2.2 times. The manufacturing sector was the city's second-largest in production value, after the service sector. The city's production value for construction has been

Table 1 Changes of the gross product and share, Yokohama City. Units: Billion yen

Year	Yokohama	Kanagawa	Japan	Share in Kanagawa	Share in Japan
1990	20,136.7	61,861.4	872,212.2	32.60%	2.30%
1995	20,662.2	60,497.6	937,100.6	34.20%	2.20%
2000	22,451.8	60,461.6	958,886.5	37.10%	2.30%
2005	22,562.9	60,082.2	972,014.6	37.60%	2.30%
2011	23,161.1	59,850.0	939,674.9	38.70%	2.50%
2015	26,541.0	65,325.1	1,017,818.4	40.60%	2.60%

Data source: Yokohama City Input–Output Report 2015

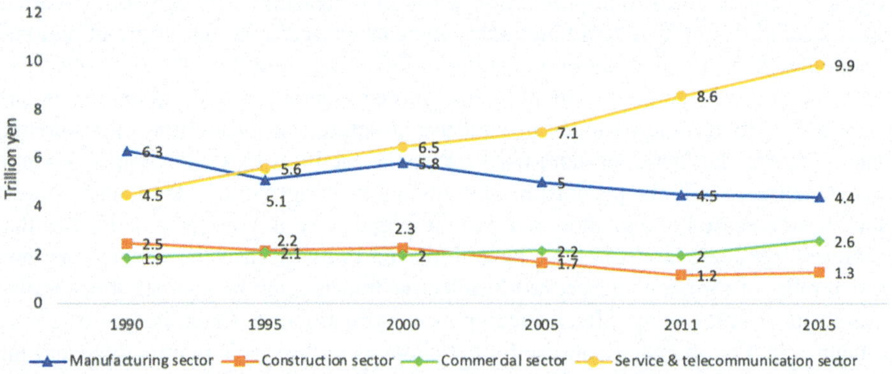

Fig. 3 Change in the city production value of major sectors. (Data source: Yokohama City Input–Output Report 2015)

declining, and the commercial sector has been declining since 2000, but showed a recovery trend in 2015. Changes in the resource consumption of a sector and the resource flows between sectors are closely related to the output value of the urban sector. The ratio of the resource input and output value of a sector reflects the resource use efficiency of a sector. Due to correlations between sectors, a decline (or increase) in the resource use efficiency of give sector may indirectly lead to an increase (or decrease) in resources used by downstream sectors.

3.2 Energy, Water, and Food Supply in Yokohama

Figure 4 shows the changes in the self-sufficiency rate of the energy, water, and food sectors in Yokohama, where the self-sufficiency rate of the electricity, gas, and heat supply sectors remained at 100% in 2011. However, the petroleum and coal products sector has a low self-sufficiency rate of about 27%. The self-sufficiency rate of the water and waste treatment sector increased from 93% in 2005 to 99.99% in 2011, basically achieving self-sufficiency. The food sector has a low rate of self-sufficiency, at 5% in 2011, an increase of 3% from 2005. The food and beverage industry's self-sufficiency rate was 16% in 2011, down from 3% in 2005.

As can be seen from Yokohama's input–output table, the production of the power, gas, and heat supply sector is heavily dependent on the mining, petroleum, and coal products sectors. The low self-sufficiency rate in the mining sector and the petroleum and coal products sectors threatens the supply security of electricity, heat, and other energy resources to cities during disasters.

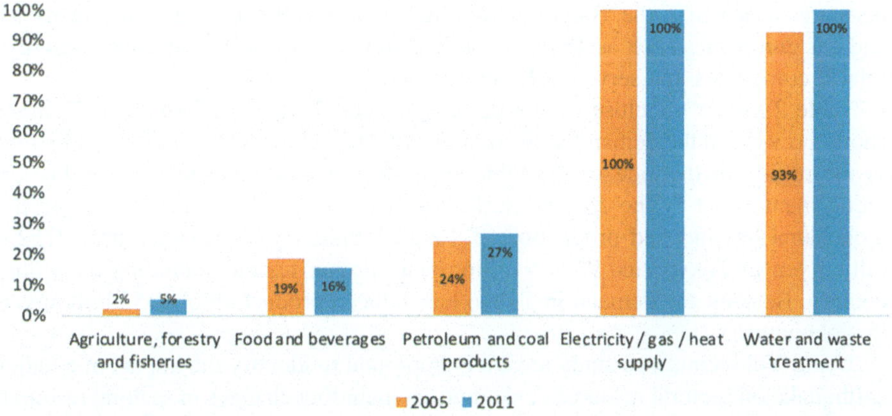

Fig. 4 Changes in self-sufficiency in Yokohama's energy, water, and food sectors. (Data source: Yokohama City Input–Output Report 2015)

3.3 *Impacts of the 2011 Great East Japan Earthquake on Yokohama*

The nuclear power plant accident caused by the earthquake in 2011 resulted in a significant reduction in power supply capacity. The proportion of thermal power generation increased, causing various problems such as increased greenhouse gases. Restrictions on electricity supply capacity put significant downward pressure on the supply side. Rolling power outages were used to limit the use of electricity in some areas, which had a great impact on the lives of citizens.

Also, the government asked large and small consumers and households to reduce their peak electricity consumption by approximately 15%. According to estimates, a 1% reduction in electricity supply would have a negative impact of approximately 0.9% on production across the sector. However, the impact on the manufacturing sector would be more pronounced than on the non-manufacturing sector. Within the manufacturing sector, Japan's major export industries, such as electronics, machinery, and equipment, transport machinery and steel, would be more affected (Koike 2012). On the other hand, the Tohoku region is an important part of not only domestic but also international supply chains. Ibaraki and Chiba prefectures, neighboring the Tohoku region, are the source of a large supply of products, including communication machinery and related equipment, computers and related equipment, general machinery, and industrial electrical equipment. Impacts on those industries would have other impacts through the supply chain, including Yokohama.

Since the earthquake, Kanagawa Prefecture, where Yokohama is located, has been committed to transition from centralized to distributed power sources (Kanagawa Prefecture Industrial and Labor Bureau 2020). To this end, Yokohama has declared its intention to realize community building with uninterrupted basic services in the event of disasters through energy decentralization and independence from the perspective of global warming countermeasures (Yokohama City 2013).

As Japan's second-largest city by population, Yokohama has a huge demand for resources. Located in the Tokyo metropolitan area, it is home to regional electricity and gas companies such as TEPCO and Tokyo Gas, as well as other large energy-related companies that serve the Kanto region.

Table 2 shows the close economic ties between Tokyo and Yokoham. Table 2 shows how much the demand in the region at the top of the table induces production in the region at the side of the table. The shaded area indicates the production inducement effect by the demand in the own region. Vertically, the demand in the Yokohama has induced production of 10925.1 billion yen in Yokohama, 2752.8 billion yen in Tokyo, and 9575.5 billion yen in other regions including the whole country. Besides, the demand in Tokyo has induced production of 1152 billion yen in Yokohama.

Using Yokohama as a study area, it is important to identify the impact of a major earthquake on sectoral resource consumption, including changes in sectoral resource use efficiency and inter-sectoral resource flows, in order to achieve resource

Table 2 Economic production links between five regions. Units: 10 billion yen

	Yokohama	Kawasaki	Other areas in Kanagawa	Tokyo	Other cities in Japan
Yokohama	1,092.51	12.30	26.94	155.20	833.63
Kawasaki	15.80	399.98	12.22	73.75	430.85
Other cities in Kana-gawa	24.87	10.10	1,160.66	118.31	661.06
Tokyo	275.28	103.46	219.16	8,322.44	7,713.06
Other areas in Japan	957.55	385.8	430.51	4,515.97	53,599.22

Data source: Ijo and Oshima 2019

conservation and reduce carbon emissions through improved energy supply and resource use efficiency in the city, and to improve the resilience of the city's sectors.

4 Data Processing and Results

4.1 Data Processing

The situation in the study area, Yokohama, was introduced in the previous section. Data required to analyze the resource consumption by Yokohama's various sectors through input–output tables includes the input–output table, energy, and water consumption data of intermediate sectors, and self-sufficiency rates of intermediate sectors. Data for 2011 and 2005 were selected in order to reflect the impacts of the 2011 Great East Japan Earthquake on Yokohama.

Input–output data for Yokohama comes from the input–output table issued by the Yokohama municipal government. Based on the input–output table, the self- sufficiency rate (I-M) of each intermediate sector can be calculated by Eq. (9).

$$\mathrm{I-M} = 1 + \frac{\text{Total imports and inflows}}{\text{Total domestic demand}} \tag{9}$$

Data for energy consumption of the intermediate sector are obtained from Agency for Natural Resources and Energy. Statistics on energy consumption are only accurate to the prefectural level, so the energy consumption of each sector in Yokohama is estimated in this study using energy consumption per million yen output by intermediate sector in Kanagawa and output by intermediate sector in Yokohama.

Data for water consumption in the intermediate sector are obtained from Yokohama City's Waterworks Bureau.

4.2 Results

4.2.1 Water and Energy Use Triggered by Intermediate Sector Production

Figure 5 shows changes in water and energy use per unit of yen arising from output in intermediate sectors in 2011 compared to 2005. Changes in water intensity arising from production of the intermediate sectors are greater for energy intensity. The water (Wa) and other manufacturing (Om) sectors experienced significant increases in production-induced water intensity, followed by the transportation (Tr) and mining (Mi) sectors, while there was a decline in water intensity caused by production in the service (Se) and food manufacturing (Fm) sectors. In terms of energy consumption, the water (Wa) and energy (El) sectors experienced significant increases in production-induced energy intensity, followed by the transportation (Tr) and mining (Mi) sectors, while there was a decline in water intensity caused by production in the service (Se) and food manufacturing (Fm) sectors. Production in the business and service sectors resulted in a decrease in the resource intensity of their own production, while production in the other manufacturing (Om), water (Wa), electricity, gas and heating (El) resulted in an increase in the resource intensity of their own production.

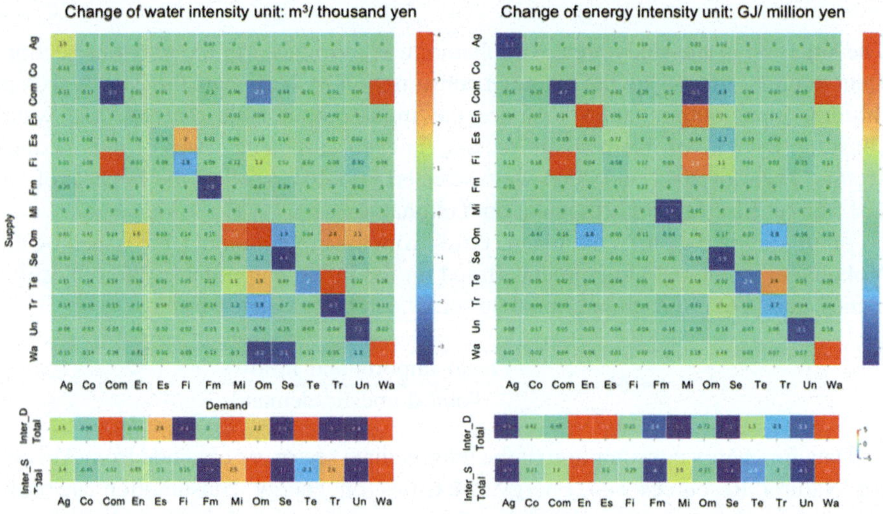

Fig. 5 Changes in intermediate sector production-induced water intensity (cu m per thousand yen) (left) and energy intensity (GJ per million yen) (right). Agriculture (Ag), Commerce (Com), Construction (Co), Electricity, gas and heating (El), Finance (Fi), Food manufacturing (Fm), Mining (Mi), Other manufacturing (Om), Real estate (Es), Service (Se), Telecommunication (Te), Transportation (Tr), Unknown classification (Un), Water and waste management (Wa), intermediate supply (Inter_S), intermediate demand (Inter_D)

Overall, in terms of water consumption in each sector, compared with 2005, the production activities of various urban sectors in 2011 led to the consumption of more water in the "other manufacturing" sector, followed by the water, telecommunication, and commercial sectors. As supplier of commodities, water consumption per thousand yen of output in the "other manufacturing" sector increased significantly, followed by the mining and transportation sectors. Conversely, water consumption associated with the service and food manufacturing sectors dropped significantly. In terms of energy consumption, production activities in various sectors of the city led to more energy consumption in the water and energy sectors. Also, the water and energy sectors significantly increased in water consumption per thousand yen of output, while energy consumption related to the service and agricultural sectors dropped significantly.

4.2.2 Energy and Water Use Triggered by the Final Sector

Commodity demand for the final sector, which is dominated by the household and export sectors, triggers the production and service activities of the intermediate sectors, which in turn leads to the consumption of resources. Through input–output analysis, it is possible to track the water and energy consumption of the intermediate sector triggered by the demand of the final sector.

The amount of energy consumption triggered by the final sectoral demand is given in Table 3. In 2011, the per capita daily energy consumption in the household sector was 4.02 kWh, a 15% decrease from 2005. From the table we can see that the energy consumption of the commercial sector triggered by the demand of the household sector was the highest, followed by the energy sector and the service

Table 3 Energy consumption triggered by the Yokohama final sector

	Household		Export	
	kWh/person/day		GJ/million yen/year	
Sector	2005	2011	2005	2011
Om	1.18	0.40	8.95	9.43
Com	0.99	0.77	0.51	0.35
Se	0.86	0.61	13.03	5.55
Re	0.62	0.54	0.27	0.26
Un	0.21	0.21	0.31	0.52
El	0.19	0.62	0.77	0.53
Fm	0.17	0.17	0.02	0.01
Tr	0.14	0.22	0.85	0.59
Te	0.13	0.10	0.69	0.80
Co	0.08	0.06	1.00	1.42
Wa	0.06	0.19	0.49	3.01
Fi	0.05	0.06	1.25	4.80
Ag	0.04	0.05	0.08	0.10
Total	4.74	4.02	28.22	27.38

Table 4 Water use triggered by the Yokohama final sector

	Household		Export	
	Liters/person/day		Liters/million yen/year	
Sector	2005	2011	2005	2011
Om	16.35	15.44	1449.87	1202.11
Com	2.52	1.80	40.97	52.10
Se	13.78	9.47	15.26	23.51
Re	0.75	1.06	0.04	0.07
Un	0.25	0.10	0.25	0.10
El	0.30	0.26	0.29	0.36
Fm	8.82	7.61	18.34	11.59
Tr	1.36	1.84	2.57	2.15
Te	0.41	0.42	0.52	0.60
Co	0.18	0.07	0.04	0.02
Wa	5.93	5.44	13.89	20.07
Fi	0.61	0.64	0.47	0.12
Ag	0.01	0.02	0.00	0.01
Total	51.26	44.17	1542.50	1312.80

sector. Compared with 2005, the energy consumption of the manufacturing sector caused by the demand of the household sector dropped significantly, while the energy demand of the energy supply and the water sector increased significantly. In terms of the energy consumption triggered by the export sector, in 2011, the annual energy consumption per million yen in the export sector was 27.38, a decrease from 2005. The energy consumption of the manufacturing sector triggered by the demand of the export sector was the highest, followed by the service sector. Compared with 2005, the energy consumption of the service sector triggered by the export sector's demand dropped significantly, while the energy demand of the water sector increased significantly.

The amount of water consumption triggered by the final sectoral demand is given in Table 4. In 2011, the per capita water consumption in the household sector was 44.17 liters per day, a 14% decrease from 2005. The manufacturing sector consumed the most water resources triggered by the demand from the household sector, followed by the service sector and the food and beverage sector. Compared with 2005, the consumption of water resources in the service industry triggered by the household sector's demand dropped significantly, while the demand for water resources in the transportation sector increased significantly. In terms of water consumption caused by the export sector, in 2011, water consumption per million yen was 1,12.80 L, down 15% from 2005. The manufacturing sector consumed the most water resources triggered by export demand, followed by the commercial and service sectors. Compared with 2005, the consumption of water resources in the manufacturing industry triggered by the export sector dropped significantly, while the demand for water resources in the water, commercial, and service sectors increased.

4.2.3 Energy Use Embodied in Production Layers

Table 5 shows the amounts and changes of energy use triggered by water and energy sectors in different production layers in 2011. Production in both the Wa and El sectors triggered energy consumption in the El sector, Pe sector. Compared with 2005, final sectoral consumption triggered more energy consumption, and production demand in the energy and water sectors triggered a substantial increase in energy consumption in the energy, petroleum, and coal sectors. The final sector's demand for products from the water sector caused a large amount of energy consumption in the water sector and commercial service sectors, and compared with 2005, this part of the energy consumption increased significantly; the production of the energy sector triggered a large amount of energy use in the construction sector and the transportation sector.

5 Discussion

In this section, we first examine Yokohama's role in securing energy for the Tokyo metropolitan area, then analyze the results of input–output analysis based on the carbon emissions of each sector in Yokohama, and how the 2011 earthquake effected resource use efficiency in each sector. Finally, we discuss the advantages and limitations of input–output analysis.

Table 5 Changes of energy use embodied in production layers (2011). Units: Terajoules (TJ)

Tier 0	Sector	Amount	Change rate			
	Wa	492.00	0.78			
	El	4940.37	2.97			
Tier 1			Tier 2		Tier 3	
Sector	Amount	Change rate	Sector	Amount	Sector	Amount
El	52.10	3.39	Pe	18.59	Pe	3.60
Pe	42.62	0.78	El	8.38	El	1.43
Wa	40.94	1.37	Un	5.24	Un	1.10
Un	14.81	−0.17	Wa	3.16	Bs	0.40
Bs	7.94	0.25	Bs	1.95	Com	0.34
Pe	841.49	0.28	Pe	128.94	Pe	20.41
El	476.75	9.98	El	54.42	El	7.55
Co	43.70	−0.20	Un	20.26	Un	4.60
Un	37.16	0.39	Met	9.76	Tr	1.65
Tr	31.02	0.92	Tr	8.12	Bs	1.64

Change rate: Changes in energy consumption compared with 2005; *Pe* petroleum and coal products, *Bs* service to business

5.1 The Role of Yokohama City in Securing Energy for the Tokyo Metropolitan Area

As shown in Fig. 6, the wide-area power system in the Tokyo metropolitan area consists of multiple 500 kV outer ring lines and transmission lines connected to the outer ring lines. A large amount of power is transmitted within the metropolitan area. There are several high-capacity power strips in the metropolitan area, the largest of which is located the Tokyo Bay area. In that area, TEPCO has capacity of 29.1 million kilowatts of thermal power stations, accounting for 70% of the total thermal power generation, while the thermal power stations located in Yokohama have capacity of 4.62 million kilowatts, accounting for 16% of the total. Yokohama also has a diverse range of power generation options inland, providing security of electricity supply to the Tokyo metropolitan area. Table 6 shows the energy supply (electricity, gas, heat) in each region of the Kanto area. The table shows that the self-sufficiency rate of Yokohama's energy sector is about 100%, and the outflow of the energy sector accounts for 65.4% of the entire Kanagawa, which fully demonstrates Yokohama's important role in ensuring the regional energy supply.

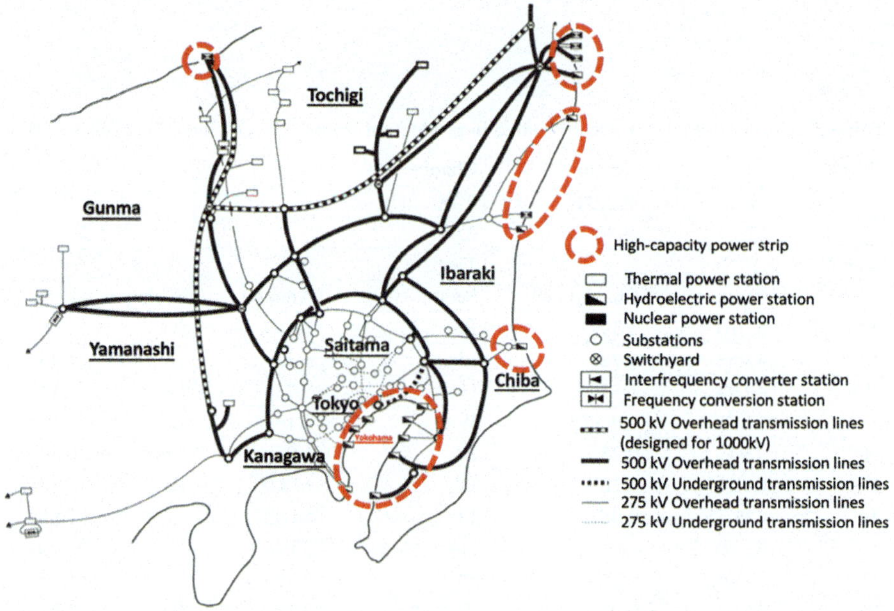

Fig. 6 Regional power system in the Tokyo metropolitan area. (Data source: Wide-area system maintenance committee secretariat 2015)

Table 6 Consumption of energy sector in Kanto region in 2011. Units: Million yen

Region	Intra-regional needs	Self-sufficiency rate	Import	Inflows	Export	Outflows
Tokyo	2,660,447	64.8%	−140	–939,696	8108	0
Saitama	942,786	59.8%	−13	–378,959	15	6103
Yamanashi	112,697	76.0%	–5	–27,086	357	0
Ibaraki	547,323	100.0%	−15	0	122	61,335
Gunma	338,266	73.5%	0	–89,479	0	1712
Tochigi	321,742	39.1%	–195,821	–195,821	0	15,757
Chiba	1,050,674	100.0%	0	0	3033	1,225,656
Kanagawa	1,456,227	66.4%	–488,531	–43	18	918,592
Yokohama	543,669	100.0%	−17	0	8	299,594

Note: Outflow refers to the sending of goods to other parts of the country; Inflow refers to the transfer of goods from one piece of land to another within a country
Made by author, data source: Input–output table for Tokyo, Saitama, Ibaraki, Gunma, Tochigi, Chiba, Kanagawa, Yokohama, 2011

5.2 *Impact of Earthquakes on CO_2 Emissions and Resource Use Efficiency*

Energy consumption is the main source of carbon emissions. According to the Japanese Ministry of the Environment, Japan's carbon dioxide emissions in 2020 were 1044 million tons, of which energy-related carbon dioxide emissions were 967 million tons (84.2%), and non-energy-related carbon dioxide emissions were 78.9 million tons (6.7%). After the 2011 earthquake, Japan's energy structure changed significantly, with the replacement of nuclear energy by natural gas, oil, and coal, which led to a significant increase in total carbon emissions and energy consumption of the energy conversion sector. The carbon emission factor of electricity also increased, thereby increasing the carbon emissions of other sectors from the consumption of electricity.

Analysis of survey of carbon emissions and energy consumption of each sector in Yokohama showed that after the 2011 earthquake, carbon emissions in all sectors increased significantly (Yokohama, 2022). Although the energy consumption of all types of sectors (except the energy conversion sector) decreased due to planned blackouts conducted by the local government, total carbon emissions increased significantly, due to a significant increase in the carbon emissions coefficient of electricity consumption in 2011 in the business, household, and industrial sectors. Figure 7 shows the trend of changes in carbon emissions per unit of electricity consumption in each sector in Yokohama before and after the earthquake. As for the

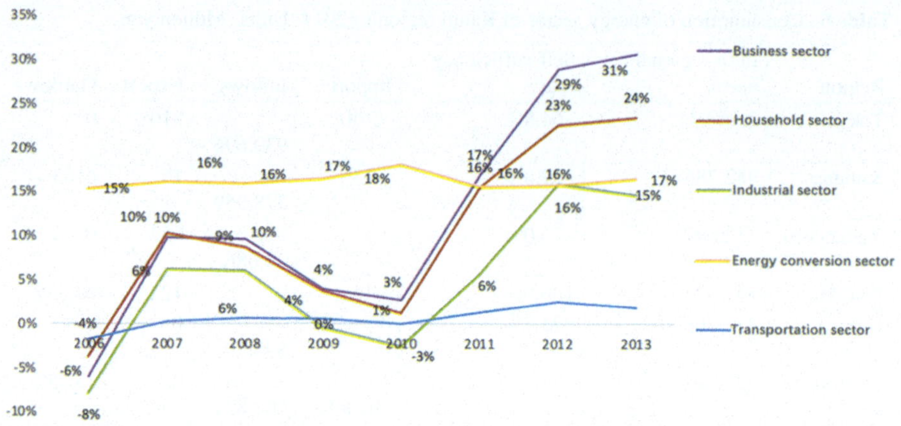

Fig. 7 Change of carbon emissions per unit of electricity use, by sector, before and after the 2011 earthquake (based on 2005). (Data source: Yokohama GHG emissions, 2019)

energy conversion sector, after the earthquake, the carbon emissions and energy consumption of this sector increased significantly, resulting in a slight change in carbon dioxide emissions per unit of energy consumption.

Through input–output analysis, we could identify the substantial increase in energy use in the energy and water sectors that occurred after the 2011 earthquake, which may indirectly increase resource consumption in other sectors. For example, from the results of changes in resource consumption triggered by the final sector, it could be seen that a decline in resource use efficiency in the energy sector increased energy consumption caused by electricity demand in the household sector. Also, by analyzing energy use embodied in production layers, we could find that the increase in energy demand in the energy sector directly led to more energy consumption in the coal and petroleum sector, leading to more carbon emissions of the energy conversion sector. On the other hand, due to damage to the power and transportation infrastructure, production of the industrial sector, including the manufacturing sector, was also greatly affected. The efficiency of water and energy use in the industrial sector declined, leading to more carbon emissions. In terms of water supply, the earthquake caused damage to the water infrastructure in the Kanto region, including both the displacement of and damage to water pipes, which also affected energy and water use efficiency of the water sector.

It is clear that the impact of changes in resource use efficiency of sectors will expand through inter-sectoral correlations and eventually affect the total resource consumption of each sector. Therefore, the resource use efficiency of the resource supply sector should be improved first, which will indirectly reduce resource consumption in other sectors. Secondly, resource consumption sectors should actively make use of natural resource systems such as solar power generation and rainwater harvesting, which can help reduce pressure on urban physical provisioning systems and enhance the ability of various sectors to deal with risks.

5.3 Advantages and Limitations of Input–Output Analysis

Through input–output analysis, the flow of material resources between sectors can be estimated based on economic linkages between urban sectors, which cannot be achieved by other methods. By linking environmental indicators with input–output tables, we can evaluate impacts on the environment in the context of a city's economic development. Meanwhile, since input–output tables have clear geographical boundaries, the connections between regions can also be clarified, so that resource transfers between regions and related impacts associated with economic activities can be further analyzed. However, as mentioned, the compilation of input–output tables is very cumbersome. Thus, not every region has an input–output table, which limits the application of regional input–output analysis. Data required for input–output analysis is difficult to obtain, and results of analysis will deviate from the actual situation depending on the accuracy and scope of data. Therefore, for input–output analysis it is necessary to clarify the specific economic activities that are to be included from the inputs and outputs of each sector into the input–output table, as well as the resource consumption caused thereby, so as to more accurately evaluate resource flows between sectors and changes in resource use efficiency.

6 Conclusion

This chapter assessed resource consumption and flows between sectors in order to analyze the impacts of the 2011 Tohoku earthquake and tsunami disaster on urban resource supply and resource use efficiency, with Yokohama as the study area. Our research found that the energy and water sectors responsible for resource generation and distribution were significantly affected by the disaster, which indirectly contributed to increased energy and water consumption triggered by some sectors due to cross-sectoral interdependencies. Also, the energy and water sectors are highly correlated, so changes on one side directly and indirectly affect the resource consumption and use efficiency on the other. Therefore, attention should be paid to how to achieve the stability of urban physical provisioning by promoting the capacity of energy and water infrastructure systems to respond to risks. Meanwhile, for sectors where resource use has declined, the resource use pattern of sectors should be improved through technological advancement or greater awareness of conservation.

Acknowledgments This article is part of the M-NEX project (m-nex.net), a grant of the Sustainable Urbanization Global Initiative (SUGI) —Food–Water–Energy Nexus by the Belmont Forum and Japan Science and Technology Agency (JST). The authors are grateful for financial support from the Taikichiro Mori Memorial Research Fund of Keio University.

References

Agency for Natural Resources and Energy; Annual Energy Report of Japan in Fiscal Year 2023, p77; https://www.enecho.meti.go.jp/about/whitepaper/2023/pdf/

Albrecht TR, Crootof A, Scott CA (2018) The water-energy-food nexus: a systematic review of methods for nexus assessment. Environ Res Lett 13(4):043002. https://doi.org/10.1088/1748-9326/aaa9c6

Arthur M, Liu G, Hao Y, Zhang L, Liang S, Asamoah EF, Lombardi GV (2019) Urban food-energy-water nexus indicators: a review. Resour Conserv Recycl 151:104481. https://doi.org/10.1016/j.resconrec.2019.104481

BEIS (2020) Energy White Paper. Energy Department 44:903–905; http://www.ret.gov.au/energy/facts/white_paper/Pages/energy_white_paper.aspx

City of Yokohama; Yokohama Economic and Trade Bureau, 2015 Yokohama City Input-Output Table; https://www.city.yokohama.lg.jp/business/kigyoshien/tokei-chosa/rennkannhyou.html

City of Yokohama, Greenhouse gas emissions in the city in FY2020; https://www.city.yokohama.lg.jp/cityinfo/koho-kocho/press/ondan/2022/0422haishutsuryou.files/0001_20220420.pdf

Dargin J, Berk A, Mostafavi A (2020) Assessment of household-level food-energy-water nexus vulnerability during disasters. Sustain Cities Soc 62:102366. https://doi.org/10.1016/j.scs.2020.102366

Feng C, Qu S, Jin Y, Tang X, Liang S, Chiu ACF, Xu M (2019) Uncovering urban food-energy-water nexus based on physical input-output analysis: the case of the Detroit Metropolitan Area. Appl Energy 252:113422. https://doi.org/10.1016/j.apenergy.2019.113422

Ghobarah A, Saatcioglu M, Nistor I (2006) The impact of the 26 December 2004 earthquake and tsunami on structures and infrastructure. Eng Struct 28(2):312–326. https://doi.org/10.1016/j.engstruct.2005.09.028

Hameed M, Moradkhani H, Ahmadalipour A, Moftakhari H, Abbaszadeh P, Alipour A (2019) A review of the 21st century challenges in the food-energy-water security in the middle east. Water (Switzerland) 11(4):682. https://doi.org/10.3390/w11040682

Haraguchi M, Kim S (2016) Critical infrastructure interdependence in New York City during Hurricane Sandy. Int J Disaster Resil Built Environ 7(2):133–143. https://doi.org/10.1108/IJDRBE-03-2015-0015

Hussien WA, Memon FA, Savic DA (2017) An integrated model to evaluate water-energy-food nexus at a household scale. Environ Model Softw 93:366–380. https://doi.org/10.1016/j.envsoft.2017.03.034

Ijo T, Oshima K (2019) Creation and analysis of an inter-industry relationship table among the seven regions of Japan, with Minato Mirai 21 and Yokohama city center as the segments. Yokohama International Social Science Research 23(3):283–295

International Energy Agency (IEA) (2021) Japan 2021—Energy Policy Review. International Energy Agency; https://www.iea.org/reports/japan-2021

Institute for Global Environmental Strategies; Yokohama in 2030.(2021). https://www.iges.or.jp/en/publication_documents/pub/policyreport/en/11892/IGES_Yokohama+Scenario_Web.pdf

Kanagawa Prefectural Industrial Labor Bureau (2020); 32 March 2018

Koike J, Ukon T (2012) Economic Impact Analysis of Electricity Supply Shortages. Collected Papers of the Civil Engineering Society d3 (Civil Engineering Design) 68(5):I_111-I_119. https://doi.org/10.2208/jscejipm.68.I_111

Koshimura S, Hayashi S, Gokon H (2014) The impact of the 2011 Tohoku earthquake tsunami disaster and implications to the reconstruction. Soils Found 54(4):560–572. https://doi.org/10.1016/j.sandf.2014.06.002

Leontief WW (1951) Input-Output Economics. 185(4):15–21

Li X, Yang L, Zheng H, Shan Y, Zhang Z, Song M, Cai B, Guan D (2019) City-level water-energy nexus in Beijing-Tianjin-Hebei region. Appl Energy 235:827–834. https://doi.org/10.1016/J.APENERGY.2018.10.097

Liu J, Hull V, Godfray HCJ, Tilman D, Gleick P, Hoff H, Pahl-Wostl C, Xu Z, Chung MG, Sun J, Li S (2018) Nexus approaches to global sustainable development. Nature Sustainability 1(9): 466–476. https://doi.org/10.1038/s41893-018-0135-8

Ministry of the Environment of Japan; Greenhouse Gas Inventory Office of Japan of NIES (2021) National Greenhouse Gas Inventory Report of Japan in Fiscal Year 2021, p 794; https://warp.da.ndl.go.jp/info:ndljp/pid/11434599/www.cger.nies.go.jp/publications/report/i144/i144.pdf

Owen A, Scott K, Barrett J (2018) Identifying critical supply chains and final products: an input-output approach to exploring the energy-water-food nexus. Appl Energy 210:632–642. https://doi.org/10.1016/j.apenergy.2017.09.069

Rose A, Benavides J, Chang SE, Szczesniak P, Lim D (1997) The regional economic impact of an earthquake: direct and indirect effects of electricity lifeline disruptions. J Reg Sci 37(3): 437–458. https://doi.org/10.1111/0022-4146.00063

Son BG, Chae S, Kocabasoglu-Hillmer C (2020) Catastrophic supply chain disruptions and supply network changes: a study of the 2011 Japanese earthquake. Int J Oper Prod Manag 41(6): 781–804. https://doi.org/10.1108/IJOPM-09-2020-0614

Wang S, Fath B, Chen B (2019) Energy–water nexus under energy mix scenarios using input–output and ecological network analyses. Appl Energy 233:827–839. https://doi.org/10.1016/J.APENERGY.2018.10.056

Yokohama City Basic Policy on Power Saving and Energy Saving Measures (2013)

Zhu X, Sun B (2017) Study on earthquake risk reduction from the perspectives of the elderly. Saf Sci 91:326–334. https://doi.org/10.1016/j.ssci.2016.08.028

The Potential of Hydrogen Energy and Innovative Diffusion Models in Japan

Baizi Zeng and Wanglin Yan

Abstract Japan's energy system is vulnerable in terms of energy security and self-sufficiency, due to the nation's scarcity of fossil fuel resources and heavy reliance on imports, but the government sees promise in hydrogen energy. Hydrogen refueling stations and fuel cells are essential components in transportation and energy sectors, but several obstacles exist. This chapter looks at two applications for hydrogen. First it compares scenarios for introducing fuel cell vehicles and buses along with the required hydrogen refueling stations in Kanagawa Prefecture. Analysis shows that uneven user distribution will hinder the deployment of hydrogen refueling stations and negatively impact service equity due to challenges providing service where hydrogen demand is low. However, fuel cell buses could play a major role in reducing the financial risks of hydrogen stations and increasing service equity geographically. The second application examined is the concept of an integrated system built around fuel cell use to provide electricity, heat, and water to individual households and buildings, seen from the perspective of the food-energy-water (FEW) nexus. Such decentralized systems could form a complete cycle of resource collection, production, distribution, and recycling based on independent home systems. Combinations of systems offer even greater prospects for the future of hydrogen.

Keywords Hydrogen energy · Fuel cell vehicle · Route bus · Ene-Farm

B. Zeng
Graduate School of Media and Governance, Keio University, Fujisawa, Kanagawa, Japan

W. Yan (✉)
Faculty of Environmental Information Studies, Keio University, Fujisawa, Kanagawa, Japan
e-mail: yan@sfc.keio.ac.jp

W. Yan et al. (eds.), *Resilient and Adaptive Tokyo*,
https://doi.org/10.1007/978-981-99-3834-6_12

1 Introduction

To achieve carbon neutrality and address domestic energy challenges, Japan has been seeking a sustainable shift in the energy mix. Hydrogen energy has received increasing attention in recent years due to multiple benefits, including being clean, renewable, and highly efficient, and capable of being produced from numerous sources. The government has proposed the idea of building a "hydrogen society" and presented a detailed road map in 2014. Since then, the development of hydrogen energy has become a fundamental national policy and a series of related strategies have been formulated.

Among the many potential applications for hydrogen energy, fuel cell vehicles (FCVs) and home-use fuel cells would be the most closely associated with changes in the lifestyles of residents. Rather than the public transportation sector, Japan has chosen private-use FCVs as one area to seek breakthroughs to expand the use of hydrogen energy in the transportation sector. However, the high cost of building hydrogen refueling stations (HRSs) and the uncertain market penetration of FCVs have created somewhat of a "chicken-and-egg" situation. The scattered and limited demand for hydrogen in the private FCV market make the economics of HRSs a major challenge, while the lack of convenient HRSs also hinders the development of the FCV market. The Japanese government envisions the development of the FCV market in parallel with the construction of new HRSs, but data show that the current market penetration of fuel cell vehicles is far below targets.

Conversely there appears to be greater potential demand for hydrogen in public transportation systems, and it would be more stable and concentrated. In contrast to Japan's development strategy, China has chosen public transportation as the breakthrough sector to expand the scale of hydrogen energy use. While building a network of HRSs, the government supports the introduction of fuel cell (FC) buses and creation of a hydrogen path to support HRS businesses in the early stages of the market.

Thus, to shine a light on the potential for such an approach in Japan, through the collection and calculation of traffic data in the study area of Kanagawa Prefecture, this chapter will compare the potential hydrogen demand from the public transportation system and private FCV market, and analyze the impacts of each on the spread of a network of hydrogen refueling stations.

On a separate but related line of pursuit, this chapter also looks at home-use fuel cell systems. Such systems use electrochemical cells designed with combined heat and power (CHP) technology, and use natural gas or hydrogen energy to produce electricity, heat, and hot water to supply homes. Due to the on-site production and co-generation technology, these systems have higher efficiency than traditional electricity production and can also reduce electrical transmission loss and heat loss. Another benefit is that home fuel cells can be configured to operate off-grid and provide continuous power and hot water for several days in the event of a natural disaster or power grid interruption. Taking this concept further, a domestic microgrid and a food-energy-water (FEW) nexus system could be built based on fuel cells, integrated with other power sources such as solar photovoltaic, as well as water tanks and storage batteries. The use of home fuel cells could transform the household of the future and have positive impacts on FEW nexus efficiency, safety, resiliency, and decarbonization.

Thus, to pursue this angle, this chapter also discusses the impacts of fuel cell technology on household power systems of the future and the FEW nexus at the household level.

2 Hydrogen Energy in Japan

2.1 Challenge of Japan's Energy Structure and Potential of Hydrogen

Energy challenges are serious for Japan because of the nation's resource scarcity. Japan's energy system is fundamentally vulnerable in terms of energy security and self-sufficiency because Japan is poor in fossil fuel resources and relies heavily on overseas imports. The energy self-sufficiency rate of Japan has remained extremely low, at about 6–7% in the past decade, due to the shutdown of nuclear power plants after the Great East Japan Earthquake in 2011, and it is the second-lowest among OECD countries. In 2020, Japan declared a goal of achieving carbon neutrality and building a decarbonized society. However, energy consumption still accounts for 84.2% of Japan's total CO_2 emissions (MOE 2021). Therefore, a structural shift and the large-scale introduction of clean and renewable alternative energy is essential to achieve these ambitious targets.

Hydrogen energy is a zero-emission energy (at least at the point of use) that can be produced from a variety of different primary resources, such as methane, natural gas, and coal. The diversity of production methods can contribute significantly to improving energy security and flexibility. Besides fossil fuel sources, hydrogen produced from renewable energy is considered a life-cycle carbon-free energy (Zhang et al. 2020). Hydrogen can also be used as an energy carrier to store and distribute an oversupply of renewable energy and play a key role in solving the issue of fluctuating output of renewable energy, and help to scale up the application of renewable energy. Combined with fuel cell plants, hydrogen energy can be used in a variety of ways, including electricity generation (hydrogen gas turbines), mobility (FCVs), and households (fuel cells) (Table 1).

2.2 Hydrogen Energy Strategies in Japan

Japan is the first country to have established a national strategy for hydrogen development, and its national energy strategies in recent years have emphasized

Table 1 Hydrogen strategies in Japan

Year	Policy	Description
2014	Fourth strategic energy plan	Introduces concept of a" hydrogen society"
	Hydrogen fuel cell road map	Describes road map to achieve a hydrogen society through three phases
2016	Hydrogen fuel cell road map (revised)	Sets clear targets for development of hydrogen use and supply
2017	Hydrogen basic strategy	Presents an action plan from the perspective of the entire energy system
2018	Fifth strategic energy basic plan	Presents fundamental reinforcement of measures for realizing a hydrogen society
2019	Hydrogen fuel cell road map (revised)	Revises targets revision and establishes a working group for follow-up in each sector
	Hydrogen and fuel cell technology development strategy	Sets targets for technology development
2020	Green growth strategy through achieving carbon neutrality in 2050	Sets hydrogen as a key method to achieve carbon neutrality

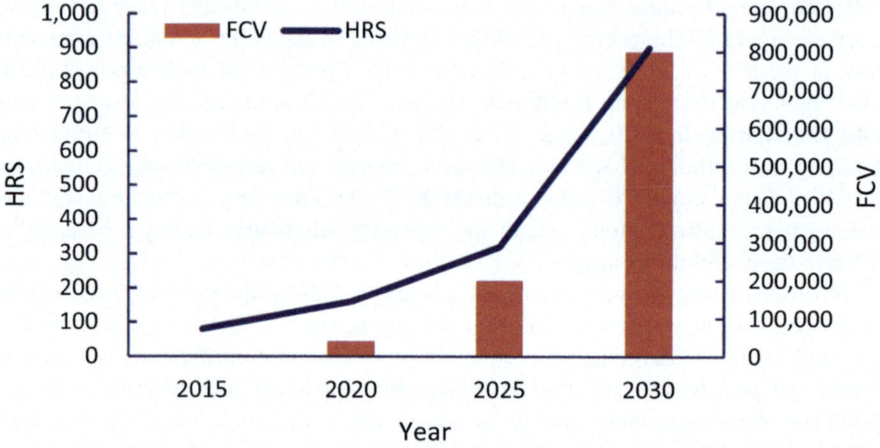

Fig. 1 Japan's hydrogen development targets (number of units) for hydrogen refueling stations and fuel cell vehicles. (Source: Ministry of Economy, Trade and Industry, 2019)

the significance of hydrogen. In 2014, the fourth Strategic Energy Plan proposed the idea of building a "hydrogen society." The same year, the Strategic Road Map for Hydrogen and Fuel Cells was released, and later revised in 2016 and 2019. It includes specific and detailed development targets for hydrogen utilization and supply, as shown in Fig. 1 and Table 2. Japan established the Basic Hydrogen Strategy in 2017 as an action plan for both the private and public sectors. In 2018, the Fifth Strategic Energy Plan again emphasized the importance of hydrogen energy and proposed the reinforcement of measures for realizing a hydrogen society. In 2019, a working group was established to evaluate targets for the road map. To

Table 2 Target achievement rates relative to hydrogen Fuel cell road map

Item and data year	HRS (2021)	Private FCV (2019)	FC bus (2019)	FC forklift (2019)
Achievement rate of 2020 target	157 (98.1%)	3,757 (9.4%)	57 (57%)	250 (50%)
Target (for 2020)	160	40,000	100	500
Target (for 2025)	320	200,000	–	–
Target (for 2030)	900	800,000	1,200	10,000

HRS hydrogen refueling station, *FCV* fuel cell vehicle, *FC* fuel cell
(Source: METI 2020a, b)

promote hydrogen energy technology innovation, the Hydrogen and Fuel Cell Technology Development Strategy was formulated, including clear targets for technological improvements. The Japanese government also established the Green Growth Strategy Through Achieving Carbon Neutrality in 2050, which focusses on three major industrial sectors that are all closely related to hydrogen energy (METI 2020a, b): energy; transportation and manufacturing; and construction. For example, in the construction sector, fuel cell systems can supply both electricity and heat for households and reduce the consumption of primary energy through co-generation. This could be an important step for next-generation housing. In the transportation and manufacturing sectors, Japan is targeting the development of hydrogen-powered fuel cell vehicles, fuel cell ships, aircraft, and machinery.

2.3 Current Development Status and Barriers

As of the most recent data, there were 157 hydrogen refueling stations in Japan (including 117 stationary and 40 mobile units), mainly in central areas of the Tokyo and Osaka metropolitan regions (Fig. 2). This number represents 98.1% of the target (2020) at that point under the Hydrogen Fuel Cell Road Map (Table 2). Generally, the supply capacity of a stationary and mobile HRS is 300 Nm^3/h and 100 Nm^3/h, respectively. More than 700 new hydrogen stations are planned for the future, which means that a rational and efficient deployment plan will be needed. Meanwhile, although the Basic Hydrogen Strategy indicates that FCVs and hydrogen station deployment are to be promoted in parallel, there is a significant gap between achievement rates for HRS and FCVs. Table 2 illustrates the achievement rates relative to target years in the road map. According to a METI report, the achievement rates in 2019 for FC bus and forklift targets were only about 50% of the 2020 targets, or 57 and 250, respectively. For private fuel cell vehicles, the number and targets achievement rate were even lower, at 3,757 units and 9.4%, respectively (METI 2019). The table suggests that development of hydrogen supply relative to market demand is extremely unbalanced.

From the perspective of investment in hydrogen energy in the mobility sector, the construction cost of an HRS has been reduced in recent years. According to the

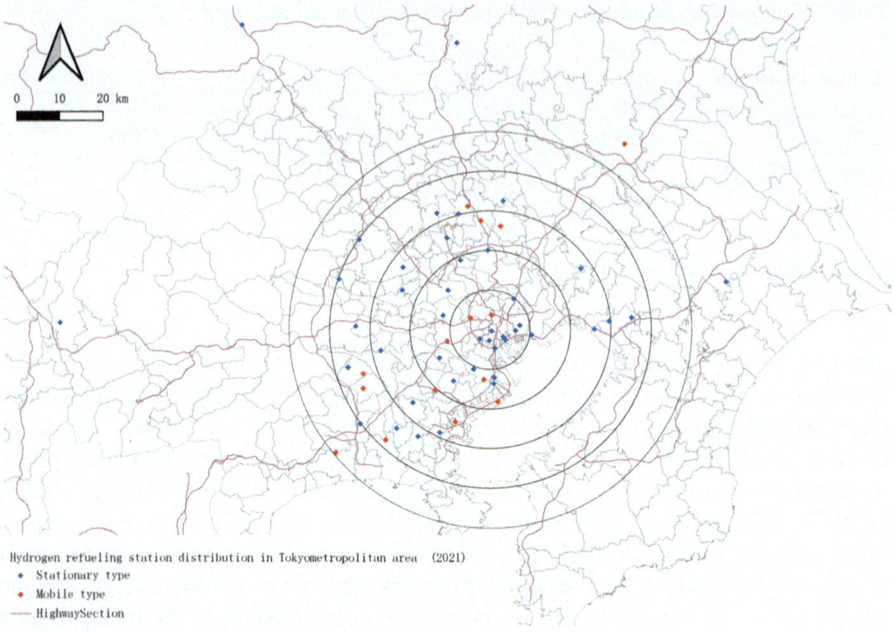

Fig 2 Hydrogen refueling station distribution in Tokyo metropolitan area (2021). (Source: Next Generation Vehicle Promotion Center: NeV, 2021)

METI report (METI 2020a, b), the average construction cost of an HRS remained at a level of 330 million yen in 2019, about 130 million yen reduced compared to the initial development period. Meanwhile, annual operating costs had decreased to about 20 million yen, down from 31 million yen per year in 2019. However, HRS construction and operating costs are still much higher than for the standard gasoline station. At the same time, low market penetration of FCVs makes it difficult for an HRS to be economically viable, and the cost of an FCV remains in the same price range as luxury cars. Taking the Toyota *Mirai* as an example, the selling price is 7.1 million to 8.6 million yen. The government offers about a three million yen subsidy per vehicle for FCV buyers, but even so, the price is still higher than most electric vehicles (EVs) or plug-in hybrid electric vehicles (PHEVs).

Hydrogen refueling stations that can produce and distribute hydrogen energy to facilities and vehicles are an important part of infrastructure for the future large-scale utilization of hydrogen energy. However, there are several obstacles to the rapid spread of independent HRSs. Two obstacles are construction and operating costs, as mentioned above. Meanwhile, as mentioned, hydrogen demand is insufficient due to low market penetration and capacity utilization is not sufficient to be economically viable in the early stages of FCV market development. In addition, a mid-size stationary hydrogen station requires about 700 m^2 of space, and finding suitable land is a challenge, especially in urban areas.

The relationship between HRS promotion and FCV market development is often described as a "chicken and egg" situation. Consumers will have confidence to purchase an FCV only if adequate hydrogen stations are available to provide satisfactory service coverage, but investors will usually build hydrogen stations only if there is sufficient hydrogen demand to maintain high capacity utilization rates and achieve profitability (Li et al. 2020; Michalski et al. 2011; Zhang et al. 2020). From the point of view of Japan's strategic plan, it might be more difficult to solve this challenge since Japan's strategy is centered on the private vehicle market, which has considerable uncertainties and cannot guarantee stable and sufficient demand at the early transition stage. Also, private FCV users are sparsely distributed nationwide and cannot be centralized to support HRS demand. Hence, investors in HRS businesses would face significant financial risks, which may hinder their investment motivation. To address this challenge, a coordinated approach to the introduction of hydrogen demand and hydrogen station deployment could be a solution. This would mean that when decision makers create an HRS layout plan, they also need to give adequate consideration to promoting commensurate hydrogen demand from fuel cell vehicles (Li et al. 2020).

3 HRS Development Proposal Based on Public Route Buses

3.1 Methodology and Data Use

The private FCV market cannot guarantee the scaling up to the extent expected. Compared to the private FCV market, a strategy to introduce FC buses might make it easier to plan the layout of a network of HRSs and coordinate the hydrogen supply side with demand side, for a number of reasons. First, route buses have large potential hydrogen demand since the annual hydrogen consumption of one FC bus amounts to that of about 45 private FCVs. Second, hydrogen demand for route buses would be stable and centralized, since route length, operational frequency, and refueling locations are fixed. Third, bus transit centers, where buses assemble and are stored, typically serve as core locations for bus operations and maintenance facilities, and would likely have sufficient land area available for HRS construction. This chapter will consider transit centers as candidate locations to install HRSs and calculate potential hydrogen demand and regional distribution for both FCVs and FC buses, then compare the economic viability of HRSs in each scenario.

This analysis uses Kanagawa Prefecture as the study area because the Kanagawa government has made a detailed strategic road map of hydrogen energy and aims to achieve these development targets in 2025 (Kanagawa Next Generation Vehicle Promotion Council 2015). According to the strategic road map, the FCV targets for 2025 range from 20,000 to 100,000 vehicles, under three scenarios. Using those three market penetration scenarios (20,000, 50,000, and 100,000 vehicles), benchmarking with percentages of targets achieved, we estimate potential FCV users based on demographic data and vehicle ownership ratios. Future population projections by 1 km mesh are multiplied by household ratio to estimate regional distribution by household type in 2025. Then the numbers of households by 1 km mesh are multiplied by vehicle ratio and FCV market penetration in different scenarios to estimate the regional distribution of potential FCV users.

This chapter then uses bus route data from National Land Information Division (2017) to calculate potential annual hydrogen demand on each bus route system for 1 year using the equation provided below. As bus frequency differs on weekdays versus weekends, potential annual hydrogen demand is calculated separately for each. The calculated potential hydrogen demand is doubled to account for the fact that route buses travel back following the same route. The results of potential hydrogen demand of each bus route are summed up and associated with their respective transit centers, to calculate the total potential hydrogen demand at each transit center. Transit center data is obtained from Google Earth, bus company data, and government reports. We start with Eq. (1):

$$D_{\text{kbus}} = E_{\text{FC bus}} \times 2 \times \left\{244 \times \left(L_k \times F_{\text{kweekdays}}\right) + 121 \times \left(L_k \times F_{\text{kweekends}}\right)\right\} \tag{1}$$

where D_{kbus} represents potential hydrogen demand of bus route system $k \in K$, and L_k is the length of bus route system $k \in K$. $E_{FC\ bus}$ is the fuel economy of an FC bus, which is 0.9 km/Nm^3. $F_{\text{kweekdays}}$ and $F_{\text{kweekends}}$ represents operational frequency of bus route $k \in K$ on weekdays and weekends (including public holidays), which are 244 days and 121 days, respectively. As mentioned, total one-way demand on each route system is multiplied by 2 to account for each bus's return trip.

Future population projections by 1 km mesh and household ratios are obtained from the National Land Information Division and the National Institute of Population and Social Security Research Center. To estimate the number and regional distribution of potential FCV users, it is crucial to take full account of future demographic changes and geographic characteristics, including population, household types, vehicle ownership rates, and location types. This chapter uses published research results of vehicle ownership rates by different mesh type (distance to station; within or outside trade area; mesh population size) and household type (single; nuclear; other) (Ariga and Matsuhashi 2013). Some of the transit center data is provided by the Kanagawa Chuo Kotsu Co. bus company, and the rest of the data is obtained from the official website of famous bus company in Kanagawa Prefecture. The total number of bus transit centers data collected is 71 and the number 1–71 will refer to each of bus transit center.

3.2 *Regional Distribution and Hydrogen Demand of FCVs*

Fig. 3 shows estimated regional distribution maps of FCV users by 1 km mesh, which vary significantly under different market penetration scenarios. As shown in

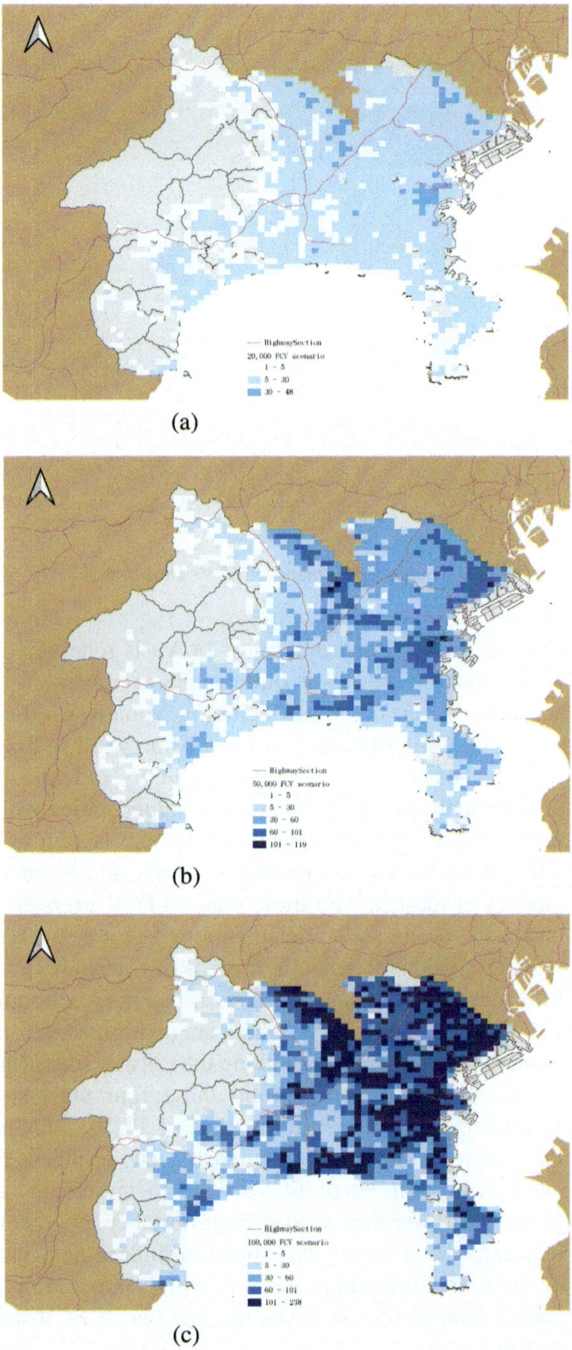

Fig. 3 Regional distribution of FCV users under three scenarios. (**a**) 20,000 FCV scenario. (**b**) 50,000 FCV scenario. (**c**) 100,000 FCV scenario

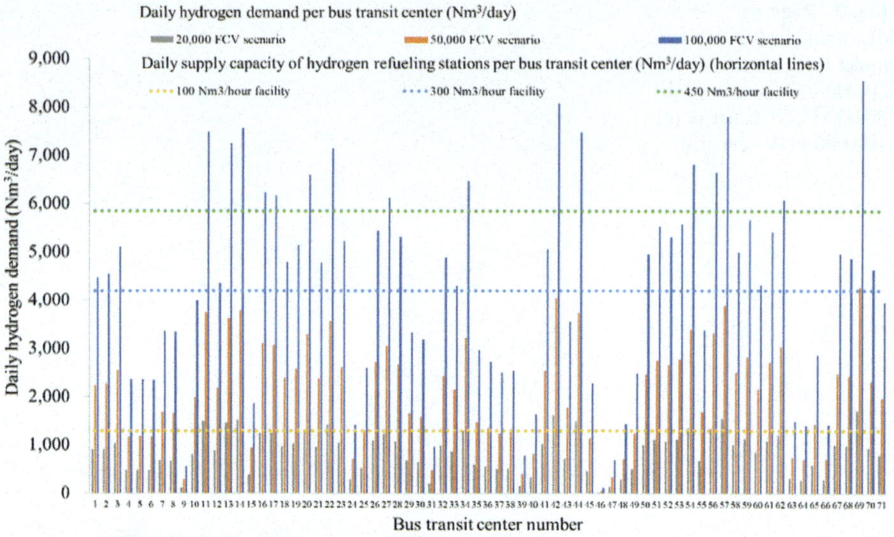

Fig. 4 Daily hydrogen demand of FCVs in different scenarios

Fig. 3a, regional distribution of FCVs is relatively even and the number of FCVs within most of meshes is quite small, at only 5–30 in the 20,000 unit development scenario. In certain areas (including Minami ward of Yokohama City, east area of Kawasaki City, Minami ward and Chuo ward of Sagamihara City), there are only a few meshes where the number of FCVs is larger than 30. These three cities are government ordinance-designated (larger) cities with populations over 500,000. In the 50,000 FCV development scenario as shown in Fig. 3b, the number of meshes with 30–60 FCV users increases rapidly and is spread over into the east area. Four clusters of meshes with more than 60 FCV users appear, mainly in the government ordinance-designated cities mentioned above, and special cities with populations of more than 200,000 in southern areas of Kanagawa Prefecture. Finally, in the 100,000 FCV development scenario shown in Fig. 3c, the number of FCV users of meshes with more than 101 FCVs which are distributed in four clusters mentioned above has increased and is spread out along the railway lines. Meshes with 30 to 101 FCV users are mainly distributed around these four clusters. In western and southeastern Kanagawa Prefecture, the number of FCVs remains extremely low, at just 1–30 FCVs, due to low population density. The results of FCV user distribution show that even if the number of FCVs were to increase to 100,000 units as projected by government, meshes with a large number of potential FCV users will be mainly concentrated in three government ordinance-designated cities and the southern area of the prefecture where there is a high population density. However, in the west, central, and southeast areas, the number of FCV users is quite small and sparsely distributed.

Fig. 4 shows the potential hydrogen demand of FCVs in three different scenarios. The red, blue, green dotted lines (horizontal) represent the supply capacity of HRSs at 100 Nm^3/h, 300 Nm^3/h, and 450 Nm^3/h, which works out to supply capacity of 1,300 Nm^3/day, 4,200 Nm^3/day, and 5,850 Nm^3/day, respectively. In the 100,000 FCV scenario, maximum and minimum hydrogen demand for FCVs is 8,554.5 Nm^3/day and 116.98 Nm^3/day, respectively. Average daily demand is 4,257 Nm^3/day, and 40 HRS sites would have enough capacity for a 100% capacity utilization rate with mid-size HRSs. However, in the two scenarios below the 100,000 FCV level in Kanagawa Prefecture, the picture changes. In the 50,000 FCV scenario, average demand is 2,128.7 Nm^3/day, only enough to justify mid-scale HRS facilities at a 50.6% utilization rate. Only one site would have enough demand to support a 100% utilization rate of a mid-sized HRS. Finally, in the 20,000 FCV scenario, hydrogen demand is extremely low and average hydrogen demand across all HRS locations is only 852.1 Nm^3/day. The hydrogen demand at most bus transit centers is below the capacity of even a small-scale HRS.

3.3 *Hydrogen Demand for Route Buses*

Figure 5 shows the potential daily hydrogen demand of FC buses and FCVs separately, and the sum of both. Hydrogen demand for FCVs is calculated using the 100,000 vehicle scenario. Comparing demand of FC buses versus FCVs, it is apparent from Fig. 5 that even if the target for FCV development was 100% achieved, which means sales of 100,000 FCVs, hydrogen demand for FC buses would still be two to four times larger. Maximum and minimum hydrogen demand for FC buses are calculated at 24,638.5 Nm^3/day and 990.2 Nm^3/day, respectively. The average daily demand for FC buses is calculated at 10,963.9 Nm^3/day, about two times larger than supply capacity of a large-scale HRS.

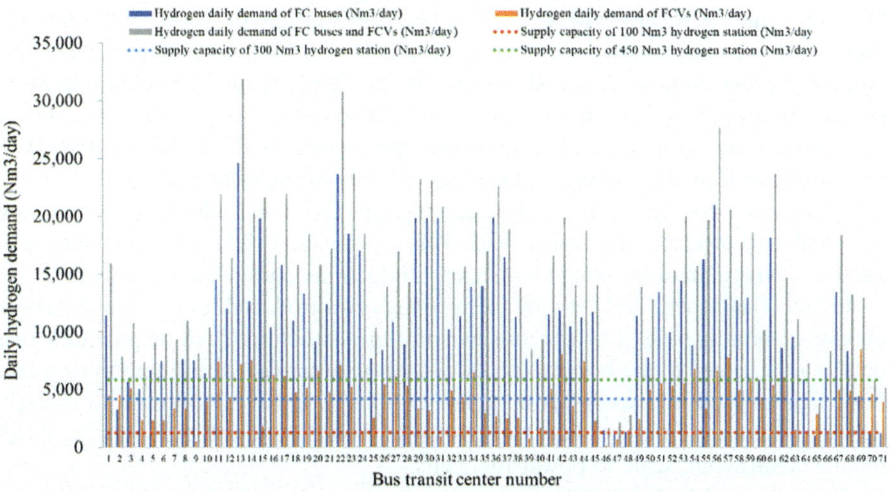

Fig. 5 Comparison of daily hydrogen demand of FC buses, FCVs, and the sum of both

For FC buses, there are 64 transit centers and 60 transit centers of which daily hydrogen demand is larger than 4,200 Nm^3/day and 5,850 Nm^3/day, which means that demand of almost all transit centers is quite enough to support a mid- or large-scale hydrogen station operating at 100% of capacity. Only three transit centers would not have enough hydrogen demand to support 100% capacity utilization rate for a mid-scale hydrogen station and four transit centers not enough hydrogen demand to justify a small HRS. For these seven transit centers (Sugita, Miyano, Yugawara, Sekimoto, Yokohama, Kugenuma, and Daishinto) the average hydrogen demand is about 1,574.5 Nm^3/day. However, if these they could also provide public refueling services for FCV users, the Sugita, Kugenuma, and Daishinto transit centers could have enough hydrogen demand to support a 100% capacity utilization rate for a mid-scale HRS. The other four transit centers could have enough demand to support a small-scale HRS at a 100% operating ratio. Another benefit of this scenario is that the estimated hydrogen demand is stable and predictable since bus routes and operational frequencies are fixed and well known.

4 Role of Fuel Cell Technology in the Food-Energy-Water Nexus

4.1 Development Status of Fuel Cells in Japan

Currently, commercial fuel cell systems can be classified into three types from the perspective of energy production method: indirect (reformed); indirect (water electrolysis); and hydrogen consuming fuel cell. Ene-Farm is the brand name in Japan for a household use indirect fuel cell system that operates through fuel (natural gas) reforming to produce hydrogen and supply power to individual households. Panasonic and Kyocera launched the Ene-Farm system in 2009, and sales for household use nationwide exceed 400,000 units so far (Fig. 6). H2One by Toshiba is a stand-alone indirect fuel cell that can be integrated with renewable energy to produce hydrogen on-site through the electrolysis of water. This system is designed to provide power for large facilities and communities, and has been installed in several Japan Rail (JR) stations, as well as office buildings and stadiums.

A hydrogen consuming fuel cell is directly charged with hydrogen gas produced separately, to produce power and heat. This type of fuel cell could be installed in locations where there is enough hydrogen production nearby, such as a factory.

Indirect (reformed) fuel cells are the type mainly introduced for use in households and small buildings as they can use existing infrastructure, such as pipelines that deliver natural gas to the home to produce hydrogen on-site, and supply power and heat. Currently, for households or small buildings, it is still difficult to directly deliver hydrogen from outside sources or to install renewable energy facilities on-site at sufficient scale to power fuel cells.

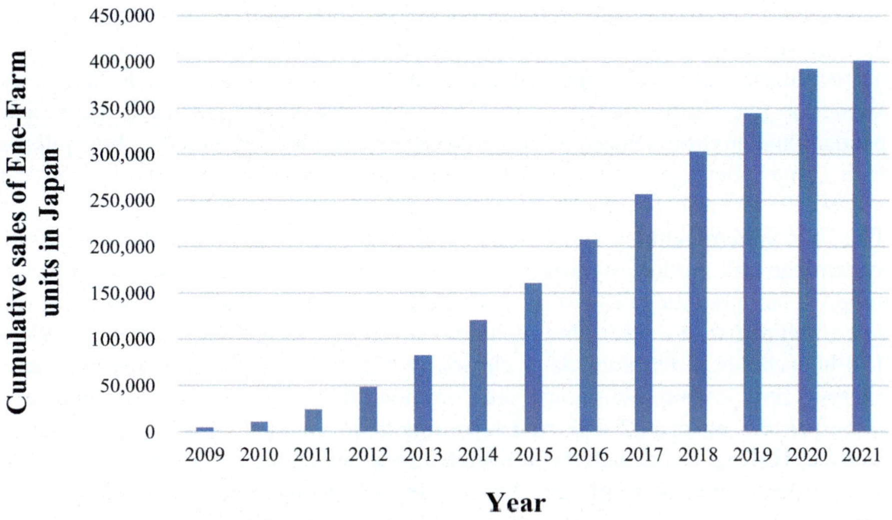

Fig. 6 Cumulative sales of Ene-Farm fuel cell units in Japan. (Source: Ene-Farm Partners)

4.2 Advantages and Significance of Introducing Fuel Cell Systems

The Ene-Farm system can extract hydrogen from liquefied petroleum gas (LPG) to generate electricity, while capturing surplus thermal energy to heat water and space in households. It has generally found applications in individual households and can be integrated with solar panels and battery to supply and storage electricity to the home. The energy management system can be configured to preferentially consume electricity from the Ene-Farm system, and users can sell the excess electricity to the power grid. In a sense, this system could also be considered as a "virtual" power plant in the future. From the perspective of energy saving, Ene-Farm can reduce power loss and heat loss in the process of production and distribution, because power is produced locally, and the co-generation technology can capture surplus heat from household components. According to data from Toshiba, this fuel cell system can reduce fossil fuel consumption in the household by more than 40%, and the savings could be even higher if the home is equipped with a solar photovoltaic system. This system can provide heat and electricity during power outages in the main grid for up to 8 days to maintain basic living (500w–700w). The system is also configured to connect to the family's fuel cell-equipped FCV to supply electricity to the home in the event of a grid interruption or natural disaster. In fact, two FCV vehicles could supply enough electricity for a small hospital or community shelters for up to 1.1 days of normal operation (Horikawa et al. 2015). The Ene-Farm is typically also equipped with a water tank as an emergency backup to supply domestic hot water. In summary, the development of fuel cell systems such as is significant for the water-energy nexus, in terms of security, efficiency, and decarbonization.

For the utilization of fuel cell technology in large facilities, Toshiba has developed a stand-alone indirect fuel cell system called H2One. Basically, this is a decentralized micro power grid that can produce, store, and distribute electricity. While the Ene-Farm still relies on a centralized gas supply, H2One is more independent and environmentally friendly, because it can be configured to be equipped with a renewable energy plant or connected to a renewable power network, and it can use excess renewable energy to manufacture hydrogen through water electrolysis. This, this system is like a micro power grid, and even without an LPG supply, this system can still produce hydrogen and distribute electricity, heat, and hot water to supply a large building or facility. According to recent simulations, this type of fuel cell combined with a solar photovoltaic micro-power grid could meet the needs of 150 households in terms of daily electricity demand, and could offer low cost of energy, zero carbon emissions, and promote the expanded use of renewable resources (Ghenai et al. 2020). And according to information from Toshiba's official website, H2One, has already been installed in Kawasaki city, can supply electricity and hot water for approximately 1 week for 300 people. It can also be used as a small-scale hydrogen station to refuel FCVs, which could reduce construction and operating costs by maintaining a relatively high capacity utilization rate, providing hydrogen for both buildings and FCVs. Actually, this kind of joint supplying system has already been put into use. Toshiba has introduced the first H2One Multi Station to fill FCVs and supply electricity to local food market in Tsuruga City, Japan.

The third type of fuel cell product directly consumes hydrogen from an external source to produce and supply electricity and heat. It could be set up near a factory or other facility that produces hydrogen (even as a by-product). Even if electricity and LP gas supplies are cut off in the event of an emergency, fuel cell systems can still generate electricity and heat by consuming hydrogen from any outside source to supply a factory, building, or single household. The H2Rex system developed by Toshiba has a typical hydrogen consuming fuel cell. More than 100 units have been sold in Japan, mainly for convenience stores, heated swimming pools, and hotels. Also, it can be used to supply hydrogen for both FCVs and fuel cell in building. Compared to conventional hydrogen energy distribution systems that only distribute hydrogen to FCVs, this type of simultaneous supply system that could have greater benefits in terms of CO_2 emission reductions and reduce primary energy consumption because of the higher operation rate and efficiency of hydrogen production. (Pak et al. 2006).

4.3 Proposal of FEW Nexus System Based on Fuel Cells

From the point of view of the FEW nexus, fuel cell can play an important role in developing the independent household FEW nexus in combination with other technologies and power sources. The M-NEX Tokyo team has proposed a FEW system which might be a feasible solution to build an independent, energy saving, low emission household FEW system, as shown in Fig. 7. This system includes

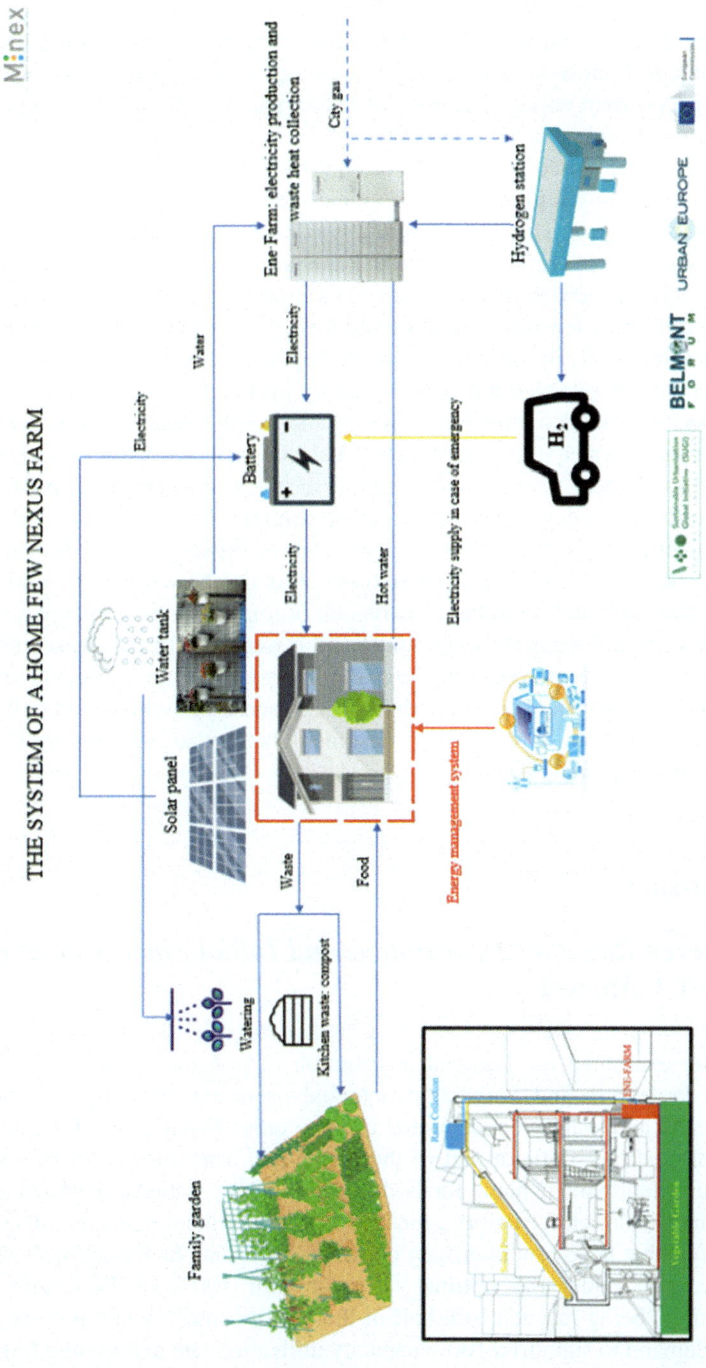

Fig. 7 Conceptual design of a home food-energy-water (FEW) nexus system. (Source: M-NEX project)

multiple processes of production, storage, distribution, consumption, and recycling, in terms of food, energy, and water. Combined with a solar panel and energy storage battery system, a fuel cell could be a core component to improve energy efficiency and reduce GHG emissions and primary energy consumption, and maintain a relatively independent energy-water nexus. Solar panels and fuel cells could generate electricity to supply to a household and reduce reliance on the power grid. Excess electricity could be stored in batteries or sold to the grid thus having this system serve as a virtual power plant on the regional grid. Using an energy management system, the whole system could preferentially consume electricity from solar panels (instead of the grid) in the daytime and use the grid to replenish insufficient power demand in night, if needed. This system would result in a smoother power supply curve and could be a feasible solution to address the problem of fluctuating output from renewable power. By-product water and heat generated in the process of power generation could be recycled and used to supply heat and hot water to a household.

Such a system can also improve resilience and safety thanks to the water storage in a tank (separate water storage tank in the fuel cell package), while electricity from the fuel cell and battery system could continue to sustain normal daily functioning of a household for serval days. The system can be configured for the fuel cell in an FCV to supply electricity for the household during an emergency, as mentioned above. Hydrogen from an external hydrogen station could also be supplied to a household fuel cell system, and this could have the benefit of improving the capacity utilization rate of hydrogen stations in the community and lower risks for the household. Also, according to the current research, from the energy consumption point of view, hydrogen provision system for a combined FCV and a household fuel cell system with combined heat and power (CHP) designs is better than distributing hydrogen only for FCVs (Pak et al. 2005).

5 Discussion

5.1 Uneven Regional Distribution and Initial Small Demand in FCV Market

This chapter examined the regional distribution of potential future FCV users in Kanagawa Prefecture and found that potential users are unevenly distributed in geographically and mainly concentrated in three cities (Yokohama, Kawasaki, and Sagamihara) and the southern part of the prefecture. This uneven distribution will hinder future deployment plans for hydrogen refueling stations (HRSs) and have negative impacts on service equity, since it will be difficult to maintain HRS services in the regions that have low hydrogen demand. According to Kanagawa's strategic road map, even under the optimal scenario with 100% of FCV unit targets (100,000 units) being reached, still half of the transit centers would not see enough hydrogen demand to support a 100% capacity utilization rate at mid- and large-scale

HRS facilities. In the 50,000 FCV scenario, almost all sites would not see enough hydrogen demand to keep a mid-scale HRS going at a 100% operating rate. In the 20,000 FCV scenario, more than 60 HRS sites would not even support a small HRS. In conclusion, even in the best-case FCV deployment scenario there is still a gap between targets and HRS deployment based on Kanagawa's road map, and this means that it is essential to introduce new and additional hydrogen demand if the prefecture wishes to promote the development of the hydrogen energy infrastructure.

5.2 *Potential of Fuel Cell Buses to Induce Hydrogen Infrastructure Development*

We have seen that demand for FCV cars is small and unevenly distributed in the study region, but hydrogen demand to power fuel cell route buses in Kanagawa would be much larger, and sufficient to justify mid- or large-scale hydrogen refueling stations at most transit centers. Hydrogen demand for FC buses would be more balanced than for FCV passenger cars alone, meaning that HRSs could be deployed even in areas with relatively low population density, and still maintain sufficient operation rates to be economically viable. The introduction of FC buses could play a significant role in reducing financial risks of hydrogen station operations and help increase service equity in specific areas of the region that have lower hydrogen demand. Our calculations showed only seven transit centers to have relatively low hydrogen demand. However, if those transit centers could be configured to provide hydrogen refueling services for private FCV users, hydrogen demand would be sufficient to justify small- or mid-sized HRS facilities.

5.3 *Impacts of Fuel Cell Plants on Future Household FEW Nexus*

Systems designed to incorporate fuel cell equipment, solar panels and co-generation technology could reduce primary energy consumption, energy loss, and CO_2 emissions arising in the process of electricity production and distribution compared to conventional electricity supplies that depend entirely on a centralized power grid. For individual households or buildings, this chapter has proposed the concept of the FEW nexus system based on fuel cells configured with other equipment. Such a system could incorporate the whole flow of FEW processes, from resource collection, production, storage, and management to final use, and independently supply electricity, hot water, and heat for several days, even during an emergency. Such a design could reduce reliance on the centralized power grid, improve resilience to disasters, and boost adaptive capacity.

Using hydrogen as a clean and potentially abundant source of renewable energy, Japan's concept of creating a hydrogen society, if realized on a large scale, could exert a significant influence on the FEW nexus. Hydrogen could play a major role in Japan's future energy structure if hydrogen-based energy consumption and demand were both to increase dramatically. In order to achieve carbon neutrality, green hydrogen produced by water electrolysis technology could become an important resource for hydrogen production. Water itself could become an energy production resource. Fuel cell systems can supply electricity, heat, and water to buildings and backup the energy-water nexus even in an emergency. As we have seen, fuel cell technology and hydrogen energy could efficiently lower risks in the FEW nexus. Decentralized energy systems like on-site hydrogen production could more fully utilize local resources to increase local energy supply potential through multiple overlapping hydrogen production methods. Scaling up hydrogen energy utilization could be an effective method to reduce dependence on a single type of energy and diversify the energy structure.

6 Conclusion

This chapter first examined the potential regional distribution of future FCV users in the study area of Kanagawa Prefecture and showed that potential users would be unevenly distributed geographically and demand relatively small. The uneven distribution of FCV users will hinder the deployment of hydrogen refueling stations and have negative impacts on service equity, since stations would not be economically viable in areas that have low demand for hydrogen. A gap still exists between FCV development targets and HRS deployment based on the prefecture's hydrogen road map.

While FCV passenger car demand is still small and unevenly distributed in the region, demand for FC buses has the potential to be considerably greater, enough to justify mid- or large-scale HRS facilities and have them operating at high utilization rates. With adequate deployment of FC buses, hydrogen demand at each refueling site would be more balanced which, means that HRSs could be deployed even in areas with low population density. Hence, the introduction of FC buses can play a significant role in reducing financial risks inherent in hydrogen station operations and help increase service equity in regions with small hydrogen demand. Some bus transit centers that have low hydrogen demand could also have extra capacity to provide refueling services for private FCV users. By doing so, those centers could have enough demand to justify small- or mid-scale HRS facilities.

In terms of fuel cell technology other than for vehicles and buses, this chapter mainly introduced two types of fuel cell products designed for household and commercial building use. Fuel cells could have positive effects on reducing the consumption of primary energy, increasing energy efficiency, and improving resilience and energy security. This chapter then proposed a possible scenario for a FEW circulation system that combines fuel cell technologies with other components, and discussed how such systems could improve the FEW nexus and increase security and resilience. Combined with other technologies, a fuel cell based decentralized FEW nexus system could cover the entire process from resource collection and recycling, production, distribution, management to final use, functioning as independent FEW-based plants for individual households or buildings. Though further advances are still needed, these concepts show the great potential for such systems to improve safety, efficiency, independence and sustainability of the FEW nexus.

Acknowledgments First of all, I would like to extend my sincere gratitude to my supervisor, professor Wanglin Yan, for his instructive advice and useful suggestions on this chapter. And I am deeply grateful to him for his kindness help in my study and life.

Second, I would like to express my heartfelt gratitude to my Co-Supervisor, professor SHAW Rajib, professor KANIE Norichika, for their kindness advice and inspiration for this chapter. And their class and lecture I have benefited a lot. I also feel grateful to professor OMAE Manabu for his help on data collecting and my fieldwork. And I want to express my sincere thanks to RH for his help and suggestions for revising this chapter.

Finally, I would like to thank Kanachu bus company for providing related data which helped a lot in my research.

References

Ariga T, Matsuhashi K (2013) Estimation of car ownership and use by online questionnaire and mesh-based population and households. J City Plan Inst Jpn 48(3):537–542

Ghenai C, Salameh T, Merabet A (2020) Technico-economic analysis of off grid solar PV/fuel cell energy system for residential community in desert region. Int J Hydrog Energy 45(20): 11460–11470. https://doi.org/10.1016/j.ijhydene.2018.05.110

Horikawa K et al (2015) New development of introducing biogasification at sewage treatment plants for promoting hydrogen stations. J Jpn Soc Civ Eng 71(6):II_43-II_52

Kanagawa Next Generation Vehicle Promotion Council (2015) "Road map for realizing a hydrogen Society in Kanagawa"

Li L, Manier H, Manier MA (2020) Integrated optimization model for hydrogen supply chain network design and hydrogen fueling station planning. Comput Chem Eng 134:106683. https://doi.org/10.1016/j.compchemeng.2019.106683

METI (2019) "The strategic road map for hydrogen and fuel cells–industry-academia-government action plan to realize a hydrogen society" https://www.meti.go.jp/english/press/2019/pdf/0312_002b.pdf. Accessed 24 Feb 2022

METI (2020a) "Status of Actions to Achieve the Strategic Roadmap for Hydrogen and Fuel Cells." https://www.meti.go.jp/shingikai/energy_environment/suiso_nenryo/roadmap_hyoka_wg/pdf/002_01_00.pdf. Accessed 18 July 2021

METI (2020b) "Green Growth Strategy Through Achieving Carbon Neutrality in 2050." https://www.meti.go.jp/english/press/2020/1225_001.html

Michalski J, Bünger U, Stiller C (2011) Business analysis of the hydrogen Refueling Station infrastructure and the role of the pricing system. Int J Hydrog Energy 36(13):8152–8157

MOE (2021) "Japan's National Greenhouse Gas Emissions in Fiscal Year 2020 (Preliminary Figures)" https://www.env.go.jp/content/900505938.pdf. Accessed 22 Feb 2022

Pak H et al (2005) Study on hydrogen infrastructure model for fuel cell vehicles and residential areas at introductory stage. Environ Sci 40:31–36

Pak H, Kobayashi N, Kato T, Suzuoki Y, Morikawa T, Katsurayama K (2006) Feasibility study on hydrogen infrastructure model for FC vehicles and commercial establishments at introductory stage. J City Plan Inst Jpn 41–3:827–832

Zhang G, Zhang J, Xie T (2020) A solution to renewable hydrogen economy for fuel cell buses—a case study for Zhangjiakou in North China. Int J Hydrog Energy 45(29):14603–14613. https://doi.org/10.1016/j.ijhydene.2020.03.206

Hydrogen Refueling Station Siting and Development Planning in the Delivery Industry

Lu Ziyu and Wanglin Yan

Abstract Hydrogen can play a vital role in energy transition as the world works to reduce greenhouse gas emissions, achieve the UN Sustainable Development Goals, and promote energy security. Recent research has looked at the siting of hydrogen refueling stations for fuel cell vehicles based on top-down blueprints or roadmaps, but low market penetration and uncertainty in the household vehicle market mean insufficient demand to make hydrogen infrastructure viable. Instead, this chapter looks at the viability of installing refueling stations at logistics service centers and converting delivery fleets to hydrogen-powered fuel cell trucks, using one company in Kanagawa Prefecture as a case study. Potential hydrogen consumption is estimated based on population data, service area, and usage of delivery services. Candidate locations are selected based on break-even analysis, zoning, and parking space criteria. Finally, environmental impacts, including GHG emission reductions, are evaluated based on well-to-wheel analysis. Hydrogen refueling stations were found to be viable at six of the company's 24 logistics centers. Relative to gasoline trucks, annual GHG emissions would be reduced by 48.3% by introducing hydrogen fuel cell trucks and hydrogen refueling stations at those centers. Future research could delve deeper into practical business models based on these findings.

Keywords Hydrogen refueling stations · Express delivery industry · Well-to-wheel analysis · Energy transition

---

L. Ziyu
Graduate School of Media and Governance, Keio University, Fujisawa, Kanagawa, Japan
e-mail: roshiu@keio.jp

W. Yan (✉)
Faculty of Environmental Information Studies, Keio University, Fujisawa, Kanagawa, Japan
e-mail: w_yan@keio.jp; yan@sfc.keio.ac.jp

W. Yan et al. (eds.), *Resilient and Adaptive Tokyo*,
https://doi.org/10.1007/978-981-99-3834-6_13

1 Introduction

Increasing economic and industrial inter-connectedness, the evolution of modern lifestyles, and the need for transition of energy sources necessitates policy change, while growing emissions of greenhouse gases (GHGs) have become a global concern, resulting in climate change and environmental pollution. A total of 196 parties adopted the Paris Agreement (United Nations 2021) and the United Nations adopted the Sustainable Development Goals (SDGs), which include environmental and energy targets, to protect the planet, limit global warming, and achieve a global peaking of GHGs as soon as possible.

Hydrogen energy is a secondary energy resource and often considered a clean fuel because it only emits water at the final stage of use. Hydrogen energy can be also produced from a variety of primary energy sources, such as natural gas, nuclear power, biomass, and renewable energy such as solar and wind. The advantages of hydrogen and fuel cells include energy storage, energy efficiency, and zero-carbon emissions (METI 2013). The core technology of hydrogen is the fuel cell, and according to Ishihara (2018), the energy efficiency of this technology ranges from about 30% to 65%, depending on the type of electrolyte used as well as temperature during operation.

Since 2014, Japan's government has promulgated a policy to establish the foundations of what it refers to as a "hydrogen society." In that context, the hydrogen refueling station (HRS) and fuel cell vehicle (FCV) play vital roles in introducing hydrogen energy to the transportation sector. Recent research has mainly emphasized the spread of HRSs, hydrogen consumption, optimal distribution of facilities, and HRS facility design, but such work has been based on assumptions about market penetration and market growth in the numbers of HRSs and FCVs, based on the blueprint presented by the national hydrogen road map.

However, Japan has been mired in a "chicken-and-egg" situation. FVC use will not increase without more HRSs, and vice versa. National policies have targeted the household vehicle market, but low market penetration and market uncertainty persist. The result has been low HRS coverage and operational rates, leading to low consumer confidence, resulting in low market penetration for FCVs. Uchida and Minamigata (2016) found that hydrogen production equipment needs sufficient utilization levels to maintain efficiency and keep operating costs down. For a business model to be viable for HRSs, the demand for hydrogen must be high enough. New ways are needed to achieve that. Most initiatives have adopted a central base and substation approach to locating refueling stations. In contrast to other research focusing on the individual household market, the research described in this chapter examines the concept of installing hydrogen stations at multiple delivery service centers (including services such as courier, package delivery, and express mail, etc.) and introducing fuel cell trucks to delivery fleets as an effective way to ensure intensive hydrogen consumption in an area. We first estimate the potential hydrogen consumption in the local delivery industry based on population data, the service area of each service center (referred to as "logistics centers" in this chapter), and the amount of delivery usage. Break-even analysis is conducted using hypothetical service centers selected for having adequate parking space to install an HRS. The study area for this research is Kanagawa Prefecture, located in the southern part of the Tokyo metropolitan area, but the concepts and approaches could be applied to any city or region.

2 Hydrogen Energy Policy in Japan

2.1 *Issues in Japan's Energy Policy*

From the perspective of energy policy security, hydrogen energy contributes a variety of benefits to the nation's energy policy in Japan. After the 2011 Fukushima Daiichi nuclear accident, the share of nuclear in the energy mix dropped to zero and still remains low. Japan has long relied on fossil fuel imports, and its energy self-sufficiency dropped from 20.3% in 2010 to 6.4% in 2014. Energy issues in Japan include low self-sufficiency, high dependency on fossil fuel energy (coal, oil, and liquefied natural gas accounted for 87.4% of the energy mix in 2017), high dependence on energy inputs from the Middle East, and trade deficits dominated by energy imports. To improve energy security and safety, Cabinet adopted the Fifth Energy Basic Plan in April 2018, emphasizing the importance of hydrogen energy in achieving the goals of reducing carbon emissions and increasing Japan's energy self-sufficiency. As transportation accounts for 20% of energy consumption and most vehicles rely on imported fossil fuels, transportation was selected as a major sector to introduce hydrogen energy. Hydrogen energy plays a vital role in long term goals to achieve GHG emission reductions, especially in the transportation sector. New Energy and Industrial Technology Development Organization (2014) reported that transportation in Japan accounted for 23.2% of total energy consumption in 2015, of which 89.1% was from vehicles, which accounted for 18.6% of Japan's total GHG emissions.

2.2 *Current Hydrogen Energy Strategy in Japan*

To achieve its hydrogen mission and vision, Japan has facilitated the introduction of hydrogen energy through the release of several policy documents (Table 1). In 2014, the Fourth Energy Basic Plan proposed the concept of a "hydrogen society," followed by a Hydrogen Fuel Cell Road Map at the end of 2014, which divided future development of hydrogen energy into three phases: expanding hydrogen energy usage, setting the foundation of a hydrogen supply chain, and finalizing a carbon-free hydrogen society. In 2016, the road map was revised and targets clarified. In 2017, a Ministerial Council enacted the Basic Hydrogen Strategy introducing hydrogen stations and fuel cell vehicles as core components of Japan's

Table 1 Policy documents related to hydrogen energy in Japan

2014	Fourth Strategic Energy Plan	Proposes concept of a "hydrogen society"
	Hydrogen Fuel Cell Road Map	Divides future development into three phases
2016	Hydrogen fuel cell road map (revised)	Clarifies targets for development
2017	Basic hydrogen strategy	Sets the foundations for hydrogen energy infrastructure
2018	Fifth strategic Energy plan	Reinforces measures for realizing a hydrogen society
2019	Hydrogen fuel cell road map (revised)	Reviews targets and achievements, and conducts follow-up in each sector
	Hydrogen and fuel cell technology development strategy	Lists actions and sets targets for technological aspects of hydrogen energy

Source: (METI 2021)

hydrogen promotion policy. Hydrogen refueling stations are at the center of the hydrogen supply chain, recognized as a part of core infrastructure to serve for the distribution of hydrogen energy. In 2019, the Ministerial Council revised the hydrogen fuel cell road map once again. This time, targets and achievements in introducing hydrogen energy were reviewed, and follow-up was conducted in each sector. The three phases of the introduction period were defined as expanding FCV numbers to 40,000 units by 2020, 200,000 units by 2025, and 800,000 units by 2030. As key components in the future supply chain, HRS targets were set at 160 locations by 2020, 320 by 2025, and 900 by 2030. By the end of August 2021, Japan had built 154 stations, and 12 of them are now still in construction.

Hasegawa (2014), Martin et al. (2017), and Miura (2018) wrote that expanding the market for hydrogen energy based on HRSs is the key to increasing demand for hydrogen energy, they assumed that transportation is the first market to expect to see market penetration and in actual the Japan government has taken efforts to introduce household FCVs. However, there are barriers to building HRSs. The cost of constructing an HRS is about 450 million yen, and even with subsidies of 200 million yen, the cost remains at 250 million yen (New Energy and Industrial Technology Development Organization 2014). The fixed portion of annual operating costs is estimated at about 40 million yen, and even with subsidies of five million yen, the total annual fixed cost would be 35 million yen, resulting in a break-even point of 70,000 kg/year. There are three types of HRS, each with a different capacity: onsite, offsite, and mobile. To achieve the optimal distribution of hydrogen energy, it is crucial to consider HRS business models in the transportation sector. It is also important to understand potential hydrogen demand. One approach to do this would be to examine existing examples in industry.

The actual number of FCVs in Japan at the end of 2019 is around 2000 (Ministry of Economy, Trade antholog Industry Agency for Natural Resource and Energy, 2019) there is currently a gap between supply and demand in the HRS business, but limited research has examined the current issues of hydrogen consumption. The general approach has been to assume that hydrogen consumption can reach the levels projected in policy targets. However, if there are problems with the assumptions, the chance of going to the next step is reduced.

Because the private vehicle (household) market for FCVs remains uncertain, alternative sectors such as transportation have been considered in hydrogen plans in both Japan and China. The rapid growth of global trade and e-commerce has been a boon for delivery services. The location-based business model and intensive use of vehicles for delivery in a defined territory are hints of potential demand for hydrogen energy. One key to success will be a proper understanding of break-even points for the large investments and financial risks involved.

2.3 Study Area

This study uses Kanagawa Prefecture as the study area, for three reasons. First, it ranges from metropolitan to rural areas, so the population can be considered to be representative of differences of total route distances and other parameters in express delivery services, which can be useful in estimating potential hydrogen consumption. For decision-makers, the results of such data analysis can help develop a road map for future development patterns. Second, Kanagawa also includes the Kawasaki area, a core manufacturing area for the automobile industry, which means that Kanagawa contains the industrial foundations to support hydrogen-related policies and is also a perfect area for trials in the research and development of fuel cell trucks. Finally, Kanagawa plays a vital role in Japan's hydrogen road map, with Kawasaki area as a core hydrogen manufacturing area that is home to a variety of hydrogen-related enterprises such as Iwatani.

3 Methodology

3.1 Assumptions for the Hydrogen Refueling Station Business Model

This study examines a scenario of introducing hydrogen demand by installing hydrogen refueling stations for fuel cell vehicle trucks at logistics centers. Such a scenario could break the chicken-and-egg dilemma by promoting market penetration for hydrogen. With the development of e-commerce and the rising trend in global trading, the delivery industry is stable and growing. Logistics centers are strategically situated as assembling places for trucks, likely to have adequate parking space to install an HRS, while adding no access costs for the trucks and meeting the need for fuel supply. Regulation and policy could also be favorable for this scenario, as it might be possible for HRSs to be installed in business and industrial districts, in contrast to residentially zoned areas where such installations may be restricted. To ascertain the potential consumption of hydrogen, it is essential to estimate total truck distance.

The route of each truck is flexible, but the distance traveled will approach the shortest route and be constrained by the service area and number of destinations. To calculate total route distance, one alternative would be to use average route distances in the industry. Such an approach, however, may result in reduced accuracy due to large differences in route distance between urban and rural areas. Thus, a model could be used to increase accuracy. According to Miyatake et al. (2016), pick-up and delivery and team pick-up and delivery are two widely delivery types in the delivery industry, but both of these methods, the number of delivery destination numbers and size of service area are the key elements that define the final route distance. Little research has focused on the relationship between route distance and delivery destinations in specific service areas, but Miyatake et al. (2016) posited that the relationship between destination density and route distance for one truck should be a square relation as shown in Eq. (1):

$$D = a\sqrt{\frac{V_i}{T} \times S_i} \tag{1}$$

where

- D refers to distance while a is a coefficient. We assume a coefficient of 1 in this research (for more accuracy in future analysis, the coefficient can be adjusted based on empirical surveys),
- V_i is the number of express deliveries which has been calculated in the previous part,
- T is the number of trucks, and this aim to find out the delivery capacity of the logistic center in 1 day.
- S_i is service area.

Therefore, the total route distance as the core of this research can be calculated from data in two parts: capacity of delivery service and consumer demand. For the former, we assume that the capacity of a logistics center is a function of the trucks operating from that center. A regular logistics center should have a reasonable amount of parking space to park the trucks. This implies a potential to accommodate an HRS. For consumer demand in a certain service area, the number of delivery destinations per day is assumed to be proportional to the population in the area. Thus, the total route distance data can be estimated as in Eq. (1).

As for service area, corporate confidentiality and regulations pose some challenges in ascertaining service area covered by each logistics center. Geographical information systems (GIS), however, can play a vital role in estimating the service area. A variety of mathematical methods can be applied. GIS applications can attempt to provide solutions or address certain issues, for instance, maximization or minimization problems, as well as planning and design. Finally, GIS enables hypothesis testing. Various forms of spatial analysis employ inferential statistics to

Fig. 1 Voronoi diagram of Sagawa logistics centers in Kanagawa Prefecture

generalize samples to populations. Hence, in this chapter, GIS is applied to analyze the optimal logistics center area coverage using a GIS Voronoi diagram tool.

Often used in mathematics, a Voronoi diagram partitions a plane into regions that are close to each of a given set of objects. In the simplest case, these objects are a finite number of points in a plane. Each object has a corresponding region (a Voronoi cell), consisting of all points of the plane closer to that object than to any other. For the present analysis, this approach can be used to estimate the optimal service area for the delivery service, and also for the HRS service. The final maps based on this analysis are presented in Figs. 1 and 3.

Once the service area and number of trucks have been determined, the next step is to determine route distance based on service area and the number of delivery destinations, using an approach based on Miyatake et al. (2016). To do so, it is necessary to estimate the usage of delivery services within a given service area. The best method to do so would be to collect primary data from the relevant enterprise, but due to corporate confidentiality and regulation constraints mentioned above, the only method is to estimate. Some secondary data is available about delivery service usage in Japan. It is possible to calculate several destinations of delivery service usage based on population data for Kanagawa Prefecture released by the Ministry of Land, Infrastructure, Transport and Tourism (2021). For the frequency of delivery service usage, this chapter refers to a survey by MyVoice (2015) based on random sampling, with 10,964 responses, which ensures a degree of universality and accuracy. Combined with the population data, delivery service usage can be calculated as shown in Eq. (2):

$$V_i = \sum_{i=1}^{n} P_i \times U_i + \sum_{i=1}^{n} P_i \times C_i, \tag{2}$$

where

- P_i is the age group in population in a given area,
- U_i is the frequency of using a delivery service to send (by age group, per period of time, based on survey),
- C_i is the frequency of using a delivery service to receive (by age group), and
- V_i is the total number of deliveries performed in the service area in 1 day.

After collecting the data on delivery numbers and service area as well as the number of trucks, which can also be defined as the capacity of the delivery services, it is possible to apply the method used by Miyatake et al. (2016) to Eq. (1) to calculate the total annual route distance of each logistics center, as shown in Eq. (3):

$$\text{Total } D = \sum\nolimits_{i=1}^{n} D \times T, \tag{3}$$

where

- D is the route distance per truck per day, and
- T is the number of trucks operating out of a logistics center.

After obtaining the total route distance, it is possible to calculate the final hydrogen consumption. To do so, it is important to know a vehicle's performance: how far can the vehicle drive on a full fuel charge? This can also be defined as fuel consumed per unit of distance. The only fuel cell vehicle currently on the market in the study area is Toyota's Mirai (a sedan), but trucks and sedan cars will differ in fuel consumption due to differences in the number of cells and other components. Thus, it is not possible to accurately estimate vehicle performance for this analysis based on data for the Mirai. Meanwhile, fuel cell trucks are in operation in China and some have data open to the public, but they are heavy-duty trucks and not being used in the express delivery industry. The best way to estimate fuel consumption would be with data based on a similar industry.

In this section, we have described the process of estimating delivery usage and total route distance of trucks, based on secondary data. Combining delivery usage from a survey and population data from the government, we can estimate total delivery numbers. Voronoi diagrams have been created to portray service areas of logistics centers in the study area, using GIS technology, based on the locations of logistics centers. Finally, the total annual route distance can be calculated based on service areas, numbers of deliveries, and numbers of truck, using a method proposed by Miyatake et al. (2016). Further below, we calculate the demand for hydrogen and potential to reduce carbon emissions by using hydrogen to fuel delivery trucks in the study area.

3.2 Data Collection and Processing

The best way to obtain data for this analysis would be to get annual data directly from the enterprise, but as mentioned, such data is not directly available. The only alternative is to make estimates. Truck fleet data is also not available to the public. Considering that zoning regulations designate certain parking space for commercial trucks and parking is regulated in certain areas, it is possible to estimate truck fleet numbers by calculations based on parking space data per size of truck (Table 2). For this study data were compiled regarding service areas, delivery destinations, truck fleet numbers, and parking space. The Ministry of Land, Infrastructure, Transport and Tourism (2021) releases data on logistics centers in each prefecture in Japan. The data can be used directly to mark each logistics center using GIS software. Google Earth is then used to calculate the amount of parking space at each logistics center.

The next task is to calculate the number of trucks per logistics center based on data on the regulation of parking space. Before a parking area can be built, the proponent must submit documentation to show there will be enough space for the expected number of trucks using a logistics center, and generally one or two additional parking places are to be allocated for emergency use. Thus, this approach can be expected to produce a reliable estimate of the number of trucks based out of each logistics center. The weight of the most widely used truck for deliveries in Japan is around 2 to 3 tons (DHL 2021). Therefore, this analysis assumes one truck per 28 m^2 of parking space per truck as a benchmark. Data sources for this analysis are as follows.

1. Administrative area: Information on the administrative area of Kanagawa Prefecture is published by the National Land Information Division. Data features include cities, villages, streets, and landforms. The map was created by QGIS.
2. Logistics center data: This analysis identified 51 logistic centers operated by five delivery businesses in Kanagawa. The data features only include each center's location. Parking space data was estimated using Google Earth. Numbers of trucks in fleets operating out of each logistics center are estimated based on parking regulations.

Table 2 Regulations pertaining to vehicle parking spaces in Japan

Vehicle size (tons)	Parking space per vehicle
Over 7.5	38 m^2
2 to 7.5	28 m^2
2	20 m^2
Less than 2	15 m^2

Source: Shift Up 2021

3. Demographic data: Data was obtained from the National Land Information Division (2018) and contains present and projected population data as well as the number of households, for municipalities of all sizes, from villages to cities. Voronoi diagram analysis based on logistics center locations was used via ArcGIS to calculate demographic data based on area proportion of streets or villages.
4. Delivery service usage data: This study utilized secondary research from the MyVoice Internet Community Service, specifically, its eleventh survey of delivery service usage. The data included 11,286 respondents during the period January 1–5, 2015. The online survey contained nine questions, including frequency of sending or receiving a delivery, choice of delivery service provider, satisfaction level, purpose of using a delivery service, preferences when choosing a service, and comments.
5. Vehicle driving performance data: Three types of trucks were considered for performance analysis based on power source: fuel cell, gasoline, and electric. For fuel cell trucks, data from NEDO (2015) on the government's Hydrogen Road Map was used. For gasoline trucks, this analysis referred to performance data for the Isuzu ELF (JC-10 model). For electric trucks, this analysis referred to data from the Yamato Transport (TA-Q-BIN) website for the Mitsubishi Fuso (eCanter model).

3.3 GHG Emission Reductions Based on Well-to-Wheel Analysis

This analysis now looks at GHG emission reductions in order to evaluate the environmental impacts of introducing hydrogen energy to the delivery industry in the study area. The only tailpipe emission from a fuel cell truck is water, but one cannot say that emissions are zero. It is important to recognize that hydrogen energy is a kind of second-hand energy produced from a variety of other resources. In the production of hydrogen, the release GHGs is inevitable. Thus, for any estimate it is important to determine a method to define and evaluate the environmental impacts.

Hondo (2005) described the use of lifecycle analysis (LCA) to analyze greenhouse emissions in a variety of industries around the world. Ahmadi and Kjeang (2017) found that LCA as an assessment strategy can provide comprehensive information about how each stage of each process in the use of an alternative fuel contributes to emissions. Today, LCA is seen as a reliable method widely applied in a variety of industries. LCA covers everything from the start of the production chain to final consumption and disposal, which makes it useful for policymakers to evaluate policies vis-à-vis sustainable development goals. LCA results can also assist enterprises that aim to develop corporate social responsibility (CSR) projects intended to contribute to GHG emission reductions.

For this study, another part of LCA analysis must involve the vehicles, in this case, delivery trucks. However, many details about fuel cell trucks are not currently available to the public. It is difficult to estimate GHG emissions without knowing details of the components of the trucks, such as the hydrogen container and fuel cell. For this study, to compare the emissions at each stage of the product lifecycle, it was decided to use a "well-to-wheel" analysis, as this approach has been widely applied in energy analysis in Japan, although it does not count GHG emissions associated with the vehicle lifecycle.

Well-to-wheel analysis includes two stages: well-to-tank focuses on the fuel itself, and tank-to-wheel focuses on emissions during driving. The former looks mainly at final emissions and energy consumption involved in producing the fuel. The latter looks mainly at the type of fuel used and the driving performance of the vehicle itself. The analysis can be express as shown in Eq. (5):

$$\text{Total } CO_2 = (\text{WtV/km } CO_2 + \text{VtW/km } CO_2) \times \text{Total D}, \tag{5}$$

where

- D is the total route distancc Total CO_2 mcans total emission of CO_2
- WtV/km CO_2 means emission of CO_2 in WtV period.
- VtW/km CO_2 means emission of CO_2 in VtW period.

In the following, we calculate GHG emissions of the three types of trucks selected for this comparison (gasoline, fuel cell, and electric).

Pouria et al. (2020) support the idea that new energy vehicles such as electric vehicles (EVs) and hydrogen fuel cell vehicles (FCVs) can eventually be the alternatives to fossil fuel powered vehicles as they emit no tailpipe emissions. Nevertheless, it must be recognized that the production of electricity and hydrogen may release GHG emissions. Hydrogen is the fuel of FCVs, but pure hydrogen is scarce in the world. Seven methods could be used to produce hydrogen (NEDO 2021): as a by-product of other processes, fossil fuel reforming, electrolysis of water-based on thermal power generation, electrolysis of water-based on renewable energy, biomass method, thermal decomposition method, and photocatalytic method. Biomass, thermal decomposition, and photocatalyst methods are currently still in the research and development phase, and a variety of issues remain, such as cost and technical constraints. Thus, these three methods are not currently considered to be practical. The electrolysis of water may see significant progress in future manufacturing advances, but cost and technical issues will prevent large-scale production in the short term. Meanwhile, over 80% of hydrogen sold in Japan is currently made from petroleum refining and petroleum chemical products. This method is also referred to as steam reforming and is in practical use. Thus, this method was selected as the main one for this study. The chemical process follows three steps as shown below, starting with Eq. (6):

$$CH_4 + H_2O \rightarrow 3H_2 + CO \tag{6}$$

For the first step, methane gas reacts with water vapor to generate hydrogen and carbon monoxide. The carbon monoxide then reacts with water vapor to finally generate hydrogen and carbon dioxide, as shown in Eq. (7):

$$CO + H_2O \rightarrow H_2 + CO_2 \tag{7}$$

Therefore, the overall process can be summarized as in Eq. (8):

$$CH_4 + 2H_2O \rightarrow 4H_2 + CO_2 \tag{8}$$

As that formula shows, producing $4H_2$ with a molecular weight of 8 generates CO_2 with a molecular weight of 44. Calculation shows that manufacturing 1 kg hydrogen will generate 5.5 kg of carbon dioxide. This will be the benchmark when applying WtW analysis of fuel cell trucks. According to NEDO (2015), the driving performance (hydrogen fuel consumption) of a small fuel cell truck is 3.1 Nm^3km (note that Nm^3 is a normal cubic meter). With this data, it is now possible to calculate potential hydrogen consumption of fuel cell trucks based on total route distance.

For gasoline-powered trucks, it is important to know the driving performance of the trucks. To compare trucks powered by different fuels, the driving performance and the sizes of the trucks of each type should be homogenous in order to improve data accuracy. Meanwhile, CO_2 emissions from the manufacturing of gasoline must be considered. Gasoline is a liquid generated at the lowest boiling point of crude oil. Energy consumption and CO_2 emissions during oil refining are shown in Table 3.

As given in Table 3, at the manufacturing stage, the energy consumption for gasoline is 0.175 MJ/MJ, while the GHG emission coefficient is 11.42 g eq-CO2/MJ and the calorific value of gasoline is 33.46 L/MJ. For this analysis, the Isuzu ELF (widely used by Sagawa Express Co.) is used as a typical example of a gasoline-powered truck. Its driving performance is 9.70 L/100 km, which works out to CO_2 emissions of 267 g-CO2/100 km.

For electric trucks, this chapter uses the eCanter model (developed by Mitsubishi Fuso, used by Yamato Takyubin), which has a battery capacity of 66 kWh and a range of about 100 km. These figures enable us to estimate the driving performance of this electric truck. According to the Tokyo Electric Power Company (2020), the CO_2 emission coefficient of generation is 0.434 kg-CO_2/kWh. With this data, we can evaluate the environmental impacts of electric vehicles.

This section has examined three types of delivery trucks to analyze CO_2 emission based on route distance. Figure 7 summarizes the potential CO_2 emission reductions based on well-to-wheel analysis.

Table 3 Gasoline consumption intensity

Energy consumption in fuel manufacturing MJ/MJ	0.175
GHG emissions during manufacturing g eq-CO2/MJ	11.42
Calorific value L/MJ	33.36

Source: Mizuho Research and Technologies 2004

4 Results

4.1 Estimates of Delivery Service Usage, Truck Fleet Route Distances, and Potential Hydrogen Consumption

Figure 1 shows the logistics centers (with identifying number and names) and areas they cover in the administrative area of Kanagawa Prefecture. Sagawa Express Co., chosen for this case study, has 24 logistics centers in Kanagawa. The eastern part of the prefecture has lower delivery coverage than the western part. One logistics center located in the harbor area of Yokohama City was eliminated from this analysis due to data error because it is a stock for airport. Most of the eastern part of the map is urban while the western part is more rural, with landforms including forests, mountains, valleys, and hills. The eastern part of the study area is mostly level plain. The Voronoi diagram has divided Kanagawa into several service areas based on the locations of logistics centers.

Figures 2 and 3 are maps created by GIS technology to portray population data, showing that most of the population is concentrated in the middle and eastern parts of Kanagawa, including the cities of Yokohama and Kawasaki. In the Yokohama area, the population density is much more intensive than any other area. The population is over 15,000 in most of the streets in Yokohama City. In the north-central area, the population is concentrated in Sagamihara, the third-largest city in Kanagawa Prefecture. In the west, most of the population is concentrated in Hadano City.

In the cities of Odawara and Minamiashigara, most Streets are around 3000 population.

The above approach reveals the population in each delivery service area. Figures 2 and 3 show that the population is concentrated in the northeast, mainly in Kawasaki and Yokohama. The Yokohama North logistics center covers the most populated service area (over 800,000) within the entire study area. Other service centers that cover a population over 500,000 include Totsuka (over 690,000), Kawasaikitama (610,000), Kamakura (600,000), Shonan (600,000), Yokohama East (590,000), and Kawasaki (580,000). In western Kanagawa, a few logistics centers cover populations of over 300,000. Others range from 150,000 to 300,000. The logistics center covering the lowest population is Ashigara (120,000).

Figure 4 shows the parking area of logistics centers (as numbered in Fig. 1). Yokohama Tsurumi logistics center has the largest parking area (9674 m^2), followed by Yokohama (9109 m^2). Logistics centers with parking area over 3000 m^2 include

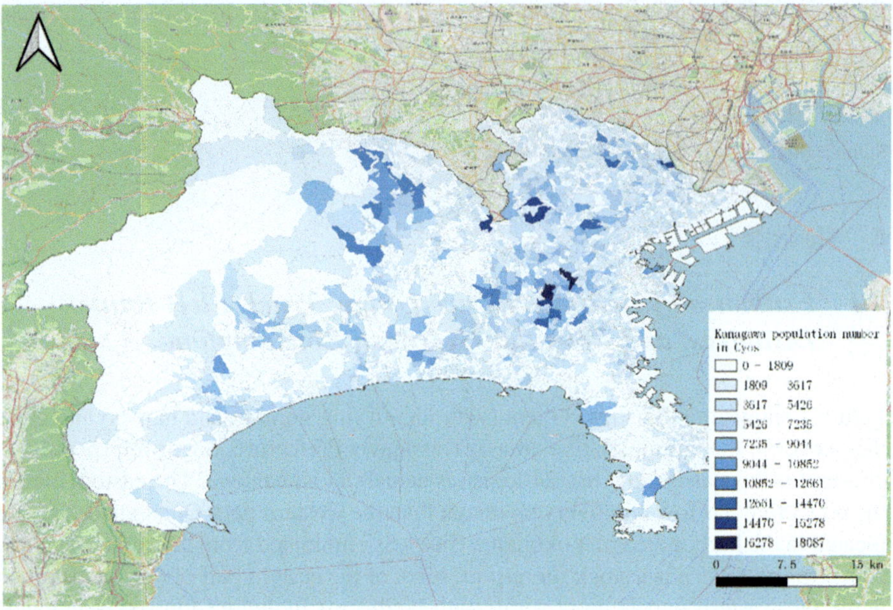

Fig. 2 Population data of Kanagawa Prefecture by census unit: small area (Data source: National Land Information Division, 2018.)

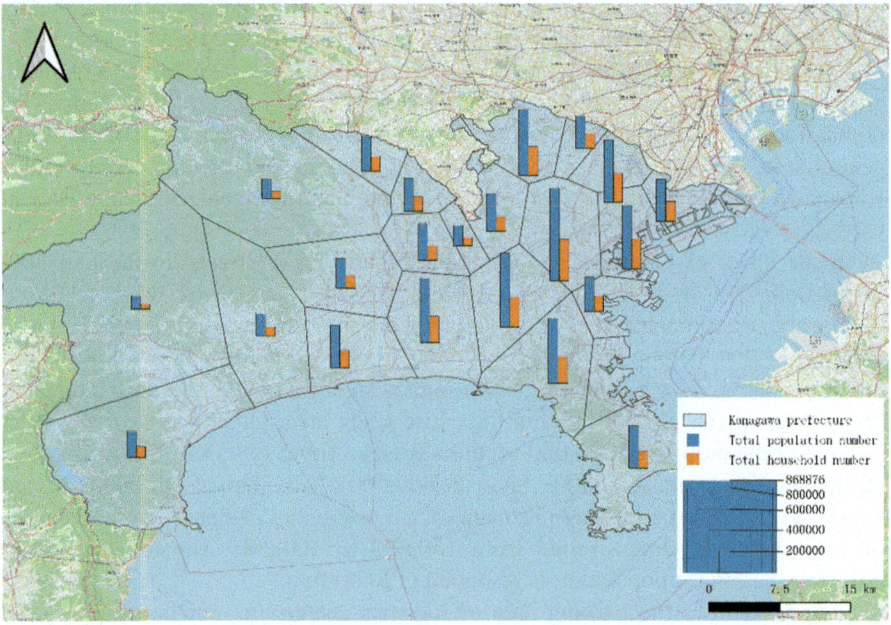

Fig. 3 Population data of Kanagawa Prefecture in Voronoi a diagram (persons, households). (Data source: National Land Information Division, 2018)

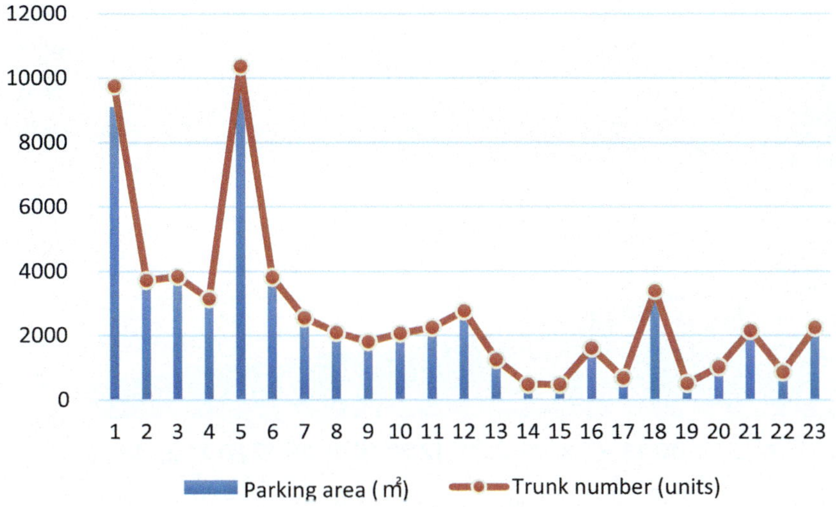

Fig. 4 Parking space and truck numbers of Sagawa logistics center in Kanagawa

Totsuka (3583 m^2), Hadano (3561 m^2), Yokohama North (3470 m^2), and Yokohama South (3161 m^2). Most of the rest range from 1000 m^2 to 3000 m^2. Logistics centers in Yokohama City have much larger parking areas than in other parts of Kanagawa, although Hadano and Shonan are also relatively large. Kanagawa prefectural government (2021) regulations governing the construction of hydrogen refueling stations require at least 1600 m^2 of space, and installations are prohibited in residence zones. Most logistics centers are in commercial or industrial zones, so for them the zoning is not a constraint. In terms of meeting minimum parking area requirements, eligible logistics centers would include Yokohama East, Yokohama Tsurumi, Yokohama North, Yokohama, Totsuka, Setagaya, Yokohama Seya, Shonan, Zama, Kawasaki Tama, Hadano, Atsugi, Sagamihara, and Ashigara. In terms of the number of potential trucks in the fleet based on parking space, according to the regulations, with 28 m^2 per truck as a benchmark, Yokohama Tsurumi and Yokohama East would accommodate the most trucks, at 346 and 325 trucks, respectively.

Figure 5 shows that Ashigara and Sagamihara have the largest delivery usage numbers. Young and middle age business users account for most the frequent delivery usage, and the usage is proportionate to population size. Also, in proportion to population, the number of deliveries is concentrated in middle and eastern Kanagawa.

Figure 6 shows the potential hydrogen consumption and total annual fleet distance of each logistics center. The ranking of hydrogen consumption is consistent with total distance. The Yokohama East logistics center's annual fleet distance is over 4,300,000 km and hydrogen consumption around 125,500 kg, followed by Yokohama North (over 3,900,000 km, around 114,500 kg), followed by Totsuka (3,400,000 km, around 99,000 kg), Shonan (3,100,000 km, around 90,200 kg),

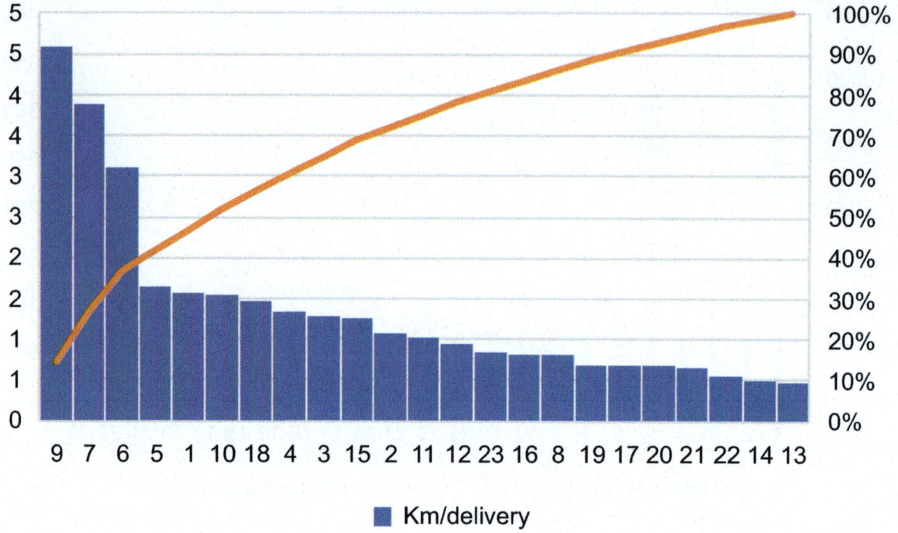

Fig. 5 Estimated delivery numbers of Sagawa logistics centers in Kanagawa

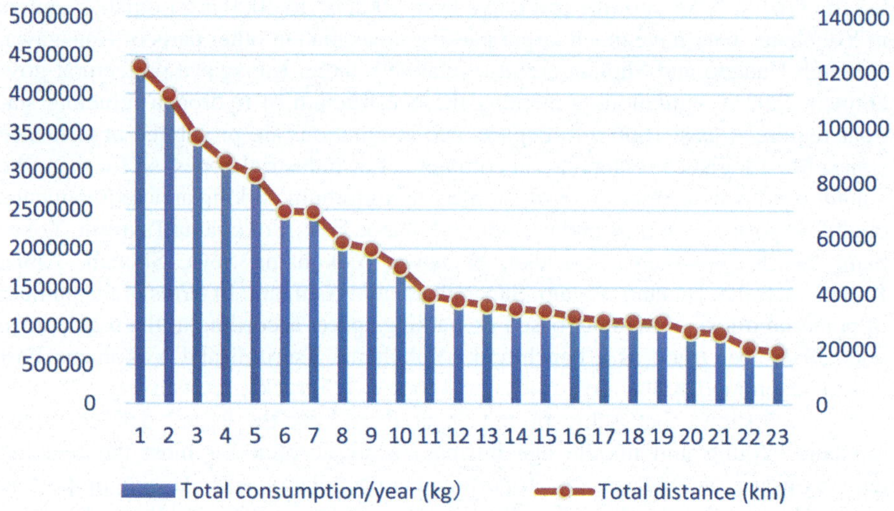

Fig. 6 Estimates of annual truck fleet distances and potential H_2 consumption by Sagawa logistics centers in Kanagawa

Yokohama Tsurumi (2,900,000 km, around 845,000 kg), Hadano (2,470,000 km, around 84,900 kg), Sagamihara (2,460,000 km, around 71,100 kg), and Kawasaki Tama (2000,000 km, around60,000 kg). At the lower end are Ashigara (1,900,000 km, around 57,200 kg), and Atsugi (1,750,000 km, around50,600 kg).

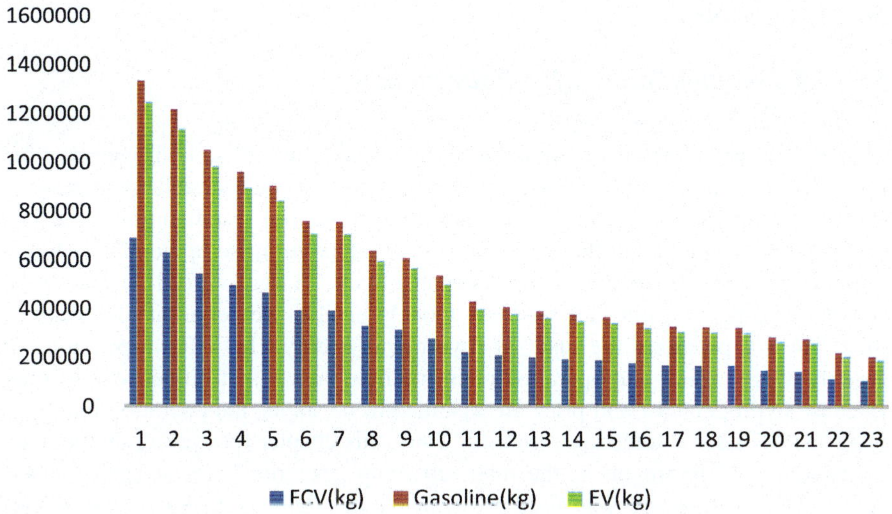

Fig. 7 Environmental evaluation of Sagawa logistics centers in Kanagawa

Below that, Zama, Yokohama, Kawasaki, Kamakura, Odawara, Sagamihara Midori, Hiratsuka, Yokohama South, and Yokosuka are around 1,000,000 km. The lowest annual fleet distance is Yokohama Seya (670,000 km, around19,300 kg) (Fig. 1 and Fig.7).

4.2 GHG Emission Reductions Based on Well-to-Wheel Analysis

For total potential GHG emission reductions relative to gasoline-powered trucks, electric vehicle trucks have 6.5% lower GHG emissions, while fuel cell vehicles (hydrogen) have 48.3% lower GHG emissions, based on total fuel consumption in a well-to-wheel analysis.

To conclude these findings, the top five Sagawa Express logistics centers in Kanagawa, in terms of annual CO_2 emissions reduced by using hydrogen instead of gasoline, are Yokohama East, Yokohama North, Totsuka, Shonan, and Yokohama Tsurumi (643,875 kg, 588,168 kg, 508,564 kg, 463,284 kg, and 435,714 kg, respectively).

5 Discussion

5.1 Recommendations for Policymakers

This chapter has examined a potential business model to utilize hydrogen as a fuel for a fleet of Sagawa Express delivery trucks powered by fuel cells in Kanagawa Prefecture, Japan. We looked at regulatory constraints for installation of hydrogen fueling stations (HRSs) at the firm's logistics centers, estimated potential hydrogen consumption, and evaluated environmental effects. Hasegawa (2014) and Miura (2018) have pointed out that business models need to be economically viable if they are to promote the use of hydrogen energy. In HRS operations in Japanhydrogen price, and operation rates are three factors that influence feasibility. Because hydrogen cost and price are assumed to be stable, the operation rate is the main factor under a company's control. Thus, this chapter focuses on the potential consumption of hydrogen as the main driver of operation rates at each logistics center. GIS technology was used to evaluate the delivery service area of each Sagawa logistics center, and a method proposed by Miyatake et al. (2016) was used to calculate the total annual fleet distance of trucks at each logistics center. EVsmartBlog (2021) reported a total average distance per day of 80 km for electric delivery trucks introduced by Sagawa. This chapter uses that number.

Our analysis has illustrated the feasibility of installing HRSs in some of the logistics centers operated by Sagawa in Kanagawa. Operation rates play an important role in the viability of HRSs, and the individual market household is unlikely to have sufficient demand in the near future, but the delivery industry has both the conditions to permit HRS installation and sufficient demand for hydrogen. Thus, logistics centers appear to be a good option to promote both HRSs and FCVs. From the regional perspective, this strategy can help reduce carbon emissions and thus contribute to achievement of the UN SDGs.

For policymakers, it is worth noting the potential benefits of enabling the installation of HRSs in logistics centers. They can create sufficient hydrogen demand to maintain HRS operations at suitable levels during the early stages of hydrogen infrastructure roll-out and set the stage for a gradual increase in household use of FCVs.

Nevertheless, policymakers should still take some additional issues into account. First, in the example of Sagawa Express in Kanagawa, in this chapter it was assumed that the entire fleet operating out of the selected logistics centers will transition to fuel cell trucks. However, the costs of a rapid and complete transition could be prohibitive, so a gradual introduction of fuel cell trucks may be more realistic. Thus, further analysis is needed to determine whether hydrogen consumption during the transition would be sufficient to support the required operation rates. Second, even though this analysis identified those sites that have sufficient parking space to install HRSs, each logistics center will have a its own unique design and structural issues to address. For example, to store high-pressure hydrogen or liquid hydrogen, safety issues need to be taken into account. These matters require further consideration.

5.2 *Contribution for Energy Society*

This research has provided policymakers a new perspective in promoting hydrogen energy. From a local perspective, this chapter presented a business model for one company in Kanagawa Prefecture to install hydrogen refueling stations at its own logistics centers. It could be part of much broader efforts to introduce hydrogen energy in the prefecture, and help to build the early foundations of hydrogen infrastructure, leading ultimately to what proponents refer to as a "hydrogen society." From an energy policy perspective, this scenario can lead to solutions to Japan's energy issues by increasing energy self-sufficiency, reducing dependency on fossil fuels imported from the Middle East, reducing trade deficits, and diversifying the energy supply. From an environmental perspective, this scenario could contribute to the achievement of carbon emission reduction targets and a transition to clean energy, by promoting hydrogen energy applications through an expanded geographical range of HRSs and FCVs. Meanwhile, this chapter demonstrated the potential of fuel cell trucks to help reduce GHG emissions. Although both electric trucks and fuel cell trucks have zero emissions while being driven, as it stands today, electricity generation in Japan results in much greater GHG emissions than hydrogen manufacturing.

6 Conclusion

There are three main findings in this chapter. First, fuel cell trucks and hydrogen refueling stations can play an important role in promoting the use of hydrogen. This chapter examined the potential consumption of hydrogen by a fleet of delivery trucks of one company in Kanagawa Prefecture, and demonstrated the scenario's feasibility as a business model. The analysis showed that 15 of 24 of Sagawa Express's logistics centers have the required space and meet zoning requirements to install HRSs. For several of Sagawa's logistics centers (including Yokohama East, Yokohama Tsurumi, Yokohama North, Totsuka, Shonan, and Hadano), break-even analysis found that the use of hydrogen is viable if the entire fleet is switched to fuel cell trucks. Considering also delivery costs, Yokohama North is a good candidate to be the first location to introduce an HRS and fuel cell truck fleet.

Second, in more rural areas of the prefecture, Sagawa's potential hydrogen demand would still be below the break-even point due to the low population density and huge service area, especially in the western part of the prefecture. Among logistics centers in the west, only Hadano can reach the break-even point, but even then the delivery performance would be inefficient. Other than logistics centers in Kawasaki, Yokohama, and the Shonan area, few other locations would have sufficient demand to justify installation of an HRS.

Third, this analysis showed in detail the potential for fuel cell trucks to reduce GHG emissions in the specific case of Sagawa Express's operations in Kanagawa

Prefecture. By well-to-wheel analysis, the use of FCVs would decrease GHG emissions by 48.3% relative to gasoline trucks, versus 6.5% in the case of EVs. In Japan, average annual per capita GHG emissions are 9000–10,000 kg. The Yokohama East logistics center has potential annual GHG emission reductions of 643,875 kg, which would be equivalent to the total emissions of about 70 people. Five logistics centers have potential annual GHG emission reductions of over 400,000 kg.

With the rapid growth of e-commerce, the use of delivery services could continue to increase. The analysis done here for one company in one prefecture in Japan suggests that, given the sufficient population density and demand for delivery services, the installation of hydrogen refueling stations in logistics centers could help the delivery industry contribute to SDGs goals and GHG emission reduction targets.

Acknowledgments The authors express their heartfelt gratitude to Yoshiaki Miyamoto and Norichika Kanie, who provided advice and encouragement for data collection, and to DHL Express for providing knowledge, information, and data regarding the express delivery industry.

References

Ahmadi P, Kjeang E (2017) Realistic simulation of fuel economy and life cycle metrics for hydrogen fuel cell vehicles. Int J Energy Res 41(5):714–727

DHL (2021) [online] Source: https://www.dhl.com/jp-ja/home/logistics-solutions.html

EVsmartBlog (2021) "Sagawa Express unveils a light EV delivery vehicle under development-Japan will change if it is realized as planned!" [online] Source:https://blog.evsmart.net/ev-news/sagawa-express-electric-mini-truck-prototype/. (in Japanese)

Hasegawa T (2014) Creation of FCEV market: deployment of independent hydrogen refueling infrastructures. Environ Inform Sci 28:119–124. (in Japanese)

Hondo H (2005) Life cycle GHG emission analysis of power generation systems: Japanese case. Energy 30(11–12):2042–2056

Akimitsu Ishihara (2018) "Role of fuel cells in hydrogen energy society and future prospects" hydrogen Energy society formation study group 2018, 3rd seminar [online]. Source: https://www.pref.aichi.jp/sankagi/shinene/suisozone/src/suisosyakai/suisosyakai20181122_yokohamadai.pdf (in Japanese)

Martin R, Lara WDS, Jesus S, Robert R, Rupert G, Solène V (2017) "Power-to-Hydrogen and Hydrogen-to-X: Which Markets? Which economic potential?" Answers from the literature. 2017 14th International conference on the European energy market

Ministry of Economy, Trade and industry (2013) "Comprehensive Resources and Energy Study Group" [online]. Source: https://www.meti.go.jp/committee/kenkyukai/energy/suiso_nenryodenchi/pdf/001_04_00.pdf (in Japanese)

Ministry of Economy, Trade antholog Industry Agency for Natural Resource and Energy (2019) "Energy issues in Japan"[online]. Source:https://www.enecho.meti.go.jp/about/special/johoteikyo/energyissue2020_1.html (in Japanese)

Shinichi Miura (2018) "Feasibility study for hydrogen Refueling Station business" the 27th annual meeting of the Energy Society of Japan. pp. 208–209 (in Japanese)

Kosuke Miyatake, Radical Minori, Link (2016) 宮武宏輔, 根本敏則, & 林克彦. 宅配便ネットワークにおける 「チーム集配」 導入のための配送密度条件. 交通学研究. *59*:205–212 (in Japanese)

Mizuho Research & Technologies (2004) "Well-to-Wheel evaluation of transportation fuel" [online]. https://www.mizuhoir.co.jp/solution/improvement/csr/lca/pdf/jisseki02_wtwghg2004.pdf (in Japanese)

New Energy and Industrial Technology Development Organization (2014) "NEDO Hydrogen Energy White Paper 2014". pp. 3–4 (in Japanese)

Pouria A, Seyed HT, Hadi A, Yousef S, Hadi G, Mehdi A (2020) The effects of driving patterns and PEM fuel cell degradation on the lifecycle assessment of hydrogen fuel cell vehicles. Int J Hydrog Energy 45(5):3595–3608

Shift Up (2021) "Regulation of parking area" [online]. https://xn%2D%2D7orpdv3oz3dptovpcm82g.com/ (in Japanese)

Uchida S, Minamigata H (2016) Tokyo gas commercial hydrogen Station. J Electr Equip Soc 46(4): 246–249

United Nations (2021) "The Paris agreement" [online]. Source: https://unfccc.int/process-and-meetings/the-paris-agreement/the-paris-agreement

Visualizing Social Capital and Actor Networks for Sustainable Suburban Areas

Sumiko Asanoumi, Ayase Yonomine, Shun Nakayama, and Wanglin Yan

Abstract In this chapter, the authors describe the Next-Generation Suburban Community Development Project, implemented collaboratively with Yokohama City and Tokyu Corporation, and activities at the WISE Living Lab, a space for experimenting with new ideas to address challenges and issues facing communities and to create value in Yokohama, Japan. The study examines the human networks and social impacts of the project and considers the role of living labs in the context of Sustainable Cities and Communities (SDG 11).

Keywords Co-creation · Sustainable city · Suburban area · Actor network · Visualization

1 Introduction

Japan is facing unprecedented low birthrates and an aging of society, leading to the emergence of issues that are here now and expected to become more prominent in future years in the suburban areas that have been supplying housing for workforces that grew apace with economic development in urban areas. Examples of such issues include aging infrastructure, an increase in numbers of older residents and decrease

S. Asanoumi
SFC Institute, Keio University, Fujisawa, Kanagawa, Japan

A. Yonomine · S. Nakayama
Graduate School of Media and Governance, Keio University, Fujisawa, Kanagawa, Japan
e-mail: t19860ay@sfc.keio.ac.jp; Shun-Nakayama@keio.jp

W. Yan (✉)
Faculty of Environmental Information Studies, Keio University, Fujisawa, Kanagawa, Japan
e-mail: yan@sfc.keio.ac.jp

W. Yan et al. (eds.), *Resilient and Adaptive Tokyo*,
https://doi.org/10.1007/978-981-99-3834-6_14

in young people, a lack of people to support community associations due to the depopulation, and the lack of financial resources.

These issues need to be treated as one integrated and interconnected whole across all areas, and to do so, it is essential to have the participation and collaboration of diverse actors, such as companies, research institutions, local residents not to mention governments. In addition, as existing systems, services, and social structures need to be reconsidered and rebuilt, experimental efforts and new approaches must be tried from new perspectives. The "living lab" concept is seen as a method of doing so.

Living labs are places where governments, private companies, local residents, and researchers can work together to develop solutions to social issues and test them in real life (McCormick and Hartmann 2017). In the context for living labs in Japan, according to a report by Koji Nishio of the Fujitsu Research Institute, living labs first originated during the 1990s with initiatives in the USA such as at MIT and the University of Chicago. In the late 1990s, the concept crossed over to Europe, and since the year 2000 living labs have been established as venues for activities in universities and public research institutions (Nishio 2012, 2016). According to reports by Kimura and Akasaka of NTT Service Evolution Laboratories on their impacts and issues, living labs are attracting attention around the world as places to create social networks and visions.

In Japan, where living labs were established to address local issues and create livable communities, and various initiatives are being undertaken, with the participation of governments, municipalities, universities, and other actors (Atsunobu and Fumiya 2018). Previous studies on living labs have included classification of functions and features of examples of living labs, and evaluating their value through interviews with participants in related activities. These studies have revealed their key elements and functions, and their value in co-creation activities. In terms of functions, living labs are reported to have created networks of links between stakeholders to engage governments, businesses, and citizens (Atsunobu and Fumiya 2018).

In this context, Yokohama City and Tokyu Corporation have been engaged in the Next-Generation Suburban Community Development project since 2012. In 2015 they opened the "Sankaku BASE" WISE Living Lab (WLL) as a base of activities for living lab initiatives. They have been working in a number of areas to create sustainable cities.

The Yan Lab at Keio University has been conducting research activities with the cooperation of the WISE Living Lab at Tama Plaza since 2018. In the absence of established methods for evaluating human networks and community connections in collaborative and co-creation projects, we made this attempt to reveal or "visualize" connections, asking this question: How did the projects and the WISE Living Lab contribute to the discovery of human resources for local community building and the creation of sustainable systems? As a result, what may have been just subjective observations can be made more concrete and potentially help in the development of future projects.

2 Project Outline: Next-Generation Suburban Community Development Project

2.1 Issues Along the Tokyu Denentoshi Line, and Signing of Comprehensive Agreement

Yokohama City and Tokyu Electric Railway (which later became Tokyu Corporation) signed an agreement on the promotion of Next-Generation Urban Community Development in April 2012, and began a collaborative effort in Aoba-ku Utsukushigaoka 1, 2, and 3-chome, north of Tama Plaza Station on the Tokyu Denentoshi Line (approx. 6,300 households) for community revitalization, featuring sustainable residential neighborhoods as model areas.

Tokyu Tama Denentoshi (Tama Garden City), where Tama Plaza Station is located, is one of the largest suburban residential areas developed through private sector initiatives as the economy grew in postwar Japan, and it is always a popular area, among the top choices of neighborhoods to live (Fig. 1). However, the average age of residents is steadily rising and there are concerns about depopulation and empty homes left behind in Aoba Ward, a major district of Yokohama, and there are concerns about the aging of apartment complexes and infrastructure around stations, inconveniences in daily living for older people in this hilly community, an increase

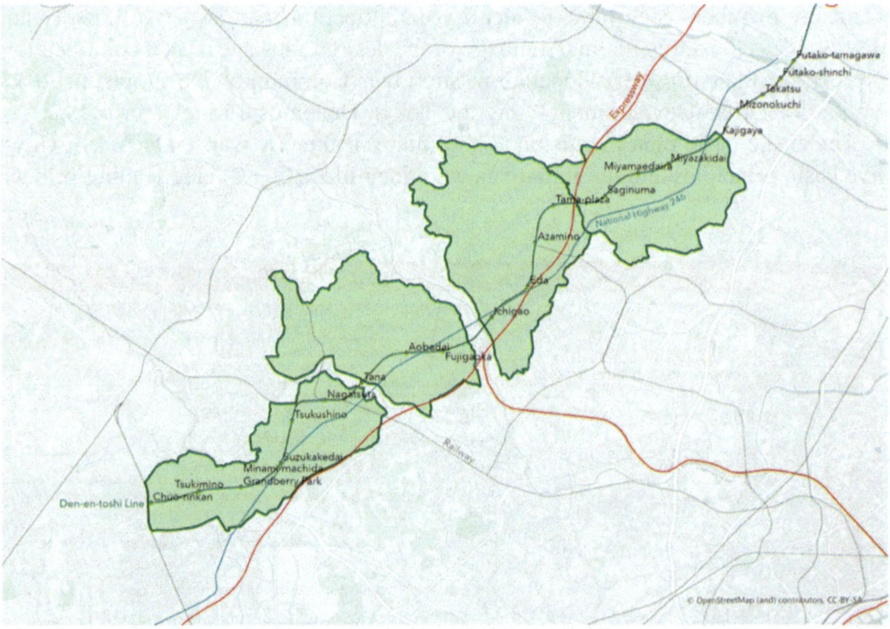

Fig. 1 Tokyu Denentoshi Line population and area (Basic Plan for Next-Generation Suburban Community Development 2013)

in households with both partners working, and the departure of young people from the suburbs due to changing lifestyles. In addition, while public and private services such as childcare and nursing care are still needed, reduced communication between old and new residents could lead to a shortage of local activities, volunteerism, and mutual assistance. There was an awareness of the need to initiate efforts to ensure this would be a sustainable suburban residential area, before these problems worsened, so the aim was to experiment in using this as a model area and use the experience and knowledge gained all along the rail line.

2.2 Basic Plan for Next-Generation Suburban Community Development and Leading Projects

2.2.1 Basic Plan for Next-Generation Suburban Community Development 2013

At the beginning of the Next-Generation Suburban Community Development Project, a full-scale survey of residents was conducted with every household to clarify the future that people envisioned for the region. In parallel, the secretariat also launched a working group for each specialty area and proceeded with discussions. In addition, five Next-Generation Suburban Community Development workshops were held with the participation of nearly 100 local residents, government staff and company members each time, to identify the attractions and issues of the city and discuss ideas to address them. An image of the desired future was then compiled into the "Basic Plan for Next-Generation Suburban Community Development 2013: Vision for Urban Development along the Tokyu Denentoshi Line" (Basic Plan).

Under the basic plan the vision of the future of the city was set a "WISE City," five basic principles and ten initiatives were identified (Fig. 2), and leading projects

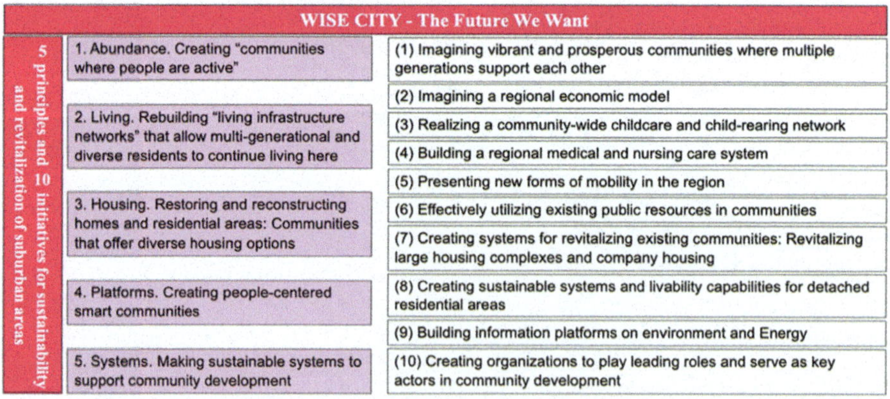

Fig. 2 WISE CITY Vision (from Next-Generation Suburban Community Development website)

aligned with priority items were launched each year. The themes cover a wide range of areas, including the environment, nursing care, child-rearing, economy, and community. Study meetings and discussion meetings, events, and social trials were held for each theme.

Among them, No. 5 (Systems)—Making sustainable system to support community development—is about developing people and organizations who can be key actors in realizing the future vision of the community and sustain community development, and to promote community development in which governments, businesses, local residents and local organizations collaborate together.

Note: The “WISE” in “WISE CITY” is an acronym for “Wellness, Walkable & Working,” “Intelligence & ICT” (ICT = information and communication technology), “Smart Sustainable & Safety,” and”Ecology, Energy, Economy.”

2.2.2 Leading Projects

The leading projects are implementation plans based on the Basic Plan for Next-Generation Suburban Community Development. Starting in fiscal 2012, six to eight themes were covered each year on a variety of topics, including health and welfare, child-raising, the environment, energy, urban infrastructure services, and regional economic revitalization, with the secretariat playing a central role in promoting collaboration among experts, companies, government, and local organizations.

Among them,“Initiatives on Area Management, etc.” is an effort to expand the support of human resources for community building and to create sustainable systems. Efforts have continued every year, including Citizen-Initiated Projects to Develop Civic Pride which were implemented between 2013 and 2014.

The “Support Project” involves community building utilizing a co-creation space, community building in the local community, study sessions and experiment projects in the Living Lab that are gaining attention as new ways to build community, and in the “Co-Creation Project,” residents are primarily engaged in collaborative efforts to address local issues and create value. The Utsukushigaoka Next-Generation Network Information Council holds regular meetings between the secretariat and community associations, shopping districts, commercial facilities, schools, etc., to discuss seasonal events and festivals, report, and coordinate activities (as of 2020).

In addition, the “Children’s and Child-raising Town Meeting,” an initiative for promoting “Machigurumi nursery and child-raising,” is an exchange of information among local nursery and educational organizations. At the twice-yearly “Family Resource Project” events children-raising related organizations and nursery-related organizations have exhibited at recycling bazaars, workshops, and information desks. The “Aoba Model” pilot project involves the Yokohama Aoba-ku Medical Association, regional medical institutions and nursing care providers to establish a network of comprehensive care in the region, and to realize a smart community as a mechanism for energy management in the region, consider the promotion of energy conservation in model areas such as household electricity conservation projects and eco-diagnostics, as well as the installation of energy-creation facilities. A seminar series on the theme of health and social experiments on regional mobility in suburban residential areas was conducted with car manufacturers and citizen monitors as an initiative to consider new forms of mobility in the community. In terms of new topics, the "Project to live and work in Denentoshi," has been developed along the rail line to create an economic model for the area.

2.3 Citizen-Initiated Projects

The "Civic Pride Project" (one of the Citizen-Initiated Projects) was launched in 2013 as one of the leading projects. It recruited a wide range of participants, including residents and action-oriented groups and organizations, to act with love and pride in their own communities. Selected projects were officially approved as Next-Generation Suburban Community Development Projects (Table 1), grants were given, and matching services were provided to bring together residents and private sector companies. Members worked on ideas for things that would not be possible by the activities of existing local organizations working alone, but spring up from people who actually live in the community. They looked at how to work across

Table 1 Approved Citizen-Initiated Projects

	Group name	Project outline
1	Utsukushigaoka Café	Connecting child-raising families with the community
2	Tama plaza Machinaka performance project	Multi-generational learning together and street performances
3	Tama plaza connect	Creating a local network of people, information, and activities
4	Tama plaza shopping street + AOBA + ART	Event using light to connect the shopping district and the park
5	3-chome Café company	Creating a community café
6	AOBA + ART	Art activities in residential areas
7	Aoba friends	Postings for health and checking in on each other
8	Utsukushigaoka diamonds	Effective use of school facilities to connect elementary schools and the local community
9	Health and community development for all at Tama plaza	Multi-generational health and community development
10	"Tama Rakushoku" food project	Creating a multi-generational eating place
11	Tama plaza friends	Creating a local network of people, information, and activities
12	Tama plaza Bunbun Denryoku co.	Generating energy locally for local consumption
13	Mori no Oto (Forest sounds)	Publication of a book that brings together the key persons of Tama plaza
14	Loco-working council Tama plaza project team	A project to live and work in a city you love
15	Tama plaza oil field development project	Recovering household cooking oil for use as energy

the boundaries between organizations and make use of resources in the community, such as parks, roads, schools and train stations. These groups then join other organizations, cooperate with each other in projects and events, and collaborate with local others in the community, such as shopping street and community associations. This project has also led to the emergence and expansion of community building personnel exchanges between the train stations along the rail line. This project attracted people around Tama Plaza who want to try something new, and created a community that supports it and a flow that expands the circle of people.

3 Activities at the WISE Living Lab

3.1 Overview of WISE Living Lab and Co-Creation Space

A central location for activities was needed in order to implement leading projects aligned with the basic plan. It needed to be located somewhere right within the community, not away in a government or corporate office.

To answer the need, the WISE Living Lab was opened in May 2015 as a node of activities for the Next-Generation Suburban Community Development Project, as well as a place to try out ideas from the Living Lab. The WISE Living Lab, located almost in the center of the Next-Generation Suburban Community Development model district, adjacent to Utsukushigaoka Park and on land owned by Tokyu Corporation, consists of three buildings: the Higashi BASE "Community and Residential Consulting Building" (east building), the Nishi BASE "Living IoT Lab" (west building), and the Mannaka BASE "Urban Planning and Residential Gallery Building" (central building) (Fig. 3).

The "PEOPLE WISE Café" community café, located in the east building, is a community café with a function of disseminating information about the Next-Generation Suburban Community Development Project and is available to be used as a workshop space, for small group presentations, and to support community activities.

The "Co-Creation Space" located in the central building is a center for the Next-Generation Suburban Community Development Project office (secretariat). It provides a venue for leading-project meetings and workshops, as well as community-building activities for residents and the Living Lab, as described below. This space is 42 square meters in size and can accommodate up to 50 people, and comes with equipment for meetings, such as 22 movable tables, white boards, a projector, screens and speakers. It is also made available for a fee and is used for corporate seminars and as a classroom, etc. Tokyu Corporation is responsible for the overall management of the facilities, while management of the Next-Generation Suburban Community Development Project is the responsibility of Yokohama City, which serves as the secretariat, as well as Tokyu Corporation, and company under contract, Ishi Planning and Design Co. (as of 2021).

Fig. 3 WISE Living Lab's exterior and Co-Creation Space. (**a**) Location of the WISE Living Lab (**b**) Exterior of the WISE Living Lab. (**c**) Co-Creation Space of WISE Living Lab

3.2 Community Development Using the WISE Living Lab and Co-Creation Space

The Co-Creation Space has been used as a base of activities for the Next-Generation Suburban Community Development Project, for the "Support Project," which invites and supports activities that lead to community building, for the "Co-Creation Project," which addresses local issues and create value through co-creation by companies and residents, and as a place for the Living Lab to test ideas.

3.2.1 Support Project

The Support Project aims to make it more accessible and easier for people to participate in the local community and community-building activities. To open up thought processes, people are invited to propose ideas, with the question "How would you like to make the most of your unique skills and hobbies and work in the Co-Creation Space?" The space is made available free of charge for activities. Ishi Planning and Design Co. personnel were entrusted with the WISE Living Lab operations and served as the local contact point and consultation desk. The Next-Generation Suburban Community Development Project secretariat reviewed applications/proposals, of which 15 were approved in fiscal 2017 and 26 in fiscal 2018.

The Co-Creation Space was used to conduct activities relating to lifestyles and hobbies that people had had previously done among friends at home, including workshops on Japanese paper dyes, embroidery, calligraphy, knitting, making accessories, miso (bean paste), and nutritional education. They were well attended, including by local housewives' groups, and were a trigger for people to get involved in the community.

In addition, there was a wide variety of activities by a diversity in the organizations and people who came together, such as from universities, non-profit citizen groups, artists, community associations, shopping districts, parent-teacher associations, government, entrepreneurs, and private business owners. Examples of activities included workshops on the environment and child-raising, non-profit citizen-writer classes, art activities that collectively archive the memories of the city by artists and local residents, presentations on walking-tours and research on local resources by universities, and a wide variety of projects, including a project to think about the street views, "Double Dutch" style outdoor skipping events, a video creator's course, an entrepreneurial small business course for women, and a wide variety of other projects. The WISE Living Lab staff regularly provided information, introduced people, and organized exchange meetings to encourage connections between groups. Gradually, cooperation flourished among the people who were connected to the WISE Living Lab, including participating in and collaborating with each other's events.

Later, the PEOPLE WISE Café took over as the access point for the community, and a variety of people, including mothers with children, entrepreneurs and remote workers held events and presentations, and this led to market exhibits organized by the café.

3.2.2 Co-Creation Project

The "Co-Creation Project," which was launched in fiscal 2019 following the Support Project, is a framework to support community-building activities by residents. The Co-Creation Space is made available free of charge to support (1) activities that contribute to the promotion of the Next-Generation Suburban Community

Development Project through industry-academia-public cooperation, (2) activities that lead to the resolution of issues in the Tama Plaza region, and (3) activities to create new things, ideas, and services that are not confined by traditional schemes and established concepts.

Projects that were approved and implemented under the Co-Creation Project include participatory street art scenery activities, compiling "memories of the city" booklets based on resident interviews, entrepreneurial classes at local high schools, sustainable city research at university labs, workshops and events for "children's town," projects to monitor/protect older people and children with apps, and SDGs card game events.

People who have been involved in various initiatives in the Next-Generation Suburban Community Development Project, including local residents who were active in the Citizen-Initiated Projects and those who participated in the Support Project, have created new activities and projects, and proposed new plans with members of other organizations.

In addition, youth were attracted to get involved, with local high school students publishing newsletters, developing desserts at the café, conducting research on foot in cooperation with university labs, and presenting their findings.

3.2.3 Living Lab Project

The Living Lab Project is a method employed by users, residents, and corporate developers, researchers, and government to create services, products, and frameworks (Fig. 4a). The Living Lab Project relating to urban development and community development is based on a flow from problem identification, information sharing, idea consideration, prototype creation, practical trials, and then verification.

This poster asks mothers what time of day they are busiest and what challenges them, and invites them to use the "shared" community fridge any time, all day long, over three days of the trial at the WISE Living Lab and PEOPLE WISE Cafe. It offers the opportunity to meet and talk together.

(a) (b)

Fig. 4 Living Lab approaches and project example. (**a**) Design of the WISE Living Lab. (**b**) Poster for the shared refrigerator project

In the Tama Plaza WISE Living Lab, the secretariat served as a matchmaker to bring together companies seeking consultations with compatible community groups on specific themes, and coordinating progress of the projects. The process of mutual understanding and sharing of issues between local residents and companies is conducted conscientiously, ideas are shared starting from community issues, trial models are created, test events are conducted, and outcomes verified.

One of the activities of the Living Lab, the Tama Plaza Living Lab, was launched in 2017 by the secretariat and a telecommunications services company, in cooperation with a local organization (Tama Plaza Connect) and with the cooperation of the University of Tokyo. "Tama Pla Bot" a service provided by the Tama Plaza Living Lab, is a chat bot service that provides local information, with each function designed and developed through workshops with residents. Services to support community activities such as community walking activities (health care, notice posting, info sharing, etc.) are also available. In addition, the local currency Machi-No-Coin (equivalent to "town coins") is a system that allows people to earn units of the "Pla" local currency in return for community contributions and community participation.

Another communications services research institute and the Living Lab did "Local Mothers' Project to Make Child-raising Easier" project under the institute's R&D division to creatively consider ideas about how information tools such as smartphones could be used in daily life in the future. Tama Plaza area mothers who are raising children wrote down their thoughts as they went through their day-to-day routines, shared insights and issues, and discussed ways to make child-raising easier for mothers.

One idea that arose was to have a shared refrigerator, which was set up in the WISE Living Lab for a week. Information was gathered on how it was used and what communication and information was exchanged with others via social media, including the opinions of the core members and event participants (Fig. 4b). Other ideas were also tried out, such as having presentations by a wine sommelier from a local wine shop and a popular local bread maker. These attracted the participation of people who previously had not been involved in community activities, such as parents with children, and men and women, both young and old.

4 Social Network Data Collection and Analysis

As mentioned in previous sections, this project fostered many activities by promoting collaboration across traditional boundaries between organizations, non-profit groups, companies and universities, by bringing together people who had been involved in the activities of community associations, shopping districts, and schools, etc.

These activities led the discovery of community-builders, an accumulation of practical experience, and the realization that the networks formed were social capital in terms addressing local issues and creating value. There was some recognition that this could all serve as a foundation of a framework for building a sustainable community, but there was no established method for understanding the links between people and organizations and for visualizing the actual situation. However, there was a growing need to be able to visualize all of this, because that could lead to verification and evaluation of the project and to future progress.

How are people, organizations, and resources connected by such needs and awareness of problems? We decided to try to compile and visualize data using the methodologies described below.

4.1 Data Collection Methods

4.1.1 Features of Data Collection Methods

To date, various data collection methods have been used in qualitative surveys for social networking research. After classifying the types and characteristics of such data collected, Yamaguchi (2003) conducted research to classify data collection methods relating to social networks with data from various data sources and summarized how they are used depending on their characteristics. According to the research, data collected for use in social network analysis can be divided into three major categories.

1. Reported data: Questionnaires are used to gather data from the target audience.
2. Observed data: The investigator observes the target audience's behavior in-situ and collects the data.
3. Recorded data: Data is compiled from existing literature and records.

With reported data approach (1) individuals in the target audience are often asked to list their people connections, and the information is used to understand their relationships. However, this method has problems with the target audience's accuracy of memory and judgment. The observed data approach (2) involves direct observation of interactions between the investigator and the subject, but this method has a problem in that continuous observation over a period of time is difficult, plus there limits on the number of subjects that can be included. Finally, the recorded data approach (3) is used to understand relationships between companies and organizations rather than individuals. This approach allows the investigator to retrace the past, but the problem is that some organizations do not retain records (Yamaguchi 2003).

Based on the advantages and disadvantages of the above three methods, it was decided to combine all three, using reported data, observed data, and recorded data.

4.2 Interview Details

To collect reported data, three interviews were conducted between July and September 2020. The method of interview was to send a questionnaire in advance and then collect data by oral interviews and recording responses (by completing a form). Three people were selected for interviews, one from the Next-Generation Suburban Community Development secretariat who was responsible for base operations, and two persons who had managed the two community cafés. The reason for the selection is that they were managers of the base of activities that was established for the Next-Generation Suburban Community Development Project. Since many initiatives related to the project were implemented there, and the space served as a base of activities such as exchanges and events among local people, we felt that these interviewees served in networking functions and roles.

We also decided to try to understand inter-relationships by looking at events in order to understand linkages amidst a wide diversity of connections. There are many things that can be called "connections" in everyday life, for example, "people I recognize" "our children are at the same school" and "we live in the same apartment complex," but for this study we decided it would be appropriate to define "actors" as those persons who plan, organize, or cooperate and participate in events to actually solve problems or create value to enrich lives, and "connections" as the relationships among those actors as revealed through "events."

In interviews, we aimed to discover "who" was cooperating with "whom," "where," and in "what kind of activity," by establishing seven items to ask about: (1) representative/leader of the organization that hosted the event, (2) the organization that hosted the event, (3) event name, (4) organizations running the event and participants of the event, (5) location, (6) frequency, and (7) purpose of the activity. "Space" for hosting events and "SDGs" aimed for by the activities were added as categories.

4.2.1 Research Results

For collection of reported data, the target audience for interviews was selected, and interviews were conducted. About 340 items of activity data were collected. To collect observed data, the research was conducted by selecting the target audience based on this reporting data, and surveys were conducted by in-person visit on the day of the activity. The collection of recorded data was an effort complementary to the collection of reported and observed data, by gathering data from event organizers in parallel, including an event's participating organizations, an event's overall description, and data posted on the Internet. In addition to interviews and observed data, we were able to use local media articles on the Internet to investigate the activities of actors with whom we could not conduct interviews, and about 80 data items were collected. As a result, a grand total of about 420 items of activity data were collected.

4.3 Visualizing the Actor Network

Using the data collected, we defined the start and end points of "connections" so that they could be used for social network analysis, and defined the frequency (strength, intensity) of "connections" by including the number of times the event occurred (Table 2). The data collected was organized into two types: "people and event" connections and "event and location" connections. The organized data was then transformed into visual form via social network analysis using Python (Fig. 5). The result was approximately 700 connections made out of approximately 420 items of activity data collected. This network analysis includes an indicator for "centricity,"

Table 2 Definition of data items in social network analysis

Item	Definitions
ID	Identification number for each connection
FROM	Connection start point (organizer, host organization, participant, participating organization, event name)
TO	Connection end point (event name, event location)
NUM	Event frequency
TYPE	Attribute: Organizer, owner, participant, or participating organization
PLACE	Name of activity location
SDGs	Which of 17 sustainable development goals

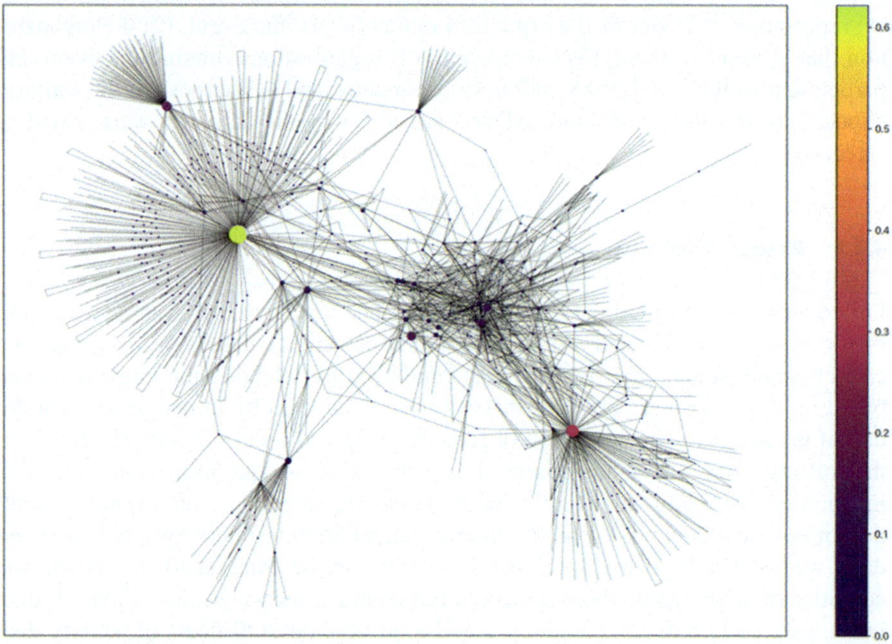

Fig. 5 Social network analysis using Python

the importance of certain actors in the network. To understand the network diagram, yellow indicates an element that has a high degree of centricity and blue means low. The yellow and pink nodes have a high degree of centricity. These are the community café, from which networks expand radially. One of the purple nodes represents the Next-Generation Suburban Community Development secretariat. Its centricity is not as high as the two community café, but one can see that this image portrays the web-like connections between actors, forming a complex network.

4.4 Discussion of Network Visualization Results

4.4.1 WISE Living Lab Contributes to the Formation of Sustainable Social Networks

With Python used to analyze and visualize social networks, one can see that the networks centering on the WISE Living Lab were significantly different from those centering on community cafes. We can see that with the WISE Living Lab, connections between actors form a complex web-like network. On the other hand, the connections between actors at other community café are mainly connected with the node (operator).

However, the Python social network diagrams do not allow one to ascertain the "location," "people/organization," and "activity," and they do not reveal much about their nature. Therefore, we applied a different platform for visualizing stakeholder relationships, known as KUMU.

In examining connections between actors centered on nodes (Fig. 6), one can see was differences between the WISE Living Lab (Next-Generation Suburban Community Development secretariat) and the PEOPLE WISE Café in terms of how connections are made. Actors connected to the WISE Living Lab have

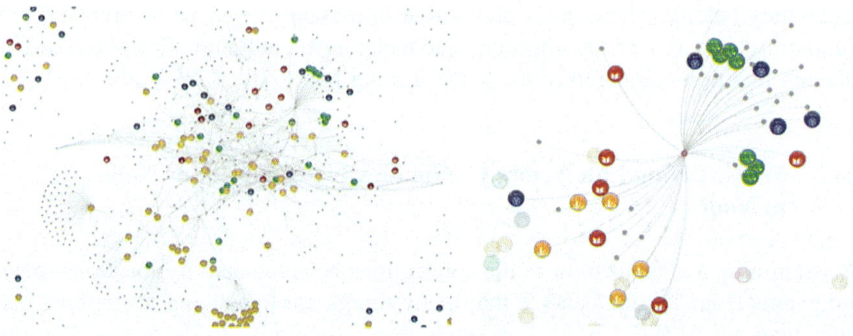

(a) (b)

Fig. 6 Actor connections centering on nodes (**a**) Actor connections (WISE Living Lab). (**b**) Actor connections (PEOPLE WISE Café)

multi-directional connections with each other, and with multiple actors. Conversely, PEOPLE WISE Café actors are connected only to the node (operator).

This shows that, when an actor conducts an event at a certain location, there is no cooperative relationship with other organizations or connections between participants. Examining these comparisons, it is believed that these differences arise from the fact that the WISE Living Lab secretariat is actively and intentionally coordinating people and organizations that are active on various issues, creating mutual relationships among organizations within the region, and companies, research institutes, and NPOs outside the region. Plus, it has been making a conscious effort to coordinate between activities. The difference with the two community cafés is that the WISE Living Lab intentionally connects people and organizations that gather there, creating new initiatives and opportunities for collaboration. It is believed that the actors connected to the WISE Living Lab play a role such in coordination, forming node-to-node (N:N) connections and helping a network evolve that does not depend on the WISE Living Lab. Complex networks, N:N type networks, and networks that are do not depend on the WISE Living Lab are, in other words, highly sustainable, because even if a certain node or actor pauses activities or disappears, other actors that form the network can carry on the function as coordinators. Considering these points, one could say that the WISE Living Lab is playing a role in the creation of a sustainable network of people and organizations that are essential to the role and function of a living lab.

On the other hand, the two community cafés, including PEOPLE WISE Café, are being used as used as places for events and individuals' activities in spaces for informal interactions among people who are not necessarily thinking about community development. This results in individual actor-to-node (1:N) connections. Based on the interviews with the operators of the two community cafés, it is evident they do not intentionally try to connect users to other users, because they are more focused on getting local people to use the community cafés for their respective hobbies and interests, group activities, and presentations. For this reason, users are probably able to do their activities without a sense of formality or feeling pressured to belong to the local community, even if they are only loosely connected to the "community" to which they belong. These cafés play a role in making it easy to be involved in the community, and as such are important and meaningful, although they play a different role and function compared to the WISE Living Lab.

4.4.2 WISE Living Lab Actors Create Flexible Connections Node to Node

By examining a visualization of the connections between activity locations (places) and events (Fig. 7), one can see that many things are happening at regional nodes other than the WISE Living Lab, including community cafés, parks, non-profit organization centers, and individual homes. It was also evident that the WISE Living Lab is being used as a place of activity just like the community cafés. Thus, one could say the WISE Living Lab is performing a node function in local activities. The

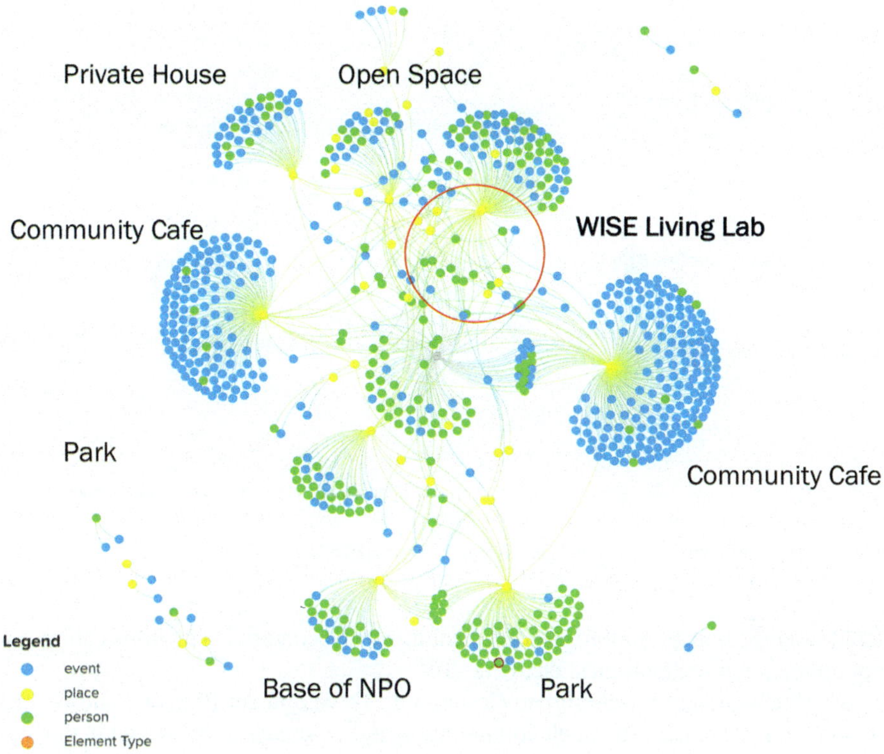

Fig. 7 Node-event connections as depicted by KUMU

activity node function is an important one, meaningful in terms of energizing local activities, addressing local issues, and enriching value creation activities.

One can also see that actors inside the red circle □ are linked to the WISE Living Lab but and, being between nodes, link multiple nodes together. The existence of such connecting individuals and organizations could be seen as enabling the formation of large networks, as they have connections of one kind of another among communities, with nodes that have some centricity. We can see that the actors connected to the WISE Living Lab also connect one place to another, and contribute to the formation of informal but extensive connections in the region.

4.4.3 Through Collaboration, Actors Were Able to Cover More SDGs

Looking at the visualization of event connections in terms of individual Sustainable Development Goals, we can see that many initiatives addressed SDG11 (Sustainable Cities and Communities), SDG17 (Partnerships), and SDG12 (Responsible Consumption and Production), as well as many other goals, such as those relating to

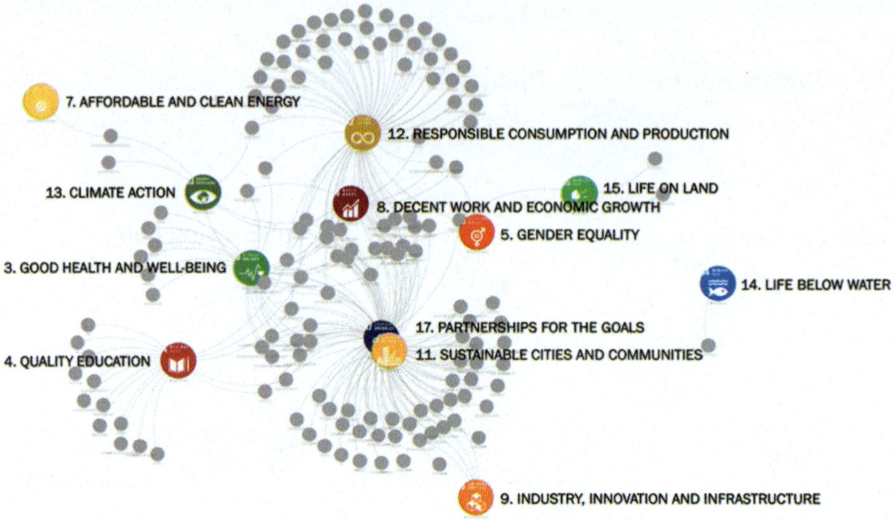

Fig. 8 Connections between SDGs and people, organizations and events

child-raising, transportation, energy, health, local economy, and information and communication technologies, etc. (Fig. 8).

The Next-Generation Suburban Community Development Project could be seen as something that arose through collaboration and co-creation, from a wide range of initiatives that are essential for the creation of sustainable suburban residential areas, the result of cross-sectoral efforts that cannot be achieved by residents, or governments, or businesses working alone.

5 Conclusion

In this study, we conducted a social network analysis for the purpose of visualizing the human networks connected by the Next-Generation Suburban Community Development Project, and considering the project's social impacts and contributions to sustainable community development.

By making it possible to visualize things this way, we were able to see at a glance that the activities were associated with many of the SDGs' targets by creating a variety of social impacts through the evolution of the next-generation suburban community development activities. In addition, with regard to the goal of the basic concepts of creating frameworks for sustainable urban development, this study suggests that the activities were creating multi-layered networks and contributing to such frameworks.

These outcomes are the result of efforts over a period of 10 years since fiscal 2012 under the Next-Generation Suburban Communities Development Project, promoting leading multi-sector projects, coordinating collaboration and co-creation among a variety of actors, and developing strategies to reach out to actors having various affiliations around the region. This ongoing effort led not only to connections between the secretariat and the local community, but also to the establishment and expansion of links between local residents and people and organizations both inside and outside the region, and to establish a network that can produce multiple projects. The project discovered new community-building actors and established relationships with them that will lead to the sustainability of community development.

The "actors" (meaning people who will take some form of action) have connections with their local communities and are the ones that carry community development forward. They are resources for local problem solving and value creation capacity. This study has made it possible to visualize the community resources that have been nurtured through the project and the Living Lab, and has clarified the effects of the initiatives and the role of the Living Lab. This approach to visualization is relevant not only to Yokohama, but could also potentially be applied to any town or city facing similar issues.

We showed that "actors" are the ones who take action as key persons in community building and form networks through events, projects, and activities. However, it is also important to note that because of differences in interviewing methods and data collection, simple comparisons of activities cannot be made across a diversity of places and locations. Over time, there can be an evolution from 1:1 to 1:N, and eventually to N:N connections. Thus, we could say that facilities such as the WISE Living Lab have a significant role to play. Going forward, the ideal situation would be have governments, companies, citizens, and universities continue working together on co-creation in the future.

Acknowledgments We would like to express our sincere appreciation to the many people who participated in the Next-Generation Suburban Development Project, including community residents who contributed to this study, the Housing Revitalization Section of Yokohama City's Housing and Architecture Bureau, Tokyu Corporation, everyone at the WISE Living Lab administrative office, staff of the PEOPLE WISE Community Café, and Mr. Ono, owner of the 3-chome Café. The study was conducted as part of the M-NEX project.

References

Atsunobu K, Fumiya A (2018) Living lab effects and challenges toward solving social issues. Servicology 5(3):4–11

Kosuke S, Yutaka M, Hideyuki N (2008) Network generation model using multiple centricity measures, Chino-to-joho. Intell Inform 20(3):410–422. (in Japanese)

McCormick K, Hartmann C (2017) The emerging landscape of urban living labs: characteristics, practices and examples. Lund University, Lund

Yamaguchi H (2003) Comparative study on data collection methods in social network analysis—published papers pf Faculty of Sociology, focusing on network data among individuals. 09189424, vol 2003-03, 36. Faculty of Sociology, Bukkyo University, Kyoto, pp 105–119. (in Japanese)

Yokohama City, Tokyu Corporation (2013) Basic plan for next generation suburban community development 2013 (in Japanese)

Yokohama City, Tokyu Corporation (2017) All participation, creating a livable city for the long term. 7 approaches to next-generation suburban development. ACTANT Co., Ltd., Ardmore, PA. (in Japanese)

Nishio Y(西尾好司). (2012). Living Lab(リビングラボ)—ユーザー・市民との共創に向けて—. 富士通総研研究レポート No.395.

Nishio Y(西尾好司). (2016). ユーザー・市民参加型共創活動としての Living Labの現状と課題. 富士通総研研究レポート No.430.

Policy Interventions for Resilience and Adaptive Cities

Rajib Shaw

Abstract Policy interventions for resilience and adaptation at the local government level is challenging for many reasons, like national policy making scheme in most cases, lack of resources, changes in risk landscape, etc. One of the key issues in recent years is the emergence of new risks such as global pandemic, digital divide and digital power concentration, energy crisis, food prices, etc. These makes deeper impacts on the existing risk landscape where disasters, extreme events, climate inactions, biodiversity losses are prominent risks in terms of likelihoods and impacts. There are different tools available for measuring resilience in the urban areas, one of them is the CDRI (Climate Disaster Resilience Index), which consists of physical, social, economic, institutional, and natural dimensions of resilience. In an increasingly complex urban area, systemic risk approach becomes more pertinent to understand the interlinkages of different systems related to urban resilience. Ten specific policy measures are suggested in this chapter: (1) Implementing RCES: resource utilization and ensuring urban rural connectivity, (2) Conducting risk assessment in terms of systemic risks, (3) Developing citizen governance interface: utilizing citizen science, (4) Supporting open data and open governance, (5) Enhancing science based adaptive governance, (6) Promoting Local Production and Consumption, (7) Using HEDRM as a common tool to enhance healthy city, (8) Enhancing 1.5 deg. lifestyle, (9) Implementing Digital *Den-en-Toshi* and ensuring digital human resource, and (10) Utilizing Disruptive technologies and Society 5.0. To implement these policies, specific entry points and change agents needs to be identified.

Keywords Resilient and adaptive cities · Systemic risk · Global risk landscape · Adaptive governance · Change Agents

R. Shaw (✉)
Graduate School of Media and Governance, Keio University, Fujisawa, Kanagawa, Japan
e-mail: shaw@sfc.keio.ac.jp

W. Yan et al. (eds.), *Resilient and Adaptive Tokyo*,
https://doi.org/10.1007/978-981-99-3834-6_15

1 Introduction

Policy interventions at the local government level is challenging for many reasons. Firstly, the policies mostly are formed at the national level, and local governments are mostly implementers of those policies. Secondly, most of the local governments (except the capital cities or large business hubs) are constrained with different types of resources, including financial, human, technological, etc. Thirdly, the risk landscape changes drastically over time, and it is always a challenge for the local governments to cope with the evolving nature of the risk landscape. However, in spite of these challenges, many local governments have made significant efforts in progressive policy making for resilience building and local adaptation.

As many of us are aware that in 2015, the world has witnessed several major global frameworks, named Sustainable Development Goals (SDGs), Paris Agreement on Climate Change and Sendai Framework for Disaster Risk Reduction (SFDRR). All these three frameworks have the same time frame from 2015 to 2030. An analysis of these three frameworks (Shaw et al. 2016) shows that the word "local" has been used extensively in all the three frameworks. "Local" is used in terms of local community, local government, local culture, local tradition, local adaptation, local materials, etc. The analysis has correctly pointed out that to make the global framework effective, local governance, local actions, and local leadership are extremely important.

In this short chapter, at first, a few issues and concepts are introduced, like global risk landscape, resilient and adaptive cities, and systemic risk approach. After that, ten specific policy measures are suggested for making cities resilient, adaptive, and sustainable.

2 Global Risk Landscape

World Economic Forum (WEF) publishes Global Risk Outlook in every year in January during the Davos Meeting. The analysis provides previous year's major risks and future potential risk. The report of 2020 (GRO 2020) shows that that environmental risk (like disaster, climate change, biodiversity losses) prevails the risk landscape globally in terms of likelihood and impacts (Fig. 1). The same analysis in 2021 (GRO 2021) puts infectious disease as one of the major risks in terms of impacts due to the impact of COVID-19 in the year 2020. However, there are two new risks arises, which are digital inequality and digital power concentration. In the year 2020, we have changed our lifestyle to work from home, and changed to digital education, digital health care, etc. However, the world does not have equal digital connectivity. There has been a north south divide in digital equality, and the impacts have been high in the rural areas, especially in developing countries. Even in developed countries, the impact is found in the age groups, where the older age groups have limited digital access, especially in the rural areas. This

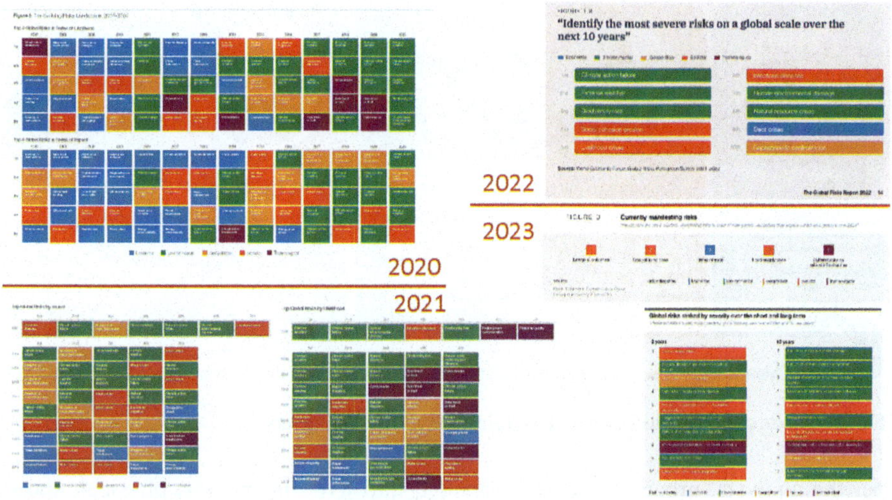

Fig. 1 Global risk landscape from 2020 to 20203

inequality has posed the challenge to access different services, and thereby is considered as a new risk.

In 2022 (GRO 2022), the same digital inequality has continued, while environmental risks have predominated other risks in spite of COVID-19 impacts, and infectious disease did not make its mark that strongly like the previous year. However, cyber security has become a new risk, while the world has been moving towards more digital services, including digital/online shopping and payment services. In 2023 (GRO 2023), the top two risks were energy crisis and cost of living, which are attributed to the Russia Ukraine war, which erupted in February 2022. Therefore, within 3 years, the global risk landscape changed drastically, with new risks such as infectious disease, digital inequality, cyber security, energy crisis, cost of living, etc. These were not perceived as major risks otherwise. The point here is that in an inter-connected world, we need to be cautious about the complex risk landscape, which changes dynamically. Although we focus on the resilience and adaptation in cities or local governments, the inter-connected nature of risks and dependency of Japanese cities and local governments to the global risk landscape, it is important to understand, analyze it, and take lessons for future preparedness.

3 Resilient and Adaptive Cities

Resilient and adaptive terms are two sides of the same coin and are closely interlinked. The more a city becomes resilient, more it becomes adaptive. Urban resilience has been a research area for many years, and there have been many ways to define urban resilience. A holistic analysis of resilience is shown by Shaw (2011). As

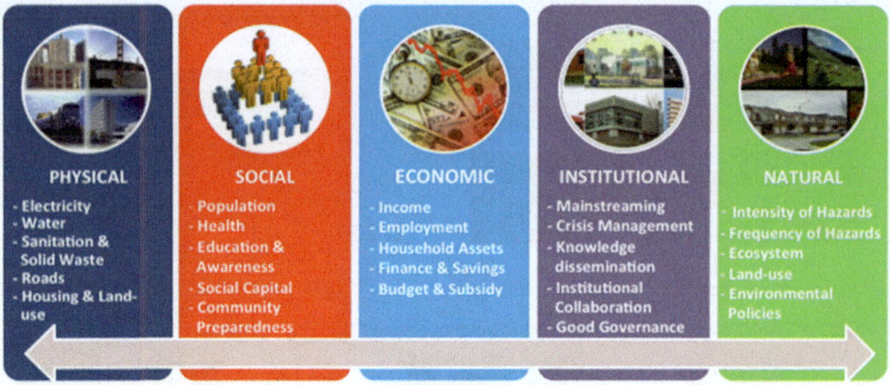

Fig. 2 Urban resilience assessment tool (dimension and parameters)

per the analysis, urban resilience is divided into five dimensions: physical, social, economic, environmental, and institutional. Each of these dimensions are divided into five parameters, and each parameter into five indicators. Thus, there are a total of 125 (5 × 5 × 5) indicators on which the data is collected. Figure 2 shows the details of each dimension. Physical resilience is divided into electricity, water, sanitation/solid waste, roads and housing/land use, etc. Electricity has status of interruption, % of city dwellers having legal access to electricity (a critical factor for developing country's cities), self-sufficiency of electricity production in city, back up service, and alternate energy sources, etc. Similar approach is done for other physical resilience parameters also, as well as for other dimensions. The key aspect of this analytical tool is to de-segregate the resilience concept to city services, so that the based on the analysis, the city officials can make their short-, medium-, and long-term plans. In most cases the cities were asked to develop their action plan and get it approved by the city senate so that the specific fund can be used for undertaking specific measures to enhance the resilience of the parameters, which are relatively lower and which gets priority in the plans.

One of the key aspects of resilience and adaptation is that the context of urban areas changes constantly. Therefore, the resilience assessment and adaptation pathways need to be continuedly updated. Thus, we recommend periodic resilience assessment using the same indicators and measure the progress of the adaptive measures. Due to climate change and related impacts, sometimes the natural resilience becomes lower since it factors severity and frequency of hazards. The physical resilience may increase with investment in the infrastructure development, but the social resilience may vary depending on the nature of population. Analysis shows that in some cities the key strength is the physical resilience, and for some other cities, it may be institutional or social resilience (Metro Manila 2010). Also, it is suggested to do the sub-city analysis where spatial variation of city's resilience can be understood, which may change periodically. Sub-city analysis (Bandung 2012) gives the local governments a better picture of the city's risk and resilience (both negative side and strength) and undertake adaptive measures.

The Asia chapter of recent IPCC report (Shaw et al. 2022) suggests that Asian cities will be exposed to increasing extreme temperatures and heatwaves, which may have adverse impacts on the health of the vulnerable population like aged population, children, pregnant mothers, etc. Similarly, the cities will also be exposed to droughts, leading to water scarcity and longer dry days, as well as extreme precipitation (short duration, heavy precipitation), which may trigger major flooding. Coastal cities are already exposed to sea level rise, which may be enhanced in due course. Tropical storms (typhoons in East and Southeast Asia, and cyclones in south Asia) will be intensified, and we will possibly see more severe storms in near future. The analysis suggests development of resilient infrastructures (like power, water, built infra, etc.) as well as focusing on nature-based solutions, which may be helpful to reduce the impacts heat waves in longer term. A combination of gray and green infrastructure policy is required for the optimum balance for urban adaptation measures.

4 Systemic Risk Approach

Systemic risk is a common framework used in the financial sectors, especially to evaluate the inter-linkages with other sectors. In recent years, several reports mention about the importance of systemic risk in disaster and climate change issues (GAR 2019, 2022). In an increasingly globalized world, we are getting strongly connected than ever before. There are different levels of connectivity. Urban rural areas connected with different resources such as food, energy, water, human resources, etc. Countries are connected with diplomatic relations, human resources, businesses, and many different systems such as education, healthcare, etc. (Mitra and Shaw 2022). Information and information technology brings another level of connectivity to all of us. Thus, where it is physical or virtual connectivity, a disruption in one system in one place affects a wider global system. And that is the concept of systemic risk. Therefore, to develop a resilient city, we need to strongly focus on different urban systems and its interdependence in terms of food system, water system, energy system, transport system, human resource system, etc. The supply chain and business continuity are core to avoid un-disrupted services during a shock or stress. Depending on the scale of shocks or stresses (i.e., how big a disaster event is and how long it continues for a stress event), it is important to customize the business continuity planning and resource management. Systemic risk approach provides a unique pathway to address different systems collectively and enhance the urban resilience during a major event.

5 Specific Policy Measures

Following section provides ten specific policy measures which help in conceptualizing resilient and adaptive city in an uncertain world.

5.1 Implementing RCES: Resource Utilization and Ensuring Urban Rural Connectivity

Regional Circular Ecological Sphere (RECS) is a concept proposed by Ministry of Environment of Government of Japan (MOEJ 2018) in its fifth Basic Environmental Plan. Later, it is also called Circular and Ecological Sphere (CES). The concept argues for enhancing plans and policies realizing the resource dependence in urban and rural area (Fig. 3). Three key issues are emphasized: (1) explore simultaneous solutions for economic, regional, and international challenges, (2) Maximize sustainable use of regional resources, and (3) Enriching and strengthening partnerships.

This specifically argues on the benefit sharing mechanism between urban and rural areas for the resource utilization, and the policy should encourage a series of agreements between relevant stakeholders to redistribute the benefits of a healthy watershed equitably so that the resource sustainability and maintaining a quality of life across the region.

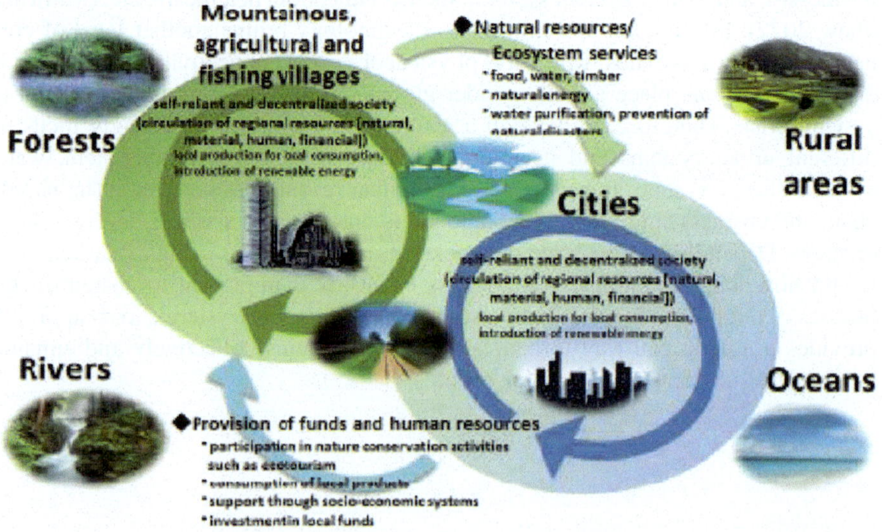

Fig. 3 Concept of RCES (Source: MOEJ 2018)

5.2 Conducting Risk Assessment in Terms of Systemic Risks

As mentioned before, the systemic risk concept is of utmost importance when we think of urban resilience. Systemic risk refers to the risk that the whole system will break down, not just the failure of individual parts. The term "systemic risks" refers to threats that have wide-reaching, cross-sectoral, or even global effects where traditional risk management and even national risk regulation are not enough (Mitra and Shaw 2022). It is critical to make proper risk assessment using systemic risk concept. The recent literature review suggests that there is hardly any tool available for assessing systemic risk in a comprehensive way. The complexity of the issue is a major challenge for developing one comprehensive assessment tool. It is suggested that the urban resilience policy should have some flexibility to allow risk assessment using systemic risk, even it is not holistic. For example, when city makes an assessment for its transport infrastructure, it needs to also take into consideration the impacts of transport failure on supply chain (goods and services) as well as health care systems and livelihood impacts of the people (business interruptions). Similarly, an interruption in the electricity system as a critical infrastructure will have deep impacts on health, education, production, transport, etc. This may not be a perfect or holistic assessment, but semi-quantitative assessment is important to understand the inter-dependency.

5.3 Developing Citizen Governance Interface: Utilizing Citizen Science

Another critical policy challenge is to develop citizen interface of governance through citizen science. Citizen science has becoming popular in the biodiversity systems, and it is now important that we utilize citizen science in the governance system to develop a resilience urban area. In a recent analysis Ozaki and Shaw (2022) have pointed out that one of the vital issues in promoting social participation of citizens could be information sharing. It also describes the cycle which citizens themselves become the main actors in generating information to promote citizen participation, and the information generated through this process leads further enhances a healthy citizen governance interface. Transparency of information sharing is a critical measure of good governance. Therefore, making specific emphasis to create citizen interface is critical for the urban resilience development program.

5.4 Supporting Open Data and Open Governance

Open data and open governance are supplementary processes to enhance citizen-based decision making in the urban area. In a recent analysis, Kanbara and Shaw

(2022) and Kanbara et al. (2022) analyzed the case of Atami landslide, exemplified that open data (here, cloud point data) helped the civic tech professional to conduct the damage assessment within 3–4 h, which was used by provincial government for the decision making in post disaster scenario. This example is a classic positive impact of open data and open governance, which brings effective decision making, as well as involve different stakeholders in providing technical advice to a resource constrained local government. However, to make it implementable, it needs different legal and higher level policy challenges. The countries which are promoting open data and open governance at the national level are prone to get the benefit of this. UNESCO has started a global campaign with different governments to make them understand the importance and benefit of open governance system. The actual global implementation may take some more time, but it is important that local initiatives starts and records success stories and good practices of open data and open governance, while the larger national policy environment for open data/open governance may take some more time.

5.5 Enhancing Science Based Adaptive Governance

In the uncertain world, the conventional scheme of governance does not work. It needs to adjust based on the local changes. While it is difficult to change the national regulations or legislative framework quite often, smaller change/adjustment in the local level is important. This is known as adaptive governance. A classic example was to respond to different types of natural hazards (like typhoon, flood, etc.) during the prolonged period of pandemic. There were adjustments of the evacuation shelter layout, shelter management, volunteer management, etc. These were not in the emergency operation manual of the local governments, and all of these can be considered as adaptive governance. Many of these were ad-hoc decisions taken by the local government during uncertainties based on the advices of the national/ prefectural governments or based on the advices of outsider stakeholders like academics or civil societies. The key issue here is how science can be used for adaptive governance at the local level. For that, scenario planning, data science can play important role, which is evident in some cases of pandemic in an early time of 2020 when vaccines were not available or it was rather difficult to understand the nature of the virus. Data science played important role in future projections of COVID-19 peaks and suggested mitigation measures to local governments. Similar use to science based adaptive governance will be useful for making cities resilience and adaptive.

5.6 *Promoting Local Production and Consumption*

In a globalized world, we often import or export food products from a far distance, where the ecological footprint becomes very high. Local Production and Consumption (LPC) model has been promoted not only to reduce the ecological footprint, but to make cities resilient in case of disruption of services, which is often the case during disasters. For local farmers selling the products in local market brings the cost of the products, as well as farmers get larger profit. The customers also get better and fresh product, close to the farm. Thus, it is a win-win situation for all. However, it is also understandable that a city or urban area cannot produce everything locally. Here, we urge that to make the cities resilient, it is important to make an assessment of the potential of LPC in the cities and make efforts to increase the percentage of LPC. That type of assessment will help the cities to make understanding of its local resilience and prioritize the supply chain which needs to be strengthened.

5.7 *Using HEDRM as a Common Tool to Enhance Healthy City*

Eco-HEDRM (Health Emergency and Disaster Risk Management) is a framework for Evidence-Based Health Policy (Tashiro and Shaw 2020) from the perspective of human security under SDG 3: Ensure healthy lives and promote well-being for all at all ages. In past different disasters, HEDRM has been used in the recovery process, especially focusing on the larger dimensions of health care services. By larger dimension, it means beyond the conventional health infrastructures like hospitals or health centers and focusing on the community well-being. Eco-HEDRM brings the ecological aspects of the community well-being, and has been used during COVID-19 in different cities globally. In recent years, IoT innovation has been attracting attention toward the realization of a society in which people and nature can coexist in harmony (One Health), based on the reflection that the negative impact of human activities on the natural environment has contributed to the spread of natural disasters, climate change, and pandemics. HEDRM integration policy, which can simultaneously realize human health and eco-health, has rarely been considered in Japan, contrary to global policy trends in disaster and health crisis management. It is important to: (1) develop a conceptual model that enables the use of simple ICT-based technologies at local sites; (2) evidence building for the use of ICT for health data at local sites in Japan, and (3) development of guidelines, including ethical guidelines, for social implementation of Eco-HEDRM.

5.8 *Enhancing 1.5 Deg. Lifestyle*

While the global negotiations in climate change always focuses on CO_2 reduction and setting up targets by 2050 or 2070, this cannot be achieved without making sincere efforts at the local level to change the lifestyle. In a major report of 1.5° lifestyle (IGES et al. 2019), one attempt was made to fill the gap between the aspiration and reality, and to begin to propose clear targets and quantifiable benefits to climate change solutions by making changes in our lifestyles. The report states: "*In terms of the gaps between actual lifestyle footprints and the targets, footprints in developed countries need to be reduced by 80–93% by 2050, assuming that actions for a 58–76% (a 8–12% reduction every year from 2019 to 2030) start immediately to achieve the 2030 target. Even developing countries need to reduce footprints by 23–84%, depending on the country and the scenario, by 2050.*" The calculation was made before the COVID-19 pandemic, and there are now new targets set up and it is almost sure that 2030 targets of SDGs and Paris Agreement cannot be achieved. It is important that the local governments, especially the cities need to start policies specifically focusing on 1.5 deg. lifestyle, make new targets and ensure that the targets are locally achieved.

5.9 *Implementing Digital Den-En-Toshi and Ensuring Digital Human Resource*

As mentioned previously, urban rural connectivity is the key to urban resilience. In an increasing digital world, urban rural digital connectivity becomes critical. GRO (2021) pointed out digital power concentration and digital divide as the potential future risk, when we are increasingly becoming dependent on the digital tools. "Digital Den-en-Toshi" is a concept of the Prime Minster Kishida Cabinet, which is launched in 2022. The objective is "to promote regional revitalization through digitalization, and furthermore, to realize bottom-up growth from the regions to the entire country." The following digital human resource development and securing are listed as important measures: (1) Develop and secure digital human resources in the public sectors and (2) implementation of online courses, etc. Local governments need to make strategic efforts to enhance digital penetration through different types of government services and reduce the urban rural digital divide.

5.10 *Utilizing Disruptive Technologies and Society 5.0*

Society 5.0 is a concept of human/people centric super smart futuristic society. Due to major demographic changes in the Japanese society, especially de-population and aging population, we need to depend more on the technologies which are inter-connected and which severs the basic needs (Kanbara et al. 2022). We now call them "disruptive" or "emerging" technologies, however, today's emerging technologies become tomorrow's essential technologies. The local governments need to develop their resilience strategies to avail different types of disruptive technologies towards Society 5.0.

6 Conclusion

To enhance urban resilience and make a city adaptive to different types of stresses and shocks, it is important to have a good policy as mentioned above. To make the policy implementable, there needs to be capable human resources as well as a strong link with the local stakeholders such as business sectors, civil society, academia, etc. It is important to note that we need specific "*Entry points*" to the local communities to make some of the policy decisions implementable at grass roots level. It can be local issues such as waste management, water conservation, social welfare, local production, etc. Each community has its own priority, and to understand it is the first step of implementation. Co-design, co-produce and co-delivery is the sustainable approach for community involvement. Apart from the entry point, "*Change Agents*" are important avenue to enter in the community. The change agent can be local elected leader, business person, or local community-based organizations and sometimes youth led innovation. There is no "one fits all" solutions for the whole city. Depending on the nature of the city's locality, specific change agent and entry points need to be identified so that some of the policy measures mentioned above can be implemented and sustained at the local level.

References

Bandung (2012) Bandung climate and disaster resilience assessment. Kyoto University, Kyoto, p 46

GAR (2019) Global assessment report on disaster risk reduction. United Nations, Geneva, p 472

GAR (2022) Global assessment report on disaster risk reduction. United Nations, Geneva, p 254

GRO (2020) Global risk report. Davos, Switzerland, p 102

GRO (2021) Global risk report. Davos, Switzerland, p 97

GRO (2022) Global risk report. Davos, Switzerland, p 117

GRO (2023) Global risk report. Davos, Switzerland, p 98

Institute for Global Environmental Strategies, Aalto University, D-Mat Ltd (2019) 1.5-degree lifestyles: targets and options for reducing lifestyle carbon footprints. Technical report. Institute for Global Environmental Strategies, Hayama

Shaw R (2011) Mapping climate and disaster resilience in cities. In: Community, Environment and Disaster Risk Management, vol 6. Emerald Group Publishing, Bingley, pp 47–61

Kanbara S, Shaw R (2022) Disaster risk reduction regime in Japan: an analysis in the perspective of open data, Open Governance. Sustainability 14:19. https://doi.org/10.3390/su14010019

Kanbara S, Shaw R, Kato N, Miyazaki H, Morita A (2022) Society 5.0, digital transformation and disasters: past, present and future. Springer, Cham, p 217

Manila M (2010) Metro Manila climate and disaster resilience assessment. Kyoto University, Kyoto, p 50

Mitra A, Shaw R (2022) Systemic risk from a disaster management perspective: a review of current research. Environ Sci Pol 140:122–133. https://www.sciencedirect.com/.../pii/S1462901122003690...

MOEJ (2018) Regional circular ecological sphere. https://www.env.go.jp/policy/kihon_keikaku/plan/plan_5/attach/ref_en-02.pdf. Accessed 12 Mar 2023

Ozaki Y, Shaw R (2022) Citizens' social participation to implement sustainable development goals (SDGs): a literature review. Sustainability 14:14471. https://doi.org/10.3390/su142114471

Shaw R, Prabhakar SVRK, Chiba Y (2016) Policy brief: SDGs, DRR and CCA: potentials for strengthening interlinkages, IGES Policy Brief, July 206, Number 34. IGES, Hayama, p 12

Shaw R, Luo Y, Cheong TS, Halim SA, Chaturvedi S, Hashizume M, Insarov GE, Ishikawa Y, Jafari M, Kitoh A, Pulhin J, Singh C, Vasant K, Zhang Z (2022) Asia. In: Pörtner H-O, Roberts DC, Tignor M, Poloczanska ES, Mintenbeck K, Alegría A, Craig M, Langsdorf S, Löschke S, Möller V, Okem A, Rama B (eds) Climate change 2022: impacts, adaptation, and vulnerability. Contribution of working group II to the sixth assessment report of the intergovernmental panel on climate change. Cambridge University Press, Cambridge, pp x–y

Tashiro A, Shaw R (2020) COVID-19 pandemic response in Japan: what is behind the initial flattening of the curve? Sustainability 12:5250. https://doi.org/10.3390/su12135250

Toward A New Resilience

Wanglin Yan and William Galloway

Abstract The urbanization of Tokyo took place in conjunction with a near continuous series of unprecedented disasters, events, and reconstructions; from the early fires in the Edo Era, to the Great Kanto Earthquake in 1923, the fire-bombing of World War II, the 1964 Tokyo Olympics, the Great East Japan Earthquake, and somewhat anti climactic, the 2020 Olympic Games. After each event history was marked and Tokyo was changed, usually improved, built back better. This chapter lightly unveils the secret behind the city's ability to recover and make use of transformational events. We conclude that Tokyo's resilient and adaptive urban system is based on three fundamentals—a strong rail network, livable communities, and convective flows. These fundamentals are supported by a complicated urban nexus, and an entangling *mesh* of sectors and people that reaches to every corner of the city and acts at multiple scales. The example of Tokyo shows that resilient and adaptive cities rely on the cooperation of both human stakeholders and multiple layers of physical factors. This insight could be valuable for growing cities in emerging countries. It provides a basis for the advancement of urban resilience and adaptivity.

Keywords Urban system · Urban nexus structure · Mesh · Urban crisis

1 The Urban Structure of Tokyo

The Tokyo Metropolitan Area is one of the most competitive cities in the world (Mori Foundation 2022), an accomplishment all the more remarkable because Tokyo is a megacity. In fact, it is currently the largest city in the world, with a population

W. Yan (✉)
Faculty of Environmental Information Studies, Keio University, Fujisawa, Kanagawa, Japan
e-mail: w_yan@keio.jp; yan@sfc.keio.ac.jp

W. Galloway
Department of Architectural Science, Toronto Metropolitan University, Toronto, ON, Canada
e-mail: will.galloway@ryerson.ca; william@frontofficetokyo.com

W. Yan et al. (eds.), *Resilient and Adaptive Tokyo*,
https://doi.org/10.1007/978-981-99-3834-6_16

nearing 40 million people. With so many people crowded together it would not be surprising to learn of any number of social or logistical problems. While those problems do exist, and historically have even been severe, the city has now become a model for sustainable urbanization for many emerging economies (ARCADIS 2022). Strolling through the streets of Tokyo we can find cultural attractions, gourmet food, vivid streetscapes and a diversity of local cultures; all supported by an efficient infrastructure, including a well-connected inner and inter city transportation network. You may wonder how a city with so many markers of success could develop in spite of a history of massive disasters. We would argue the achievement is more intriguing because the city has done so well through deliberate choice.

The chapters of this volume include stories that describe physical and institutional conditions. They outline nexus thinking methods, for observing and evaluating cross-sectoral efforts. And they propose solutions for the development of green infrastructure, renewable energy, and participatory planning. Reading through the material we began to feel that we are indirectly describing a particular code for making a resilient metropolis. That code can be generalized by four parts - a powerful city core, an advanced railway network, compact communities, and the resulting convective flow of goods and people cross a metropolitan region. Each of these are elaborated on below.

1.1 A Powerful City Core

Generally, the central area of a city is the location of its social and economic engine, and if the city is large enough, it can propel a nation. During its long history Tokyo established a central business district (CBD) around Marunouchi-Tokyo Station. Unlike some cities in the world where the business center shifted ore weakened as a result of suburbanization, Tokyo never stopped upgrading its CBD. Quite the opposite, as suburbs grew around the metropolis in the late half of the twentieth century, subcenters such as Shinjuku, Shibuya, Ikebukuro, and Ueno also grew at a remarkable pace. All of them are well known destinations in the city, some of them are world famous. Not by accident, all of them are all located on the Yamanote Line, a ring-shaped train line that encircles much of what might be called central Tokyo. Functionally, these subcenters were not a replacement of the CBD but were the containers of spillover effects. Consequently, the CBD and subcenters form an enlarged urban structure, with the main CBD and multiple subcenter cores scattered just inside and around the Yamanote Line. The huge compact region created by this ring was designated as the central core of the Tokyo Metropolitan Area (TMA) from the 1990s, a designation continuously revised in the city's strategic development plan (Tokyo 2022). Although the monolithic approach to this large area is often criticized, many people believe that when "Tokyo is strong, Japan is strong; [and when] Tokyo is weak, Japan is weak." Meanwhile, the Japanese government also decided to decentralize some divisions and institutions of the government to new cities such as the Minato Mirai 21 District in Yokohama City, Makuhari City in

Chiba, Saitama New City in Omiya, and Saitama City; all in order to improve its disaster resilience through decentralization, and to improve regional sustainability. The effort has consequently motivated the private sector to move out from the core to the suburbs as a way to balance their development.

1.2 Advanced Railway Networks

Tokyo is famous for its well-developed railway network and its use as the main public transportation instrument (Calimente 2012; Yajima et al. 2019). The railway network connects the central core, new cities, and local communities, forming a strong framework for the urban systems that support the region (Kono n.d.). With this railway network we can see that:

- The central core of Tokyo accommodates the daily movement of eight million workers, the highest rate in the cities of the world.
- The high density, high frequency, and seamless connection of trains between the central core and suburbia creates the most efficient and convenient commuting system.
- The railroad stations were built as both service and living centers for residents. This makes Tokyo one of the most advanced models of transit-oriented development (TOD) in the world (Calimente 2012).
- The partnership of railway companies, municipalities, landowners as well as citizens forms a new type of commons in the modern city that can be leveraged as a tool in urban planning and land development (Nakamura 2018). Cooperation between these groups makes the TOD possible in most of the metropolitan area and supports the establishment of both the working and living practices of its citizens (Sanders 2015).

1.3 Livable Communities

Residential communities are the basic organ of a city. Livability is the most fundamental factor under consideration when people determine where to live, an idea which includes the cost of finding or building a home (Wheeler 2004). The goal of urbanization, in an ideal world, is to continually protect communities from disasters and risks, to improve access to food, water, and energy, and to meet the needs of citizens as they progress through the varied stages of life (Ramaswami 2020). Planning in Tokyo has managed to maintain a focus on the perspective of its residents and the production of a better living environment even through its most dramatic period of rapid economic growth in the post war years (Liu et al. 2022). Several chapters in this volume make this point, as some readers may have noticed. With regards to the future, adapting to a shrinking population, the metropolitan government has developed a picture of Tokyo for 2040 where residents live in compact developments along the railway network. The plan aims to also revitalize underused land and social infrastructure in the hinterland of existing railway systems. As such it is not only a physical plan but a political movement promoted by the Ministry of Land, Infrastructure, Transport and Tourism (MLIT). If done correctly it would additionally act as a part of the city's trend toward building a zero carbon society.

1.4 Convective Flow of Goods, People, and Information

Tokyo expanded radically during much of the twentieth century, growing from a little more than one million people to over thirty-one million. In 1962 it passed into the small club of world megacities (TMG), and currently has a population of around 36 million in the greater metropolitan area, making it the largest city in the world.

All of this growth was supported by the massive flow of energy and resources, both in and out. The scale of urbanization caused severe environmental problems, while opportunities concentrated more and more in urban areas, causing a flood of population out of the countryside and into cities like Tokyo. Reversing this one-way flow from rural to urban areas was an urgent issue for the Japanese government. As a result, the central government revised its national terrestrial planning in 2008, highlighting *convective flow* (MLIT 2008). This term, "convective flow" signifies the tight communication and exchange of goods and services between urban and rural areas as well as among regions. According to the new plan, urban and regional plans and projects were to help develop local centers and nodes with unique characteristics. Ideally, the centers and nodes would not only be the locus of consumption but would additionally provide goods and services to other nodes in the system. Since its adoption pragmatic policy has been implemented under the supervision of the MLIT, making use of three paths, each aimed at creating a sustainable social infrastructure as well as public services. The three paths are as follows; first, control housing and urban stock by regulating new construction of hard infrastructure; second, upgrade incrementally in order to adapt to slow onset change; and third, manage urban facilities like parks or water treatment plants through engagement with the private sector.

Beyond investment in infrastructure, investment in ecosystems was also promoted across the country. Through the accumulation of investment in natural resources it is expected that productivity and access to public services could be improved. To give an example, a typical program was the concept of the "Circular Sphere of Forest, River, Settlement, and Sea," promoted by the Ministry of Environment since 2007 (MOE 2007). This idea is easily linked with the principles of nexus thinking, as the discussion in chapter "Design-Led Nexus Approach for Sustainable Urbanization" of this volume.

A powerful city core, an advanced railway network, livable communities, and convective flows, are all aspects of a single entity, the city of Tokyo. While there are other aspects to consider, such as good governance and social cohesion, it is our conclusion that these are the key elements that shape the essential structure of a resilient and adaptive Tokyo. They ensure the city sits at the top of the list for sustainability among the world's megacities.

2 The Urban Nexus in Tokyo

With the urban structure of Tokyo in mind, we can conceptualize the complexity of a metropolitan urban nexus system with a mesh, as illustrated in Fig. 1. The horizontal layers consist of the physical and social factors supporting the management of a city or region, and the columns are spatial scales. From bottom to top, the nexus factors are stacked in order of the ecological basement, geographic agglomeration, hard infrastructure, soft infrastructure, actors, connectors as well as people. Each layer is related to community, city, and the metropolis in consideration of the general systems of urban planning and management.

The terms within the mesh can be thought of as parts of what the philosopher Timothy Morton calls Ecological Thought (Morton 2009), where Nature is deeply entwined with the synthetic constructs of humanity. In his way of thinking the 'mesh' takes on the same meaning as a network, like the Internet or world-wide-web. As he writes, "*By extension, 'mesh' can mean 'a complex situation or series of events in which a person is entangled; a concatenation of con-straining or restricting forces or circumstances; a snare'*" (Morton 2010). From the point of view of the authors this is an idea very similar to the centra term of this volume, the Nexus. While we do not intend to look deeply into the philosophical significance of this pronouncement it is a useful point of reference. Applying the term helps us to

	Community	City	Metropolis
Vision Well-being	Compact life	Multi-centric urban structure	Global megalopolis
Connector Product/goods/services	Seamless connection	Competition/i innovation	Industry and logistics
Actors Company/organization	Developers/citizens/ landowners	Municipality/council/ NPO	Multi-level governance
Soft infrastructure Institution (Law/rule)	Neighborhood planning	City planning	Metropolitan strategy planning
Hard Infrastructure Urban infrastructure	Transit-oriented development (TOD)	Railway-led urbanisation	Convection of urban and rural
Agglomeration Building and land use	Growth/shrinkage	Adaptation	Layered development
Ecological foundations Planet/Resources	Geology/topology	Multiple circulation	Sustainability

Fig. 1 Urban nexus structure of the metropolis

grasp the withdrawn (Morton 2012) factors hidden beneath the complicated foreground.

The nodes of the mesh in Fig. 1 are localized to many characteristics in Tokyo, expressing the evolution of systems between nature and society. Through the evolutionary urbanization process, cities have attracted people, expanded buildings and built-up land, and consequently changed the interface between human and nature in the form of hard infrastructure and soft institutions. From the perspective of hard infrastructure, advanced railways and public transportation systems are the typical components that ensured the possibility of Transit-Oriented Development over the metropolitan region. Meanwhile, planning systems are the institutions created to manage both the infrastructure and urban systems.

Urban planning in Tokyo contains formal knowledge legally established in urban planning at the neighborhood level, the city level and the regional level. Various actors including government officials, politicians, and professionals are engaging in planning with the development of experience, products, projects, services, and eventually forming industries and the logistical flow of materials and energy. The top layer of the mesh indicates the ultimate objective of urban planning and management, to create and support more well-being and a high quality of life for citizens.

Historically, Tokyo has led the movement of urbanization through transit-oriented development. Residential communities are constructed with both public and private products (i.e., buildings, roads, and infrastructure), and services collected around railway stations, leading to a relatively compact life. There are always contradictions and conflicts among the multiple layers and scales because of different interests by stakeholders, as well as the timeframes connected to each. Nonetheless, the function and performance of services relies on the cooperation of sectors, both formally and informally. Nexus thinking and the nexus approach provide a pragmatic method to explore common benefits by leveraging our understanding of those interactions and interlinkages. Interactions in the urban nexus have been popularly recognized in scientific research (Newell et al. 2019) while the benefits of interlinkages are often handled in practice as a result of politics and institutional norms (Romero-lankao et al. 2017).

The design-led nexus approach by M-Nex suggests three components: an iterative design method, the FEWprint evaluation tool (ten Caat et al. 2022), and the cooperation of a participatory living lab to engage in practice (Yan et al. 2021). In this way the M-Nex approach is an integration of knowledge and action, from data management, to workshops and stakeholder engagement (Chapter "Design-Led Nexus Approach for Sustainable Urbanization"). Chapter "Calculating the Demand for Food, Energy, and Water in the Spatial Perspective" examined the effect of food-energy-water at the neighborhood scale while chapter "Assessing Urban Resource Consumption and Carbon Emissions from a Food–Energy–Water Nexus Perspective" analyzed the interlinkage of sectors. Through the chapters we recognized the impact of urbanization on land use, natural resources, and green infrastructure. We learned of the nexus effects of food-energy-water nexus at the scale of the neighborhood, the city, and the region, considering CO_2 emissions across industries.

Solutions and their nexus effects, including energy systems, participatory planning activities at living labs, and collaborative political instruments were proposed and examined for further discussion. We believe this knowledge is valuable for understanding Tokyo so far and can be used to push forwards to a more sustainable pattern of urbanization.

3 Challenges in Tokyo for More Resilience and Adaptivity

Because of its geographic location and geological conditions, the urban development of Tokyo was filled with a history of combating disasters and disruptive events. This was true from its very beginning, punctuated by large events such as the Meiji Fire in 1657, the Great Kanto Earthquake in 1923, the war in 1945, and recently, the Great East Japan Earthquake of March 11, 2011. Each disaster caused misfortune to the city, but post-disaster reconstruction also brought new opportunities for urban planning learning and practice. Consequently, the city was built with both hard and soft infrastructure grounded in resilience.

The most recent lesson comes from the Great East Japan Earthquake, the event that struck Japan on March 11, 2011. In the post-disaster reconstruction the power system controlled by Tokyo Electric Power Company was dismantled, the electric power market was fully liberalized, and new energy systems and urban resilience structures were developed (Ofuji and Tatsumi 2016; Shinkawa 2018). At the same time, new ideas and mindsets were born through the process of reconstruction. Some of the new practices have contributed to global communities on resilient construction - the combination of hard infrastructure and soft aspects of activities including public participation were adopted in the Sendai Framework (Murayama 2016; Diaz-Sarachaga and Jato-Espino 2019).

Tokyo originally planned to declare its recovery from the disaster at the 2020 Olympic games. Unfortunately, the sudden outbreak of COVD-19 got in the way.

To some extent, so too did the urban structure and systems described above. The assets accumulated in Tokyo over a century and a half of urbanization would normally be viewed as markers of affluence in an advanced global city. COVID-19 revealed a new challenge, one that flew in the face of those achievements. Around the world cities with a high density of movement and business exchange were the first to suffer. In a mega city like Tokyo the established rhythm of production and daily life was abruptly stopped during the pandemic (Aruga 2021; Dubinsky 2022; Boratinskii et al. 2023; Yabe et al. 2020; Tsuboi et al. 2022).

Thanks to the development of information technology and telecommuting, somehow the cities were able to maintain many of their essential functions. However, there have been changes. Crowded commuter trains were largely empty, and bustling nightclubs were suddenly deserted. Meanwhile, people realized that the daily commute need not be part of their life. The once proud idea of "working in the center, living in the suburbs" lost its shine, and could be seen as a social norm created out of habit as much as an actual need, at least in our time of digital connectivity. By introducing remote working, people have more time to spend in their neighborhoods. Wandering around the communities, many people noticed the poor communal environment, the narrow streets and alleys, the lack of green space, broken pedestrian paths, and so on. Positively speaking, the work-from-home movement could lead to improved livability in local neighborhoods, and may even become part of the process of adapting positively to Japan's aging society.

The Japanese government has promoted a so-called compact plus network for the next generation of urban development for years. In this scenario, people who lived in detached houses in suburbia gradually return to the urban center. However, the sudden pandemic might have broken this movement. Recent statistical data shows that the number of people moving out of Tokyo in 2020 surpassed the number moving in (Tokyo 2022). This may resolve the problem of over-concentration in Tokyo. Meanwhile, it will bring challenges for government and industrial sectors as they try to identify and predict future demands and develop plans to meet emerging trends. We imagine this could be an opportunity to rebalance the development of the city center, suburbia, and rural areas. It may also require us to rearrange large areas around railways stations in suburban regions. During the period of suburbanization, huge amounts of land was transformed to residential use and many live far enough from the stations that daily life is heavily dependent on private cars, even though commuting to the center is by train. The new social context, if it continues, could become a chance to revitalize the local environment and local communities, as residents become more involved in their decentralized lifestyle. However, this will not be realized easily because of path dependency (Romero-lankao et al. 2017), particularly the rigid political and sector-based regulating systems. Nexus thinking and the nexus approach are a useful philosophy that could shed light on the situation and help with development of a novel methodology to increase adaptation. Which is in the end what Tokyo excels at.

4 Conclusion

Entering the third decades of the twenty-first century, metropolitan regions are expected to be more prosperous and to better resist the effects of unprecedented disaster. This includes adapting to both climatic and social change.

Tokyo, one of the largest metropolitan areas in the world, has demonstrated the possibility of transformation from traditional to modern, from post-modern to resilient and adaptive. Today, Tokyo is still glorious, and the 1-year delay of the Tokyo Olympics in 2020 left a legacy for the core of the city as well as the nation. The 100-year pandemic provided an opportunity to review the path we have taken and to look with new eyes toward the future. The practice of resilience and adaptation is becoming ever more important across scales, sectors, and disciplines. Even so it also feels like it is always imminent. Perhaps that is its correct position, as resilience requires a certain amount of comfort with uncertainty and a lack of conclusion.

The current version of Tokyo, like many metropolitan areas in the world, faces a series of interlocking crises. Some are long term, some are short. Recovering from the pandemic, a shrinking population (and its partner problem, an aging population), planning for climate change and the next disaster, whether earthquake or something unforeseen. Tokyo is still in the process of building itself, coming to terms with problems so large and encompassing they cannot be comprehended and yet need to be responded to. These kind of problems are what Timothy Morton calls hyper-objects (Morton 2013)–economic, social, and climate crises. Alternatively, it might be easier to think of them as a poly-crisis, to use the term recently popularized by Adam Tooze (Wolf 2022), where multiple crises overlap and interact, creating problems larger than each one taken on its own. However, we label the challenge of taking on the enmeshed problems of our time, it is our hope that this text helps to prepare for them, and to develop better plans for our cities.

Acknowledgments This chapter is built on the achievements of the M-NEX project (https://m-nex.net) sponsored by the SUGI nexus and Japan Agency of Science and Technology (JST). The authors express their great gratitude to the contributors and partners of this publication project.

References

ARCADIS (2022) The ARCADIS Sustainable Cities Index 2022. https://www.arcadis.com/en/knowledge-hub/perspectives/global/sustainable-cities-index. Accessed 28 Jan 2023

Aruga K (2021) Changes in human mobility under the COVID-19 pandemic and the Tokyo fuel market. J Risk Financ Manag 14(4):163

Boratinskii V, Loopmans M, Osaragi T (2023) Reshuffling City life: spatial and functional dynamics of urban activity in Tokyo during COVID-19. Int J Urban Sci:1–27

Calimente J (2012) Rail integrated communities in Tokyo. J Transp Land Use 5:19–32. https://www.jstor.org/stable/26201682

Diaz-Sarachaga JM, Jato-Espino D (2019) Do sustainable community rating systems address resilience? Cities 93(May):62–71

Dubinsky Y (2022) The Olympic games, nation branding, and public diplomacy in a post-pandemic world: reflections on Tokyo 2020 and beyond. Place Brand Public Dipl. https://doi.org/10.1057/s41254-021-00255-x

Kono (n.d.) "Transit-oriented development versus car-oriented sprawl: the story of Tokyo and Adelaide." In 29th Australasian Transport Research Forum. https://australasiantransportresearchforum.org.au/wp-content/uploads/2022/03/2006_Soltani_Kono.pdf

Liu Y, Nath N, Murayama A, Manabe R (2022) Transit-oriented development with urban sprawl? Four phases of urban growth and policy intervention in Tokyo. Land Use Policy 112 (January):105854

MILT (2008) Planning for reshaping the national land. https://wwwmlitgojp/kokudoseisaku/kokudoseisaku_tk3_000043html. Accessed 28 Jan 2023

MMF (2022) Global power city index YEARBOOK 2022. Mori Memorial Foundation, Tokyo. https://www.mori-m-foundation.or.jp/english/ius2/gpci2/. Accessed 28 Jan 2023

MOE (2007) Mori-kawa-sato-umi project. https://www.env.go.jp/nature/morisatokawaumi/morisatokawaumich.html

Morton T (2009) Ecology without nature: rethinking environmental aesthetics. Harvard University Press, Cambridge, MA

Morton T (2010) Thinking ecology: the mesh, the strange stranger, and the beautiful soul. Collapse 6:265–293

Morton T (2012) Everything we need: scarcity, scale, Hyperobjects. Archit Des 82(4):78–81

Morton T (2013) Hyperobjects: philosophy and ecology after the end of the world. U of Minnesota Press, Minneapolis, MN

Murayama A (2016) Land use planning for depopulating and aging Society in Japan. In: Yamagata Y, Maruyama H (eds) Urban resilience. Springer, Cham, p 172

Nakamura, Takashi. 2018. "Maintenance of the railway user under the aging Society in Tokyo Metropolitan Suburban Area." 7009185. International Institute of Social and Economic Sciences.. https://EconPapers.repec.org/RePEc:sek:iacpro:7009185

Newell JP, Goldstein B, Foster A (2019) A 40-year review of food–energy–water nexus literature and its application to the urban scale. Environ Res Lett 14(7):073003

Ofuji K, Tatsumi N (2016) Wholesale and retail Electricity Markets in Japan: results of market revitalization measures and prospects for the current. Econ Energy Environ Policy 5(1):31–50

Ramaswami A (2020) Unpacking the urban infrastructure nexus with environment, health, livability, well-being, and equity. One Earth 2(2):120–124

Romero-lankao P, Mcphearson T, Davidson DJ (2017) The food-energy-water nexus and urban complexity. Nat Clim Chang 7(4):233–235

Sanders J (2015) Linking station node- and place functions to traffic flow: a case study of the Tokyu Den-En Toshi line in Tokyo, Japan. University of Twente, Enschede. http://essay.utwente.nl/67082/

Shinkawa T (2018) Electricity system and market in Japan. Electricity and Gas Surveillance Commission, Tokyo. https://www.emsc.meti.go.jp/english/info/public/pdf/180122.pdf

ten Caat P, Nick MJ, Tenpierik TS, Tillie NMJD, van den Dobbelsteen AAJF, Thün G, Cullen S, Nakayama S, Karanisa T, Monti S (2022) Towards fossil free cities—emission assessment of food and resources consumption with the FEWprint carbon accounting platform. Clean Environ Syst 4(100074):100074

Tokyo (2022) Future Tokyo: Tokyo's long-term strategy version up 2022. https://www.seisakukikaku.metro.tokyo.lg.jp/basic-plan/2022/10/pdf/versionup2022en.pdf

Tsuboi K, Fujiwara N, Itoh R (2022) Influence of trip distance and population density on Intra-City mobility patterns in Tokyo during COVID-19 pandemic. PLoS One 17(10):e0276741

Wheeler SM (2004) Planning for sustainability: creating livable, equitable and ecological communities, vol 9780203300. Routledge, London

Wolf, M. (2022) How to think about policy in a polycrisis. https://www.ft.com/content/a1918fec-2c8f-4051-ad78-c300b0fc9adb. Accessed 4 Apr 2023.

Yabe T, Tsubouchi K, Fujiwara N, Wada T, Sekimoto Y, Ukkusuri SV (2020) Non-compulsory measures sufficiently reduced human mobility in Tokyo during the COVID-19 epidemic. Sci Rep 10(1):18053

Yajima I, Kishii, Nakano (2019) TOD practice in Japan: Tokyo, A global city created by railways. Sciences, Tokyo. http://113.157.192.76/toshi/tosiko/content/001361990.pdf

Yan W, van Delbsteen A, Keeffe G, Thün G, Roggema R (2021) Developing and implementing the design-led nexus approach for sustainable urbanisation. In: CITIES 20.50—Creating habitats for the 3rd millennium: smart—sustainable—climate neutral. Proceedings of REAL CORP 2021, 26th international conference on urban development, regional planning and information society. CORP—Competence Center of Urban and Regional Planning, Schwechat, pp 251–260

Correction to: Design-Led Nexus Approach for Sustainable Urbanization

Wanglin Yan and Shun Nakayama

Correction to:
Chapter 2 in: W. Yan et al. (eds.), *Resilient and Adaptive Tokyo*,
https://doi.org/10.1007/978-981-99-3834-6_2

The original version of this book was inadvertently published with an incorrect image for Fig. 4 in Chap. 2. The correct image has now been updated.

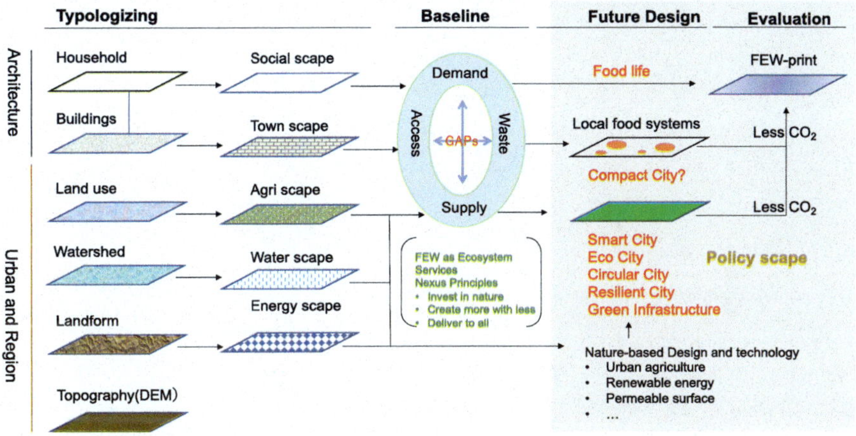

Fig. 4 Implementing the M-NEX method in urban design

The updated version of this chapter can be found at
https://doi.org/10.1007/978-981-99-3834-6_2

W. Yan et al. (eds.), *Resilient and Adaptive Tokyo*,
https://doi.org/10.1007/978-981-99-3834-6_17

The manufacturer's authorised representative in the EU is Springer Nature Customer Service Centre GmbH, Europaplatz 3, 69115 Heidelberg, Germany. If you have any concerns regarding our products, please contact ProductSafety@springernature.com

Printed and bound by CPI Group (UK) Ltd, Croydon, CR0 4YY
01/10/2026
02225866-0009